Methods in Protein Sequence Analysis

Methods in Protein Sequence Analysis

Edited by

Kazutomo Imahori

Mitsubishi Kasei Institute of Life Sciences
Tokyo, Japan

and

Fumio Sakiyama

Osaka University
Osaka, Japan

Plenum Press • New York and London

Library of Congress Cataloging-in-Publication Data

Methods in protein sequence analysis / edited by Kazutomo Imahori and Fumio Sakiyama.
p. cm.
"Proceedings of the Ninth International Conference on Methods in Protein Sequence Analysis, held September 20-24, 1992, in Otsu, Japan"--T.p. verso.
Includes bibliographical references and index.
ISBN 0-306-44488-7
1. Amino acid sequence--Congresses. I. Imahori, Kazutomo. II. Sakiyama, Fumio. III. International Conference on Methods in Protein Sequence Analysis (9th : 1992 : Otsu, Japan)
QP551.M3886 1993
547.7'5--dc20 93-7788
CIP

Proceedings of the Ninth International Conference on Methods in Protein Sequence Analysis, held September 20–24, 1992, in Otsu, Japan

ISBN 0-306-44488-7

A Division of Plenum Publishing Corporation
233 Spring Street, New York, N.Y. 10013

Printed in the United States of America

ORGANIZATION OF 9th MPSA

Organizing Committee:

Kazutomo Imahori (Chairman), Akira Tsugita (Executive Chairman), Shin-ichi Ishii (Vice Chairman), Koiti Titani (Vice Chairman), Fumio Sakiyama (Secretary General), Yasutsugu Shimonishi (Treasurer), Koichi Suzuki (Program Committee Chairman).
Members: Toshio Fukui, Gunki Funatsu, Sadaaki Iwanaga, Yukiteru Katsube, Seiichi Kawashima, Hiroshi Matsubara, Hisayuki Matsuo, Tatsuo Miyazawa, Kozo Nagano, Takahisa Ohta, Tsuneo Okuyama, Tairo Oshima, Kenji Takahashi, Susumu Tsunasawa, Tsuneko Uchida.

Scientific Advisory Committee:

Ettore Appella (U.S.A.), M. Zouhair Atassi (U.S.A.), Patrick Argos (Germany), Klaus Biemann (U.S.A.), Jui-Yoa Chang (Switzerland), Maria A. Coletti-Previero (France), Marshall Elzinga (U.S.A.), Angelo Fontana (Italy), Gerhard Frank (Switzerland), Agnes Henschen (U.S.A.), Michael W. Hunkapiller (U.S.A.), Pierre Jollés (France), Hans Jörnvall (Sweden), Richard A. Laursen (U.S.A.), Hans Neurath (U.S.A.), Richard N. Perham (U.K.), Richard J. Simpson (Australia), Joël Vandekerckhove (Belgium), John E. Walker (U.K.), Kenneth A. Walsh (U.S.A.), Brigitte Wittmann-Liebold (Germany).

Advisors:

Morio Ikehara, Tokuji Ikenaka, Masao Kotani.

Secretaries:

Hiroko Ito, Chizu Sasai, Naoko Ami.

PREFACE

The Ninth International Conference on Methods in Protein Sequence Analysis was held for the first time in Asia from September 20 to September 24, 1992 in Otsu (a city near Kyoto), Japan. Approximately 400 delegates attended the meeting. Forty papers were presented orally and 147 poster presentations were discussed. Academic sessions were held from early in the morning until late in the evening. We are confident that the Conference was successful in providing up-to-date information about methods in protein sequence analysis to all participants. Moreover, with the knowledge and understanding of the present standard of various methods of analysis that are being used and will be used, we were able to clarify areas that need to be evaluated, to be improved and be explored further.

Major topics in the Conference mostly covered areas in the methodology of protein sequence analysis, such as: micropreparation and microsequencing of proteins, mass spectrometry, post-translational modification, prediction and database analysis, and analysis of protein structures of special interests. The evolution of genetic engineering in molecular biology has greatly accelerated the accumulation of knowledge on the amino acid sequence of novel proteins regardless of whether they are expressed or not expressed in living organisms. In the early stage of accumulation of structural information, the amino acid sequence itself is worthy of notice. However, it is difficult for researchers to understand the real structures of proteins that are actually functioning in and outside cells as their nucleotide sequences are insufficient to present detailed structures of those proteins. Protein chemistry, however, can tell us much more about the detail of the covalent structure of protein molecules that what can be deduced from amino acid sequences.

With these considerations in mind, our major concern for organizing the Conference was to highlight the updated method for protein microsequencing and for the analysis of structural information collected by protein sequencing, as done in the previous conferences. In particular, we focused our attention on the development of mass spectrometry over the past two years and a special session was organized on this topic. It is predicted that mass spectrometry will become a promising tool for the microsequence analysis of proteins in the near future. We hope mass spectrometry will refine itself more and more in an effort to overcome the drawbacks such as a lack of quantitative measures and the cost of instruments, in addition to the continuing efforts being made until only several molecules of a protein are necessary for sequencing.

This book documents almost all invited papers, including several papers from poster presentations which have been selected by the Scientific Program Committee. We believe updated methodologies described for protein purification, the elucidation of amino acid sequence, the analysis of site and mode of co- and post-translational modification, molecular weight determination, the prediction of tertiary structure from amino acid sequences and the analysis of structure database, will all serve to help daily research on proteins, as well as to inspire new ideas for the advancement of protein science.

The editors wish to express special thanks to Professor Koichi Suzuki, Chairman of the Scientific Program Committee, who, with the secretarial assistance of Drs. Hisashi Hirano and Hiroshi Kawasaki, has devoted much time to the organization of the scientific program that made the 9th MPSA a successful meeting. Thanks are also due to Dr. Hisashi Hirano and his colleagues for their assistance in the editing process, especially for transcripting all manuscripts of this book to a special format in order to have them printed in a unified style with unified lettering. Furthermore, the editors wish to acknowledge the helpful suggestions made by Ms. Mary Phillips Born of Plenum Publishing Corporation throughout the preparation of this book.

Kazutomo Imahori
Fumio Sakiyama

November, 1992

ACKNOWLEDGEMENTS

Thanks are due to the following organizations for their support:

Foundation for Advancement of International Sciences
Japan Bioindustry Association
Japan Society for Bioscience, Biotechnology, and Agrochemistry
Pharmaceutical Society of Japan
Protein Research Foundation
The Biophysical Society of Japan
The Chemical Society of Japan
The Japanese Biochemical Society
The Molecular Biology Society of Japan
The Protein Engineering Society of Japan
The Society of Fermentation Technology, Japan
The Society of Synthetic Organic Chemistry, Japan

Special thanks are also due to the following sponsors:

Ajinomoto Co., Inc.
Applied Biosystems
Asahi Chemical Industries Co., Ltd.
Ciba-Geigy Foundation (Japan) for the Promotion of Science
Finnigan-Mat Instruments
Fuji Oil Co., Ltd.
Japan Health Sciences Foundation
JASCO Corporation
JEOL (Japan Electronics Laboratory)
Kanegafuchi Chemical Industry Co., Ltd.
Kao Corp.
Kirin Brewery Co., Ltd.
Life Science Foundation of Japan
M & S Instruments Trading Inc.
Mitsubishi Kasei Corp.
Mitsui Toatsu Chemicals, Inc.
Nestle K. K.
Nissei Sangyo Co., Ltd.

Nissin Flour Milling Co., Ltd.
Oriental Yeast Co., Ltd.
Osaka Pharmaceutical Manufacturers Association
Seikagaku Corp.
Seiko Instruments, Inc.
Shimadzu Corporation
Shimadzu Science Foundation
Showa Denko K. K.
Sumitomo Chemical Co., Ltd.
Suntory Ltd.
Takara Shuzo Co., Ltd.
Teijin Ltd.
The Cell Science Research Foundation
The Commemorative Association for the Japan World Exposition (1970)
The Mitsubishi Trust & Banking Corp.
The Pharmaceutical Manufacturers' Association of Tokyo
The Protein Research Foundation
Tokuyama Soda Co., Ltd.
Toray Industries, Inc.
Toyobo Co., Ltd.
Uehara Memorial Foundation
Unitika, Ltd.
Wako Pure Chemical Industry, Ltd.
Yoshida Foundation for Science and Technology

THE EDMAN AWARD 1992

This year two scientists were honored as recipients of the 1992 Edman Award sponsored by Millipore Corporation. The Edman Award Committee (Agnes Henschen, Adam Inglis, Hans Jörnvall, Fumio Sakiyama and Johann Salnikow) carefully discussed research areas and candidates eligible for the award and decided to honor Professor Klaus Biemann and Professor Donald F. Hunt. Both of these scientists have been pioneers of the development and application of mass spectrometry techniques for contemporary protein sequence analysis.

Professor Klaus Biemann received his Ph. D. degree in organic chemistry from the University of Innsbruck, Austria, in 1951. In 1955, he moved to the Department of Chemistry at MIT in Cambridge, MA, as a postdoctoral fellow, was appointed to the faculty in 1957, and promoted to full professor in 1963. He began his work in the structure determination of natural products by mass spectrometry in 1958. One year later, he published his first paper on mass spectrometric amino acid sequencing of peptides. The methodology, which required reduction to polyamines and interpretation of the electron-impact spectra obtained with a gas-chromatography/mass spectrometer system, was continuously improved over the next two decades. One example of its utility was the determination (in collaboration with H. G. Khorana) of the amino acid sequence of bacteriorhodopsin, a membrane protein difficult to analyze with the Edman technique alone.

Beginning in 1981, Professor Biemann developed a strategy for the verification and correction of the primary structure of proteins deduced from DNA sequences or by other means. It was based on the ability to measure the molecular weights of large, underivatized peptides, even in mixtures, by fast atom bombardment mass spectrometry, an ionization technique introduced at that time by M. Barber (Manchester, U.K.). In the mid-80s, high energy collision in a magnetic tandem mass spectrometer was explored extensively for direct peptide sequencing. This methodology was then used in the determination of the primary structure of a number of proteins, particularly thioredoxins and glutaredoxins, as well as modified peptides.

Professor Donald F. Hunt received his Ph. D. degree in organometallic chemistry at the University of Massachusetts, did a year of postdoctoral research in mass spectrometry with Professor Klaus Biemann at MIT, and then joined the faculty at the University of Virginia in 1968. In a landmark paper published in 1981, Professor Hunt proposed and implemented a new strategy for protein sequence analysis based on tandem mass spectrometry. The first unknown peptide and protein sequences solved by this technique were described by him in 1984 and 1987, respectively. Much of Professor Hunt's recent efforts have focused on extending the above approach by developing methods for the isolation, manipulation, and sequence analysis of protein sample at the low and subpicomole level. The combination of microcapillary HPLC and electrospray ionization/tandem mass spectrometry on a triple quadropole instrument was developed for this purpose in 1991. Proteins at the 1-5 pmole level can now be enzymatically digested within 50-75 micron fused silica columns, and the resulting peptides then washed directly onto a similar column packed with C-18 particles, separated, and eluted directly into the tandem mass spectrometer for sequential analysis. The power of this approach was further demonstrated by sequence analysis of peptides presented to the immune system on the surface of antigen presenting cells. Mixtures of peptides containing more than 1000 components were isolated in this work and individual components were sequenced at the 100-300 fmole level by the above methodology. Identification of peptide antigens associated with viral infection, cellular transformation, and autoimmunity should follow as a result of this analytical breakthrough.

So, as you can see from the contributions made by these two outstanding scientists, they have truly been major pioneers in the development and application of mass spectrometry techniques for contemporary protein sequence analysis. We are now anxiously awaiting the day when only a single protein molecule will be required to complete its structure.

Fumio Sakiyama *

* The author deeply appreciates all help extended by Dr. Ettore Appella in the preparation of this manuscript for laudation.

CONTENTS

MICROPREPARATION AND MICROSEQUENCING

MASS SPECTROMETRY

POST-TRANSLATIONAL MODIFICATIONS

PREDICTION AND DATABASE ANALYSIS

PROTEIN STRUCTURES OF SPECIAL INTERESTS

MICROPREPARATION AND MICROSEQUENCING

CAPILLARY LIQUID CHROMATOGRAPHY: A TOOL FOR PROTEIN STRUCTURAL ANALYSIS

Robert L. Moritz and Richard J. Simpson

Joint Protein Structure Laboratory, Ludwig Institute for Cancer Research (Melbourne Branch) and the Walter and Eliza Hall Institute for Medical Research
Parkville, Victoria 3050 Australia

INTRODUCTION

Microbore column (1-2.1mm I.D.) liquid chromatography is, without question, one of the most important and widely used techniques for the recovery and isolation of proteins and peptides for the purpose of structural analysis. It is also the method of choice for the identification and quantitation of phenylthiohydantoin (PTH) derivatives, the products of the Edman degradation procedure employed for amino acid sequence analysis. The major advantage of microbore columns over their conventional column counterparts (4.6 mm I.D.) is that they offer enhanced mass sensitivity (5-20 fold) and decreased peak volumes (40-60 μl) without any compromise of column resolution (for reviews see Refs. Simpson and Nice, 1989; Simpson *et al.*, 1991; Simpson *et al.*, 1989).

Although the potential advantages for further column miniaturization have long been recognized (Ishii *et al.*, 1977; Yang, 1982; Novotny, 1988, Novotny, 1990; Verzele *et al.*, 1988), progress with this technology has been constrained by the limited availability of packed capillary columns (<0.5-mm I.D.) and instrumentation designed to facilitate the operation of such columns. Concerning the latter, the ability to operate and form reliable solvent gradients at very low flow rates (1-5 μl/min) as well as UV detector design are of the utmost importance. In a previous report we described the design of a liquid chromatographic system that allows continuous gradient elution at low flow rates of 1-5 μl/min, from reversed-phase capillary columns (<0.320 mm I.D.) (Moritz and Simpson, 1992; Moritz and Simpson, 1992; Moritz *et al.*, 1992). A simple procedure for packing fused silica capillary (FSC) columns was developed and a U-shaped 'longitudinal' capillary flow cell (6-mm pathlength) for measuring the UV absorbance of the column eluate was assessed.

In this paper we have extended these studies to include several microscale applications of this technology for protein structural analysis. These include rapid trace enrichment

Methods in Protein Sequence Analysis, Edited by K. Imahori and F. Sakiyama, Plenum Press, New York, 1993

(concentration) of low nanogram levels of protein from large volumes, peptide mapping and PTH-amino acid analysis.

EXPERIMENTAL

Capillary HPLC System

Solvent delivery was supplied by either an Applied Biosystems Model 120A or a Hewlett-Packard Model HP1090A liquid chromatograph. Accurate flow rates of 1-5 µl/min and reproducible gradient formation were achieved by installing a preinjection solvent split that diverted most of the solvent flow through *ca.* 100 cm of 0.100 mm I.D. x 0.260 mm O.D. FSC tubing at a 1/16-in tee. Capillary columns were directly connected to either a Rheodyne Model 8125 injector fitted with 0.5-5 µl injection loops or to a Rheodyne Model 7520 injector fitted with 0.2-1.0 µl internal sample rotors. Solvent flow through the column was accurately set at 1-5 µl/min by adjusting the master pump flow-rate (100-200 µl/min). Frequent monitoring of the actual flow through the capillary column is recommended since variations in flow rate can occur over time with this split-flow approach. Flow measurement is readily achieved by attaching a 10 µl chromatographic syringe to the effluent capillary of the flow cell and accurately timing the meniscus with a stop-watch.

UV Detection

Two different UV detectors were used in this study. For the Applied Biosystems Model 120A system, the convential flow cell in this instrument was replaced with a 6-mm optical pathlength U-shaped 'longitudinal' capillary flow cell (LC Packings, Amsterdam, the Netherlands). In the second system, a Hewlett-Packard solvent delivery system was coupled to a Spectra Focus forward optics scanning detector fitted with the same style U-shaped flow cell mentioned above, but configured for this detector. Data collected were either sent to a strip-chart recorder via the single wavelength mode or, in the latter case, collected on an IBM PS/2 computer (Model P70 386) with Spectra Focus software installed and operated in the multiple-wavelength mode (195-340 nm, 5 nm intervals).

Column Construction

Column Assembly. Capillary columns of 0.200 mm and 0.320 mm I.D. and varying lengths were constructed in the following manner. Lengths (50-mm) of 0.420 mm O.D. x 0.320 mm I.D. or 0.340 mm O.D. x 0.200 mm I.D. polyimide-coated FSC tubing (Polymicro, Phoenix, AZ, USA) were used as disc cutters to fabricate column frits from 0.45-µm porosity hydrophilic PVDF membrane (Millipore). For the 0.320-mm I.D. column, a length of 0.275 mm O.D. x 0.075 mm I.D. FSC tubing was used to insert the frit to a depth of *ca.* 10-mm. In the case of the 0.200 mm I.D. columns, 0.190 mm O.D. x 0.050 mm I.D. FSC tubing was used to insert the frit. A small bead *(ca.* 3-mm in length) of epoxy resin, that had been precured with a heat gun to achieve a 'tacky' consistency, was positioned at the join and further cured to hardness. *Note: (i) proper precuring of the epoxy resin is important to ensure that the resin sets quickly, thereby minimising the risk of it seeping up the capillary and blocking the frit; (ii) thick-walled FSC tubing (e.g., 0.540 mm O.D. versus 0.420 mm O.D.) was employed to facilitate column packing at high pressure (e.g., 400 bar versus 70 bar) and in doing so, reducing the column-packing time from 16h to 30 min. Typically, this column construction procedure takes ca. 5-10 min.*

Column-Packing Procedure. The slurry-packing procedure used to pack the capillary columns used in this study has been previously described (Moritz and Simpson, 1992). The steps employed to pack a 0.320 mm I.D. x 200 mm column are as follows:
• Prepare slurry (45 mg packing/500 µl n-propanol), sonicate for 15 min. • Connect capillary column to high-pressure column packer (Shandon, Runcorn, UK) via a slurry reservoir (empty 50 mm x 2.1 mm I.D. stainless steel glass-lined tube with 1/4-in column end fittings and fill column with packing solvent (n-propanol) at 400 bar. • Empty the reservoir and refill with 160 µl packing slurry (~16 mg). • Pack the column at constant pressure (400 bar) for 30 min. • Condition the column with 50% aqueous methanol (400 bar) for 30 min.

Peptide Mapping

Peptide mapping of *S. aureus* protease digest of recombinant murine interleukin-6 on a 200 mm x 0.320 mm I.D. microcolumn packed with Brownlee RP-300 has been described elsewhere (Moritz and Simpson, 1992).

RESULTS AND DISCUSSION

Chromatography of Proteins

Mass Sensitivity. The advantages of narrow-bore and capillary columns with respect to enhanced sensitivity of protein detection and reduced peak volumes are shown in Figure 1. All columns were packed with the same support (Brownlee RP-300 C8, 7-µm dimethyloctylsilica, pore size 300 Å) and operated at equivalent flow velocities. A comparison of the 4.6 and 2.1 mm I.D. columns revealed a five-fold increase in sensitivity by using the 2.1 mm I.D. column. This increase in mass sensitivity is expected since the extra-column volumes and detector cell geometry (Hewlett-Packard model 1090A diode-array cell) were identical for the operation of both columns. In the case of the 0.320 mm I.D. column, 25-fold and 125-fold inereases in sensitivity of protein detection were achieved when compared with 2.1 and 4.6-mm I.D. columns. The constant peak band widths, as a function of time, indicated that the three columns had comparable column efficiencies. Peak volumes for the 0.320, 2.1, and 4.6-mm I.D. columns were 2.5, 133.5 and 600 µl, respectively.

Detector Linearity. The linearity of the detector response for the 6-mm pathlength capillary flow cell, using our packed 0.320 mm I.D. capillary column (Figure lC), for lysozyme extended to ~2 µg, while the detection limit was in the range 50-100 pg (Moritz and Simpson, 1992). For lysozyme and ribonuclease, the minimum detectable amounts (at 215 nm) which produce a peak that is twice the baseline noise were ~ 50 pg (3.6 fmol) and ~100 pg (7.3 fmol), respectively.

Trace Enrichment/Resolution/Loadability. In the three panels shown in Figure 2, a sample mixture of proteins was loaded onto a 0.320 mm I.D. column in volumes of 5-, 50-, and 500-µ1. Resolution of the protein mixture achieved by a 5-µl injection was comparable to that achieved by 50-µ1 and 500-µ1. The close agreement between the three profiles is a good demonstration of the trace-enrichment capabilily of this capillary column.

Typically, proteins were recovered in eluent volumes of 1-4 µl. Figure 3 illustrates the separation efficiency that can be achieved for proteins on a 150 x 0.200 mm I.D. reversed-phase capillary column (Brownlee RP-300) packed by us. Using gradient elution between

0.1% aqueous TFA and acetonitrile at 1.6 µl/min, the resolution of this set of proteins is similar to that obtained on larger-bore columns packed with the same support. The effect of the mass of protein on peak width for protein loads of 50-100 ng is minimal (data not shown). For protein loads in excess of 1 µg the band width increased significantly and the band profile became distorted.

Peptide Mapping. Peptide mapping by RP-HPLC is a widely-used technique for protein structural analysis. An example of high sensitivity peptide mapping obtained with capillary chromatography is given for a 180 ng (9 pmol) *S. aureus* protease digest of recombinant

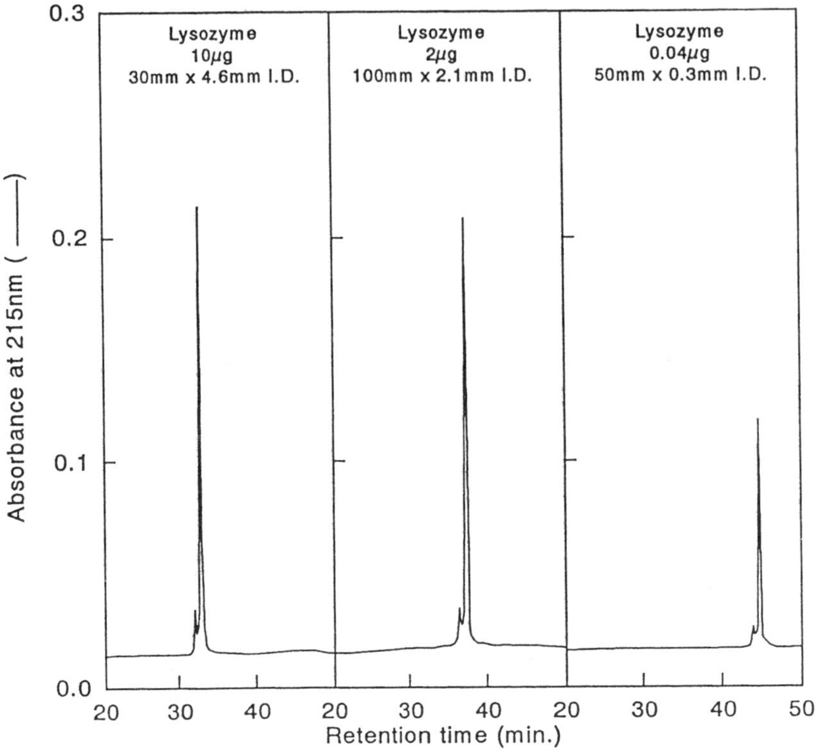

Figure 1. Effect of column internal diameter and flow rate on detector response. Column support, Brownlee RP-300 C8 (7 µm dimethyloctylsilica, 300 Å pore size, 30 x 4.6 mm I.D. and 100 x 2.1 mm I.D., Applied Biosystems, and 50 x 0.320 mm I.D. packed in-house); linear 60-min gradient from 100% B, where eluent A is 0.1% aqueous TFA and eluent B is 60% aqueous acetonitrile containing 0.09% TFA; column temperature, 45°C; detection, 215 nm; flow rates, 1 ml/min, 200 µl/min and 4 µl/min for 4.6, 2.1 and 0.320 mm I.D. columns respectively. Sample lysozyme (4.6 mm I.D. column: 10 µg; 2.1 mm I.D. column: 2 µg; 0.320 mm I.D. column: 0.04 µg). (From Moritz, R. L. and Simpson, R. J. *J. Chromatogr.* 1992. With permission).

mouse IL-6 (Figure 4). Selected peptides were recovered manually and subsequently identified by Edman degradation. Sequence data for peptides 1 and 2 (see legend to Figure 4) was obtained at low picomole levels (4-7 pmol), this corresponds to approximate yields of 42% and 77%, respectively. In this study, peptide 1 was collected directly onto a Polybrene-treated disk while peptide 2 was collected in an Eppendorf tube and then transferred, using a 2-µl Gilson pipette, onto the sequencer sample disk. If the peptides need to be stored before sequence analysis, we prefer to collect them in an Eppendorf tube containing 10-µl of 0.1% aqueous TFA/0.01% Tween 20. The sequences obtained were in perfect agreement with the published sequences.

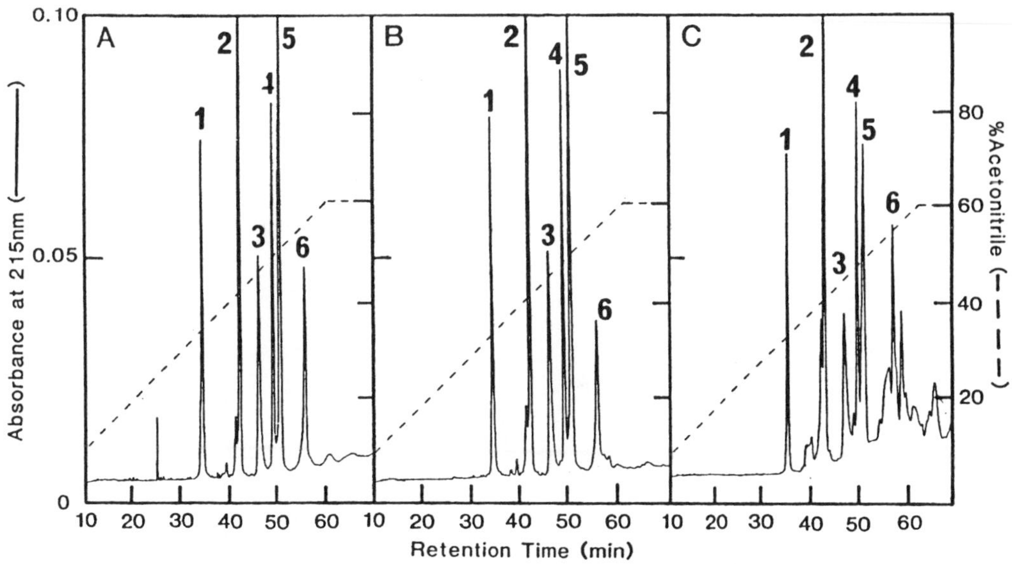

Figure 2. Trace enrichment of protein standards on a 200 mm x 0.320 mm I.D. microcolumn, slurry packed with Brownlee RP-300 silica. Chromatographic conditions are given in Figure 1. Flow rate, 3.6 µl/min. Proteins; 1, ribonuclease B; 2, lysozyme; 3, bovine serum albumin; 4, carbonic anhydrase; 5, myoglobin and 6, ovalbumin. Loading conditions: (A) 5-µl injection volume loaded via 5-µl loop; (B) 50-µ1 volume loaded via a 50-µ1 sample loop; (C) 500-µ1 volume loaded in ten 50-µl portions using a 50-µl loop. Sample: 50 ng protein standards in 0.05% aqueous TFA/0.01% Tween 20 (From Moritz, R. L and Simpson, R. J., *J. Microcol. Sep.*, 1992. With permission).

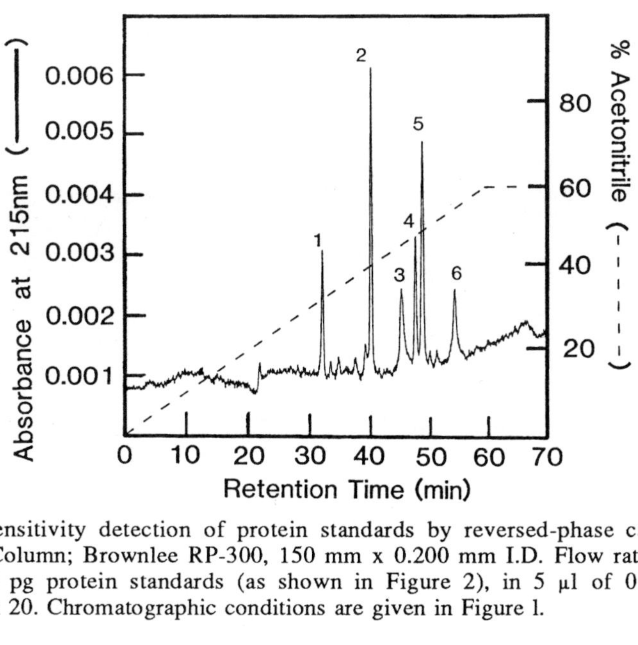

Figure 3. High sensitivity detection of protein standards by reversed-phase capillary liquid chromatography. Column; Brownlee RP-300, 150 mm x 0.200 mm I.D. Flow rate, 1.4 µl/min. Sample load: 500 pg protein standards (as shown in Figure 2), in 5 µl of 0.05% aqueous TFA/0.01% Tween 20. Chromatographic conditions are given in Figure 1.

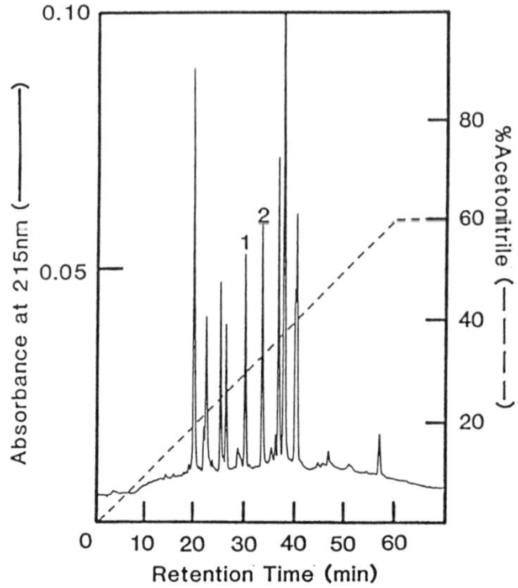

Figure 4. Peptide map of *S. aureus* V8 digestion of recombinant murine IL-6 (180 ng). Column; Brownlee RP-300, 200 mm x 0.320 mm I.D. See Figure 1 for chromatographic conditions. Selected peptides, peptides 1 and 2, were identified by sequence analysis as Ile-Gln-Leu-Asn-Lys-Ile-Ser-Ser-Gln (4 pmol) and Thr-Leu-Ile-His-Phe-Asn-Gln-Glu (7 pmol), respectively (From Moritz, R. L and Simpson, R. J., *J. Microcol. Sep.*, 1992. With permission).

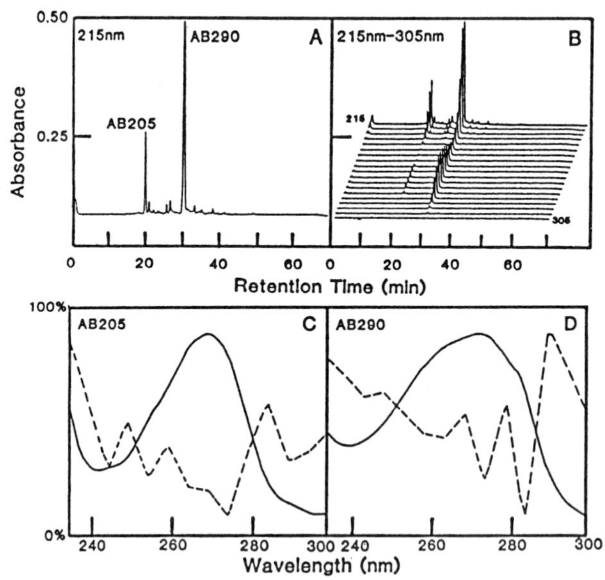

Figure 5. Spectral analysis of aromatic amino acid-containing synthetic peptides chromatographed by reversed-phase capillary liquid chromatography. Chromatographic conditions as given in Figure 1. (A) 0.320 mm I.D. capillary column profile of synthetic peptides (250 ng) AB205 and AB290, (B) multiple signal plot; (C and D) zero-order (solid lines) and second-order derivative spectra (dashed lines) of synthetic peptides AB205 and AB290, respectively (From Moritz, R. L. and Simpson, R. J. *J. Chromatogr.*, 1992. With permission).

Multiple-Wavelength Detection

We were interested in establishing whether the multiple-wavelength detector (Spectra-Physics, Spectra Focus), a forward-optics system, could be adapted for use in our capillary liquid chromatography system for identifying aromatic amino acid-containing peptides. Two synthetic peptides, a tyrosine-containing peptide AB205 (SRDMYESEGGDGY) and a tryptophan-containing peptide AB290 (DLAWWELR), were chromatographed on a reversed-phase capillary column (Figure 5A). Inspection of the multiple signal plot of the data (Figure 5B) reveals major absorption peaks in the range 270-290 nm which are indicative of the aromatic amino acids (Grego *et al.*, 1986). Second-order derivative spectroscopic analysis of these data (Figure 5C and D) reveals minima at 277 ± 5 nm for AB205 (characteristic for tyrosine) and 287 ± 5 nm for AB290 (characteristic for tryptophan).

Phenylthiohydantoin Amino Acid Analysis

Figure 6 illustrates the separation efficiency for PTH-amino acids obtained with a 150 x 0.320 mm I.D. capillary column packed with Applied Biosystems PTH-C_{18} support. The sensitivity of detection of PTH-amino acids (~25 fmol) is at least 25 times higher than that achieved with a corresponding 2.1-mm I.D. column. Although the enhanced mass sensitivity of PTH-amino acid analysis by capillary liquid chromatography offers great potential for low-femtomole protein sequence analysis, a major drawback for interfacing this technology with existing sequencer hardware is that the volume of organic solvent that can be applied to a capillary column must be kept below the critical elution percentage to allow for efficient chromatography.

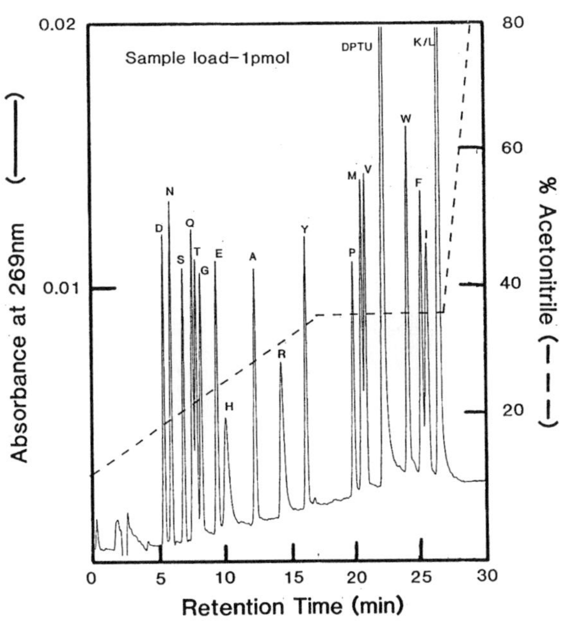

Figure 6. Separation of PTH-amino acids by reversed-phase capillary liquid chromatography. Column, 150 mm x 0.300 mm I.D. Applied Biosystems PTH-C_{18}. Solvent A. 8.3 mM sodium acetate-5% tetrahydrofuran (pH 4.1); solvent B, acetonitrile. Column temp., 55°C, Flow rate: 5 µl/min. Sample load: mixture of 1 pmol each of PTH-amino acids in 0.5 µl of 2% aqueous acetonitrile (From Moritz, R. L. and Simpson, R. J. *J. Chromatogr*. 1992. With permission).

REFERENCES

Grego, B., Nice, E. C. and Simpson, R. J., 1986 Use of scanning diode array detector with reversed-phase microbore columns for the real time spectral analysis of aromatic amino acids in peptides and proteins at the sub-microgram level: Application to peptide and protein sequencing, *J. Chromatogr.* 352:359.

Ishii, D., Asai, K., Hibi, K., Jonokuchi, T. and Nagaya, M., 1977 A study of micro-high-performance liquid chromatography. I. Development of technique for miniaturization of high-performance liquid chromatography, *J. Chromatogr.* 144:157.

Moritz, R. L.,Ward, L. D. and Simpson, R. J., 1992 High sensitivity peptide mapping utilizing reversed-phase microbore and microcolumn liquid chromatography, *in*: Techniques in Protein Chemistry III, Angeletti, H., ed., Academic Press, San Diego, pp 97-106.

Moritz, R. L. and Simpson, R. J., 1992 Application of capillary reversed-phase high performance liquid chromatography to high-sensitivity protein sequence analysis, *J. Chromatogr.* 599:119.

Moritz, R. L. and Simpson, R. J., 1992 Purification of proteins and peptides for sequence analysis using microcolumn liquid chromatography, *J. Microcol. Sep.* 4:485.

Novotny, M., 1988 Recent advances in microcolumn liquid chromatography, *Anal. Chem.* 60:500.

Novotny, M., 1990 Recent advances in the isolation and structural studies of biomacromolecules using microcolumn techniques, *J. Microcol. Sep.* 2:7.

Simpson, R. J., Begg, G. S., Reid, E., Rubira, M.R., Ward, L. D. and Moritz, R. L., 1989 Application of microbore HPLC for purifying proteins and peptides for microsequence analysis, *in*: Methods in Protein Sequence Analysis, Wittmann-Liebold, B., ed., Springer-Verlag, Berlin, pp 256-262.

Simpson, R. J., Moritz, R. L., Reid, G. E. and Ward, L. D., 1991 Current strategies for the purification of subnanomole amounts of protein and polypeptides for microsequence analysis, *in*: Methods in Protein Sequence Analysis, Jörnvall, H., Höög, J.-O. and Gustavsson, A.-M., eds., Birkhauser Verlag, Basel, pp 67-78.

Simpson, R. J. and Nice, E. C., 1989 Strategies for the purification of subnanomole amounts of protein and polypeptides for microsequence analysis, *in*: The Use of HPLC in Receptor Biochemtistry, Alan R.Liss, Inc., pp 201-244.

Verzele, M., Dewaele, C. and De Weerdt, M., 1988 Micro-LC: What are its chances? *LC-GC* 6:966.

Yang, F., 1982 Fused-silica narrow-bore microparticle-packed-column high-performance liquid chromatography, *J. Chromatogr.* 236:265.

ROUTINE AMINO ACID SEQUENCING ON 2D-GEL SEPARATED PROTEINS : A PROTEIN ELUTION AND CONCENTRATION GEL SYSTEM

J. Vandekerckhove, M. Rider, H.-H. Rasmussen, S. De Boeck, M. Puype, J. Van Damme, B. Gesser, and J. Celis

Laboratory of Physiological Chemistry, University Gent
K.L. Ledeganckstraat 35, B-9000 Gent, Belgium
Institute of Medical Biochemistry, Aarhus University
Ole Worms Allé, 170, DK-8000 Aarhus, Denmark

INTRODUCTION

2D-gel electrophoresis is routinely used to study the integral protein composition of a given cell type, often revealing more than 2,000 individual spots in a single gel. Due to its high resolution power, the technique has been employed to study alterations in cellular protein expression in response to various stimuli or as a result of differentiation and development (Celis and Bravo, 1984). Identification of the separated proteins can be achieved by comigration electrophoresis or by immunological techniques. Proteins can also be recognized by their amino acid composition, their exact molecular weight as determined by mass spectrometry or by partial amino acid sequencing. The latter approach further allows cDNA cloning from the resultant sequences.

Sequencing of 2D-gel separated proteins became possible with the introduction of protein electroblotting methods which allow efficient transfer of sample from the gel matrix onto supports suitable for gas-phase sequencing or related techniques (Vandekerckhove *et al.*, 1985; Aebersold *et al.*, 1986; Matsudaira, 1987; Bauw *et al.*, 1987 and Eckerskorn *et al.*, 1988). Direct NH_2-terminal sequencing is the most sensitive method (1-5 μg of protein), but, when gaps or ambiguous assignments are seen, verification of the sequence by other means often demands much more material. In many cases, proteins are either artefactually or naturally blocked. *In situ* membrane cleavages (Aebersold *et al.*, 1987) or in *matrix* gel digestions (Ward *et al.*, 1990), yield many peptides, most of which may be obtained in pure form and sequenced. The latter approach is less sensitive as more manipulations are involved (5-10 μg of protein), but is ideally adapted for the PCR-technology.

The success of these methods mainly depends on the ratio of protein to gel volume or blot surface. Therefore our efforts have been directed towards trying to develop a system that

Methods in Protein Sequence Analysis, Edited by K. Imahori and F. Sakiyama, Plenum Press, New York, 1993

would allow the concentration of proteins into small gel volumes, prior to electroblotting or "in gel" digestion. We have previously described a gel-elution-concentration system that permitted us to elute proteins from multiple spots collected from 2D-gels and reconcentrate them into a new gel (Rasmussen *et al.*, 1991). Using an improved system, we can now reproducibly concentrate proteins from ± 1.0 ml into spots of 5-10 mm^2. This represents a concentration factor of about 200, with yields which are higher than 80%.

We have used this technique to concentrate the proteins present in collected 2D-gel spots and have microsequenced up to 240 proteins purified from human cell lines and tissues. Here we describe in detail the gel-concentration procedure and illustrate this with the analysis of 2D-gel purified proteins.

MATERIALS AND METHODS

Bovine serum albumin, ovalbumin and dephosphorylated casein were obtained from Sigma Chemical Company, St. Louis, USA. Actin was purified from rabbit skeletal muscle as described (Spudich and Watt, 1971). Endolysine C and porcine trypsin (sequencing grade) were from Boehringer, Mannheim FRG. Coloured protein markers (rainbow markers), [γ-^{32}P]ATP and [2-^{14}C] iodoacetic acid were from Amersham International, England. PVDF membranes (Immobilon) were from Millipore Corporation, Bedford, USA. All other chemicals were Janssen Chimica, Geel, Belgium, Merck, Darmstadt, FRG or Serva, Heidelberg, FRG.

^{14}C Labelling of Proteins

Proteins were dissolved in 50 µl of 8 M urea/0.1 M Tris-HCl pH 8.8 at a concentration of 2 mg/ml and reduced with 2 µl of 5mM dithiothreitol (DTT) by heating at 50°C for 30 min. The samples were allowed to cool and the proteins precipitated with 6 volumes of absolute ethanol at -20°C for 1 h. The precipitated proteins were collected in an Eppendorf centrifuge (10,000 rpm x 5 min) and redissolved in 50 µl of 8 M urea/0.1 M Tris-HCl pH 8.8. The proteins were carboxymethylated by adding 1 µl of 22 mM [2-^{14}C] iodoacetic acid (specific radioactivity 56 µCi/µmol) and incubation at 25°C for 20 min. Finally, the labelled proteins were precipitated with ethanol, centrifuged, dissolved in 50 µl of Laemmli sample buffer and separated by SDS-PAGE as described below.

Phosphorylation of Casein

Casein (625 µg) was phosphorylated in a final volume of 0.2 ml of 50 mM Hepes, 5 mM $MgCl_2$, 0.1 mM EDTA, 1 mM DTT pH 7.5 containing 7 mUnits of cyclic AMP-dependent protein kinase catalytic subunits and 0.1 mM [γ-^{32}P] Mg-ATP (specific radioactivity 80 cpm/pmol). After 30 min at 37°C, the protein was precipitated with 1 ml of 10% (w/v) trichloroacetic acid, and dissolved in sample buffer for electrophoresis.

SDS-Polyacrylamide Gel Electrophoresis

Purified proteins (5-20 µg per track) were separated in 1.5 mm thick SDS-gels with acrylamide concentrations adjusted to the M_r of the protein : 10% acrylamide gels were used for the separation of bovine serum albumin (67,000 M_r) and 12.5% acrylamide gels were used for the separation of ovalbumin (43,000 M_r), actin (42,000 M_r) and casein (23,000 M_r). Gels were stained for 20 min in a solution of 0.25% Coomassie Brilliant Blue R250 (Serva, Heidelberg, FRG) in 45% methanol and 9% acetic acid. Destaining was done in 5%

methanol and 7.5% acetic acid. Protein bands were excised and washed by agitation in distilled water for 2 h. The washed protein bands were cut in 3 mm x 3 mm x 1.5 mm pieces and stored at -20°C. Prior to sample concentration in the second gel (see below) the gel pieces were combined and equilibrated for 1 h by submerging in 200 μl of sample buffer.

Protein Quantitation

The bands corresponding to the labelled proteins and labelled protein spots from stained concentration gels (see below) were cut out and dissolved in 0.4 ml of 30% vol. H_2O_2 for 2 h at 80°C. Radiolabel in the dissolved gel/H_2O_2-mixture was counted in 10 ml of Optiphase High Safe 3 scintillant (LKB).

Protein Recovery from 2D-Gels

Details concerning cell culture, 2-D gel electrophoresis, protein detection and computerized data analysis have been described in detail elsewhere (Celis *et al.*, 1990; Celis *et al.*, 1991). Protein spots from several (up to 40) Coomassie Brilliant Blue stained, dried 2-D gels - some of which were stored for more than one year - were cut out with a scalpel and combined. Protein extracts were routinely mixed with [^{35}S]methionine-labelled cellular proteins from AMA cells or keratinocytes in order to assign the corresponding number in the respective databases to the microsequenced polypeptide (Celis *et al.*, 1990). The dried gel pieces were rehydrated by incubation in water for 1 h, washed and then incubated at 4°C for 6-18 h in sample buffer (50mM DTT, 12 mM Tris-HCl pH 6.8, 10% glycerol, 0.1% Bromophenol Blue, 1% sodium dodecyl sulfate) and any decrease in the pH was corrected by adding 1 M Tris base. The rehydrated gel pieces and remaining sample buffer were boiled for 10 min and loaded into a gel slot of the concentration gel (see below).

Concentration Gels

Concentration gels were cast as shown in Figure 1A. The sample well was formed by a 1 cm x 2 mm thick spacer set between two parallel spacers 0.5 cm x 2 mm thick. The three slot-forming spacers were attached to the top edge of the back gel plate with adhesive tape. The separation gel, whose acrylamide concentration was determined by the M_r of the protein (see above), formed the lower gel with a height 5 cm from the bottom of the plate. Once set this was overlaid with a 5% acrylamide stacking gel whose level came up to 1 cm below the top edge of the front plate. The central spacer was removed leaving a well 1 cm x 4 cm x 2 mm, which was rinsed once with distilled water. The bottom spacer was removed and the slab gel was mounted in the electrophoresis tank. The pre-equilibrated gel pieces (cut at approximately 2 mm x 2 mm), together with sample buffer, were transferred to the slot, and overlaid with a suspension of similarly treated blank gel pieces taken from the same gel. The suspension was added so that its level was at least 5 mm above the level of the stacking gel. The well was then carefully filled up with electrophoresis buffer and the run was started at 250 V. When all the sample had entered the gel, the voltage was increased to 350 V and maintained until the tracking dye has entered the separation gel over about 1 cm. During electerophoresis the apparatus was cooled by fanning air over the glass plates.

Electroblotting and Digestion on PVDF Membranes

The separation gel containing the concentrated protein spot was electroblotted onto PVDF membranes (Bauw *et al.*, 1987) and stained with either 0.2% (w/v) Ponceau S in 1% (v/v) acetic acid or with 0.2% (w/v) Amido black in 1% (v/v) acetic acid, by shaking for 2 min with dye solution. The blots were then washed 3-4 times in distilled water and the protein spots were cut out and immediately immersed in a solution of 0.2% polyvinylpyrrolidone in water to block remaining protein adsorbing sites. *In situ* digestions on the PVDF-membranes were carried out in 0.2 ml of 0.1 M Tris-HCl pH 8.5 with 1 µg of trypsin or in 0.2 ml of 0.1 M Tris-HCl pH 8.8 with 0.3 µg of endolysine C, both at 37°C for 3-4 h. The digestion was stopped by addition of trifluoroacetic acid (2% final concentration).

Coomassie Staining and Digestion in the Gel

The separation gel was cut from the larger gel and stained with Coomassie Brilliant Blue (see above). It was further washed extensively with distilled water until neutrality. The protein spot was excised in a gel piece 3 mm x 2 mm x 2 mm and digested *in matrix* as described by Rosenfeld *et al.*, 1992.

Reverse Phase HPLC and Microsequencing

Peptides were separated by reverse phase HPLC on a C18 Vydac 2.1 x 250 mm column (Separations Group, Hesperia CA), equilibrated in 0.1% trifluoroacetic acid/5% acetonitrile (solvent A). A linear gradient program of 5-100% B in 100 min, where B = 0.1% trifluoroacetic acid/70% acetonitrile, was used to elute the peptides at a flow rate of 80 µl/min, and the eluate was monitored by absorbance at 214 nm using an Applied Biosystems model 759A Absorbance Detector. Peptides were collected in Eppendorf tubes and then directly applied on the gas-phase sequenator (Applied Biosystems model 470A), equipped with a 120A PTH-analyser.

RESULTS AND DISCUSSION

Figure 1B shows the construction of the concentration gel and describes the protein concentrating effect. In this system, proteins are electrophoretically eluted out of the gel pieces in the sample well and concentrated in a very small volume in a new gel. The concentration effect is obtained from a combination of protein stacking and a horizontal compression effect of the stacked protein band. The latter occurs when the conductivity within the sample is much lower than in the stacking gel and is augmented by keeping the sample separated from the rest of the gel by the vertical spacers.

Several points need attention in this system: (1). The gel pieces should be washed in water before applying the sample buffer. This reduces the acetic acid used in the staining and destaining procedure so that the pH should not be further adjusted or adjusted with minimal amounts of base. In addition this will also dilute the concentration of the sample buffer, lowering its conductivity. (2). The level of the piled gel pieces in the sample well should rise to ± 1 cm above the level of the stacking gel. Thus, the electrophoretic front in the sample will start with a ± 1 cm "handicap" versus the front in the stacking gel. This handicap will leave sufficient time for the electrophoretic front to move horizontally at the exit of the sample well and to meet the stacked protein front when it leaves the exit of the sample well. If there are not sufficient pieces available, then similarly treated blank gel pieces, cut from the same gel can be added until the suspension reaches the described level. (3). Gel pieces should be kept as small blocks and not be crushed because that could clog the sample slot by trapping air bubbles formed during the electrophoresis.

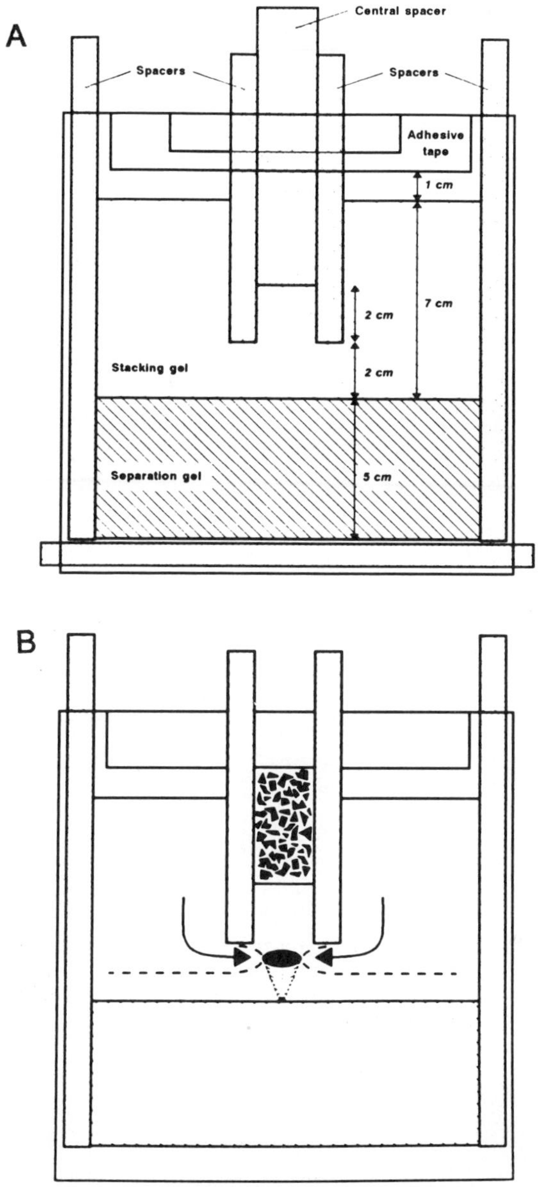

Figure 1. Construction of the concentration gel (A). The concentration gel system mounted in the slab gel apparatus (B). Note the horizontally moving electrophoresis front at the exit of the sample well causing a compression of the stacked protein band (for details see MATERIALS AND METHODS).

The volume of the well is about 0.8 ml, which is sufficient to accomodate about 20-40 gel pieces equilibrated in sample buffer. Sample concentration is not affected by the number of gel pieces, but when the pieces are heavily pressed in the well, the concentrated protein tends to streak out in a vertical line rather than forming a spot.

Figure 2 illustrates time-lapse photography of the migration of coloured proteins through the concentration gel. The coloured proteins were added to a suspension of gel pieces in the slot of the concentration gel. After running at 250 V, the proteins enter the stacking gel as a sharp band between the two vertical spacers (Figure 2a). At this point, the voltage is increased to 350 V. Note that the migration of the protein between the spacers is retarded with respect to the solvent front down the sides of the spacers (Figure 2b), due to the start handicap. The concentration of the sample begins when the solvent front on the outside passes the ends of the two vertical spacers (Figure 2c), and continues as the proteins are compressed into a narrow spot (Figure 2d). As the spot enters the separation gel, the proteins are separated into their individual components (Figure 2e, f).

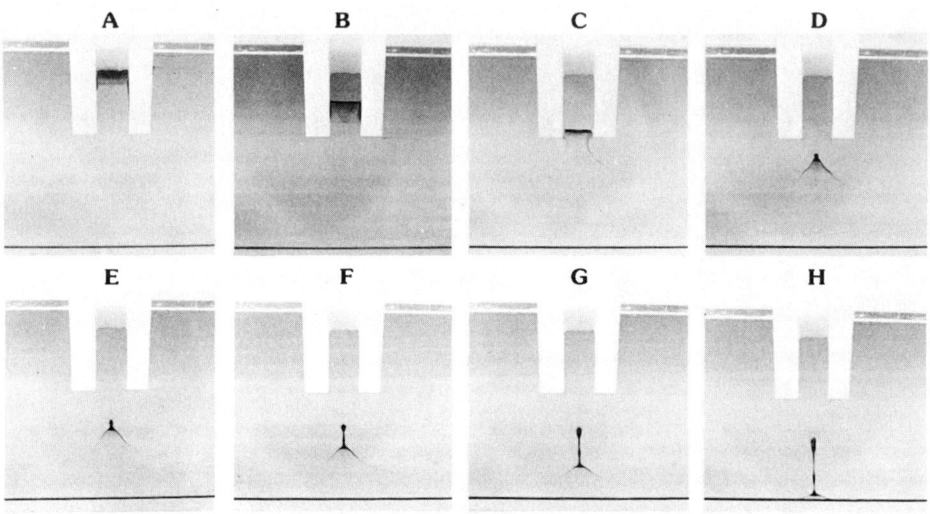

Figure 2. Time-lapse photography of the migration of coloured proteins through the concentration gel. Coloured protein markers were loaded in the sample well, together with blank polyacrylamide gel pieces. For this particular experiment, the sample buffer did not contain bromophenolblue. The concentration gel was run as described in "RESULTS AND DISCUSSION". Pictures were taken at different time intervals. Notice the concentration effect seen between C and D.

In order to measure the recovery of proteins from 1-D gels after the concentration gel, casein was phosphorylated with ^{32}P and purified in a 1-D gel with a thickness of 1.5 mm. After running the concentration gel, the recovery of ^{32}P-labelled casein was about 80%. The same recovery was obtained when casein was separated in several mini-gels rather than a 1.5 mm slab gel. In these experiments, one could argue that the yields were not higher because of the dephosphorylation of casein in the gel. Similarly the recoveries in concentration gels of ^{14}C-labelled proteins were greater than 80% for proteins having a range of molecular masses from 23,000-67,000. The recovery was about the same for 5 µg or 20 µg of the different proteins (data not shown).

The concentration gel technique was used for actin to compare the peptide yields from digests of actin on PVDF membranes after electroblotting, with digests in the gel *matrix*. After running a concentration gel with gel pieces containing 20 µg of actin, the protein was electroblotted onto PVDF and digested with endolysine C. The yields of the peptide peaks were not affected by the use of either Amido Black or Ponceau S to stain the protein on the PVDF membrane. Actin (20 µg) was also localized in the concentration gel after staining with Coomassie Blue. A gel piece about 3 mm x 3 mm x 2 mm was cut out and digested with endolysine C. The peptide yields from actin digested in the gel (Figure 3A), were greater than those obtained after digestion on PVDF (Figure 3B). This experiment illustrates the general outcome when blot *in situ* digestion yields are compared with peptide recoveries after digestion in the gel matrix. As illustrated in Figures 3A and 3B, in matrix gel digestion generally results (1) in higher peptide recoveries and (2) in a larger number of peptides. Especially, when large proteins are analysed this could cause a problem in the peptide purification procedure.

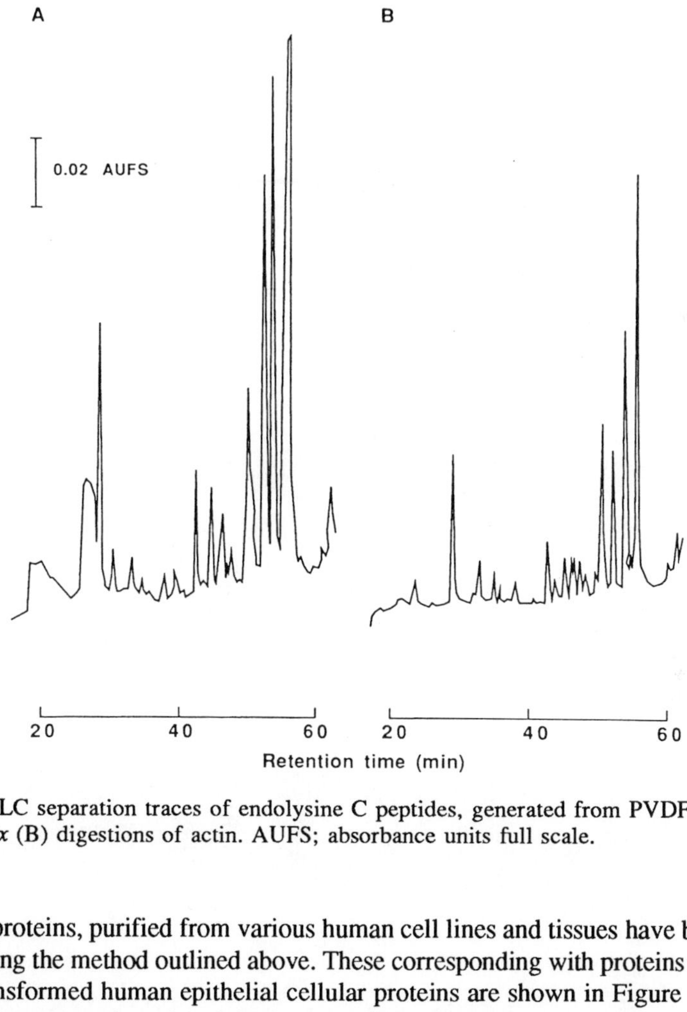

Figure 3. HPLC separation traces of endolysine C peptides, generated from PVDF -- *in situ* (A) or gel *in matrix* (B) digestions of actin. AUFS; absorbance units full scale.

Up to 240 proteins, purified from various human cell lines and tissues have been partially sequenced using the method outlined above. These corresponding with proteins in the master 2D-gel of transformed human epithelial cellular proteins are shown in Figure 4. Several of these proteins represent less than 0.01% of the total cellular protein content. About 50% of the proteins were identified on the basis of homology or identity with sequences recorded in protein sequence data bases. In combination with other identification methods, the number of characterized proteins now amounts to more than 500 (Rasmussen *et al.*, 1992). This large number of identified proteins opens the possibility to study integral biochemical pathways at the level of coordinated regulation of the participating enzymes. This 2D-gel approach has also been used to identify marker proteins for pathological conditions. The method is also particularly suited for studying the effects of pharmaceutical drugs on enzyme regulation in various tissues and cultured cells.

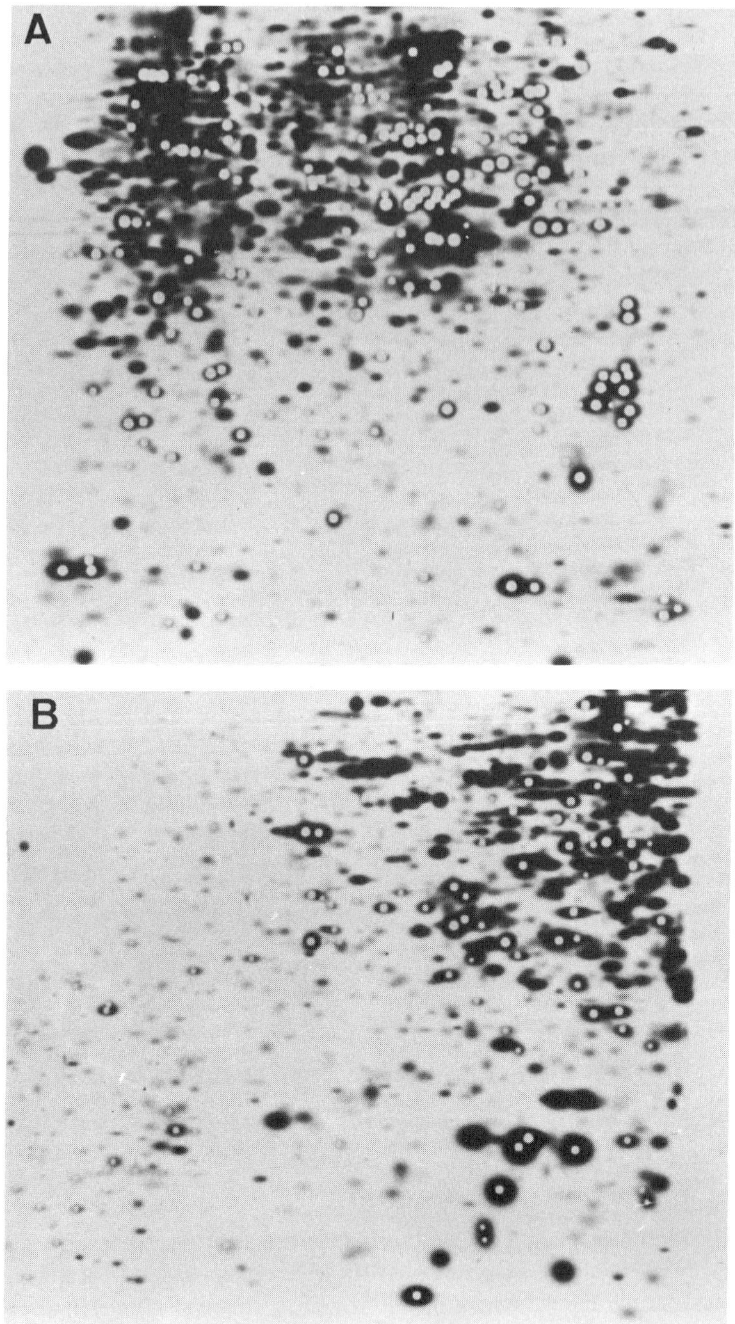

Figure 4. Images of 2D-gels for transformed human epithelial amnion cellular proteins resolved by (A) IEF and (B) NEPHGE. Microsequenced proteins are flagged with white spots.

Acknowledgements: The group in Aarhus was supported by grants from the Danish Biotechnology Programme, the Danish Cancer Society, the Danish Medical Research Council, the Danish Rheumatoid Society, the Velux Fund, NOVO Nordish and the Danish

Foundation for Advancement of Medical Sciences. The work in Ghent was supported by grants from the Belgian National Fund for Scientific Research, the Commission of the European Community and the Program Concerted Research Action of the Flemish Community.

REFERENCES

Aebersold,, R. H., Teplow, D. B., Hood, L. E. and Kent, S. B. H., 1986, Electroblotting onto activated glass, *J. Biol. Chem.* 261:4229.

Aebersold, R. H., Leavitt, J., Saavedra, R. A., Hood, L. E. and Kent, S. B. H., 1987 Internal amino acid sequence analysis of protein separated by one or two dimensional gel electrophoresis after *in situ* protease digestion on nitrocellulose, *Proc. Natl. Acad. Sci. USA* 84:6970.

Bauw, G., De Loose, M., Inzé, D., Van Montagu, M. and Vandekerckhove, J., 1987 Alterations in the phenotype of plant cells studied by NH_2-terminal amino acid-secuence analysis of proteins electroblotted from two-dimensional gel-separated total extracts, *Proc. Natl. Acad. Sci. USA* 84:4806.

Celis, J. Bravo, R., 1984 (eds.) Two Dimensional Gel Electrophoresis of Proteins: Methods and Applications, Academic Press, New York.

Celis, J. E., Gesser, B., Rasmussen, H.-H., Madsen, P., Leffers, H., Dejgaard, K., Honoré, B., Olsen, E., Ratz, G., Lauridsen, J. B., Basse, B., Mauritzen, S., Hellerup, M., Andersen, A., Walbum, E., Celis, A., Bauw, G., Puype, M., Van Damme, J. and Vandekerckhove, J., 1990 Comprehensive two-dimensional gel protein databases offer a global approach to the analysis of human cells : The transformed amnion cells (AMA) master database and its link to genome DNA sequence data, *Electrophoresis* 11:989.

Celis, J. E., Rasmussen, H.-H., Leffers, H., Madsen, P., Honoré, B., Gesser, B., Dejgaard, K. and Vandekerckhove, J., 1991 Human cellular protein patterns and their link to genome DNA sequence data : usefulness of two-dimensional gel electrophoresis and microsequencing, *FASEB J.* 5:2200.

Eckerskorn, C., Mewes, W., Goretzki, H. and Lottspeich, F., 1988 A new siliconized-glass fiber as support for protein-chemical analysis of electroblotted proteins, *Eur. J. Biochem.* 176:509.

Matsudaira, P. J., 1987 Sequence from picomole quantities of proteins electroblotted onto polyvinylidene difluoride membranes, *J. Biol. Chem.* 262:10035.

Rasmussen, H.-H., Van Damme, J., Bauw, G., Puype, M., Gesser, B., Celis, J., and Vandekerckhove, J., 1991 Protein-electroblotting and microsequencing in establishing integrated human protein databases, *in*: Jörnvall, H., Höög, J.-O. and Gustavson, A.-M., eds, Mehtods in Protein Sequencec Analysis, Birkhauser Verlag, Basel, pp. 103-114.

Rasmussen, H.-H., Van Damme, J., Puype, M., Gesser, B., Celis, J. and Vandekerckhove, J., Microsequences of 145 proteins recorded in the two-dimensional gel protein database of normal human epidermal keratinocytes, *Electrophoresis,* in press.

Rosenfeld, J., Capdevieille, J., Cuillemot, J. C. and Ferrara, P., 1992 In-gel digestion of proteins for internal sequence analysis after one- or two-dimensional gel electrophoresis, *Anal. Biochem.* 203:173.

Spudich, J. A. and Watt, S., 1971 The regulation of rabbit skeletal muscle contraction, *J. Biol. Chem.* 246:1866.

Vandekerckhove, J., Bauw, G., Puype, M., Van Damme, J. and Van Montagu, M., 1985 Protein-blotting on Polybrene-coated glass-fiber sheets. A basis for acid hydrolysis and gas-phase sequencing of picomole quantities of protein previously separated on sodium dodecyl sulfate/polyacrylamide gel, *Eur. J. Biochem.* 152:9.

Ward, L. D., Reid, G. E., Moritz, R. L. and Simpson, R. J., 1990 Peptide mapping and internal sequencing of proteins from acrylamide gels, *in*: Villafranca, J. J., ed., Current Research in Protein Chemistry: Techniques, Structure and Function, Academic Press, San Diego, pp. 179-190.

CAPILLARY ELECTROPHORESIS IN STRUCTURAL CHARACTERIZATION OF POLYPEPTIDES

Tomas Bergman

Department of Chemistry I, Karolinska Institutet
S-104 01 Stockholm, Sweden

INTRODUCTION

Capillary zone electrophoresis is efficient in peptide analysis, allowing short separation times, high resolution and high sensitivity (cf. Karger *et al.*, 1989). These features make the technique attractive also for preparative attempts in structural analyses, but the small volumes (nl) and the correspondingly small amounts (fmol) have long been considered a difficulty in preparative applications. However, recently we showed that capillary zone electrophoresis can be used to recover peptides for structural analysis (Bergman *et al.*, 1991; Bergman and Jörnvall, 1992; Bergman, 1992). Sufficient amounts for sequence analysis are collected in only a few preparative separations, while single runs yield enough material for sensitive amino acid analysis by the phenylthiocarbamyl (PTC) derivatization method (cf. Bergman *et al.*, 1986). Preparative results with similar capillary electrophoresis systems have also been reported by others (Smith *et al.*, 1992; Omori, 1992). In addition, a finding that sensitive detection of underivatized amino acids is possible by capillary zone electrophoresis, provides novel opportunities for direct analysis of results from digestions with exopeptidases (Bergman *et al.*, 1991). Apart from rapid purity checks of preparations from reverse phase HPLC or conventional chromatography (Large, 1991; Agerberth *et al.*, 1991), capillary zone electrophoresis can furthermore provide support for the identity of an isolated peptide through comigration experiments with synthetic replicas (Sillard *et al.*, 1992).

This contribution supplies results on the use of capillary zone electrophoresis in analysis of complex mixtures of small peptides, protein digests and HPLC-purified peptides. Data from amino acid and sequence analyses of polypeptides collected in a limited number of preparative runs are also given. We now demonstrate the separation of intact proteins in uncoated capillaries without the addition of detergents or polymers which, when included, make preparations incompatible with subsequent structural analyses. Finally, separation and sensitive detection of underivatized amino acids is presented.

Methods in Protein Sequence Analysis, Edited by K. Imahori
and F. Sakiyama, Plenum Press, New York, 1993

MATERIALS AND METHODS

Proteins and peptides were prepared from natural sources or by solid phase peptide synthesis (cf. Kent, 1988). Bovine insulin B chain was carboxymethylated and purified as described (Bergman *et al.*, 1991). Proteolytic digestions were performed with Glu-C endoprotease from *Staphylococcus aureus* (Boehringer) and chymotrypsin (Merck). For HPLC separation of chymotryptic peptides, a Vydac C8 column was developed at 1 ml/min with a linear gradient of acetonitrile in 0.1% aqueous trifluoroacetic acid. A calibration mixture of proteins in the range 14.4 - 94 kDa was from Pharmacia LKB.

Capillary zone electrophoresis in 50 mM phosphate buffer, pH 2.5 or 7.0, was performed with a Beckman P/ACE 2000 instrument, equipped with an IBM PS/2 Windows-based control including the System Gold (Beckman) software. Capillaries of fused silica with 75 μm inner diameters and 57 cm total lengths were used. The ultraviolet detector (200 or 214 nm) was positioned 7 cm from the outlet end. Samples were injected by pressure and electrophoresis was conducted at 20-25 kV. The temperature was thermostatically controlled at 25°C. In preparative runs, the separated polypeptides were recovered by programmed positioning at the capillary outlet of conically shaped microvials containing 10 μl phosphate buffer.

Amino acid analysis of polypeptides prepared by capillary zone electrophoresis was performed as described (Bergman *et al.*, 1986), and sequence analysis with an Applied Biosystems 470A instrument, utilizing PTH analysis by reverse phase HPLC (Kaiser *et al.*, 1988).

RESULTS AND DISCUSSION

The analytical power of capillary zone electrophoresis for small peptides is demonstrated by separation of an HPLC-purified brain extract containing di-, tri- and tetrapeptides in amounts from 2 pmol to 100 fmol (Figure 1). When analyzed by reverse phase HPLC with 500-fold more material, only one broad peak is apparent (data not shown). In contrast, capillary zone electrophoresis resolves the sample into more than ten components (Figure 1A). In order to identify a particular tripeptide, that peptide was synthetically prepared and mixed with the biological sample. The synthetic tripeptide comigrates with one of the components (Figure 1B), and this also applies at a different pH (data not shown). Thus, such comigration studies with synthetic peptides can provide strong positive evidence for the presence of a specific structure in a complex mixture of similarly sized peptides.

Protein digests are readily analyzed by capillary zone electrophoresis in amounts as small as 100 - 200 fmol, or less, and typically a complete peptide map is obtained within 20 min. Sensitivity studies with 1-2 kDa peptides reveal that 10 fmol is possible to detect. Capillary zone electrophoresis of a Glu-C endoprotease digest of insulin B chain (200 fmol) is complete within 13 min (Figure 2) which confirms the analytical strength. However, the technique is applicable also to preparative work. In this case, aliquots of protein digests are concentrated to approximately 40 pmol/μl final solution, through lyophilization and resolubilization in a lesser volume of electrolyte buffer (2 to 3 μl are sufficient for injections). As shown before (Bergman *et al.*, 1991; Bergman and Jörnvall, 1992), the digest separated in Figure 2 was concentrated to 36 pmol/μl and peptides 2 and 3 were collected in four and five preparative runs, respectively. Amounts recovered were 19 pmol (peptide 2) and 24 pmol (peptide 3), showing that 5 pmol readily can be injected without overloading the system. The preparations were directly applied to sequencers and interpretations were possible up to the C-termini of both peptides. The data for peptide 3 are given in Table 1.

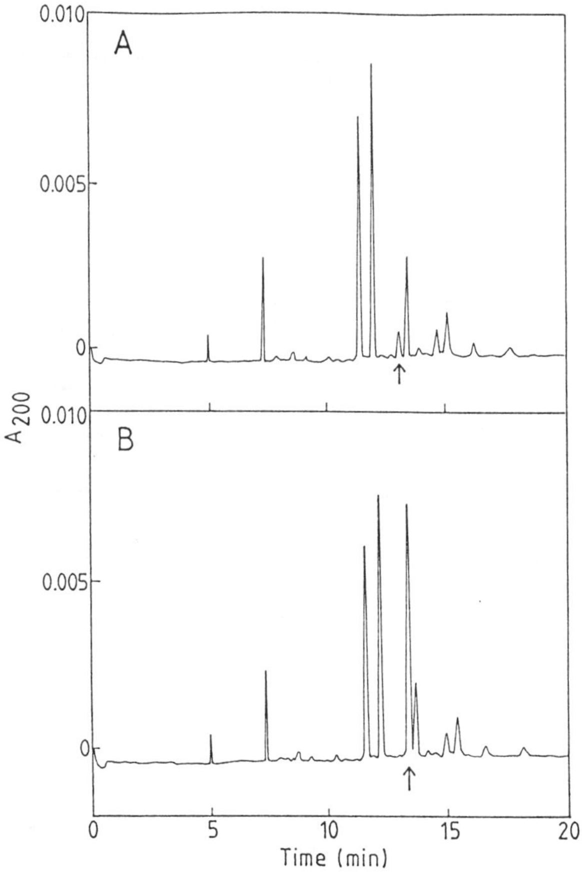

Figure 1. Capillary zone electrophoresis of di-, tri- and tetrapeptides (2 pmol - 100 fmol) from an HPLC-purified brain extract.
A: Analytical separation. B: Sample in A mixed with a synthetic tripeptide (arrow).

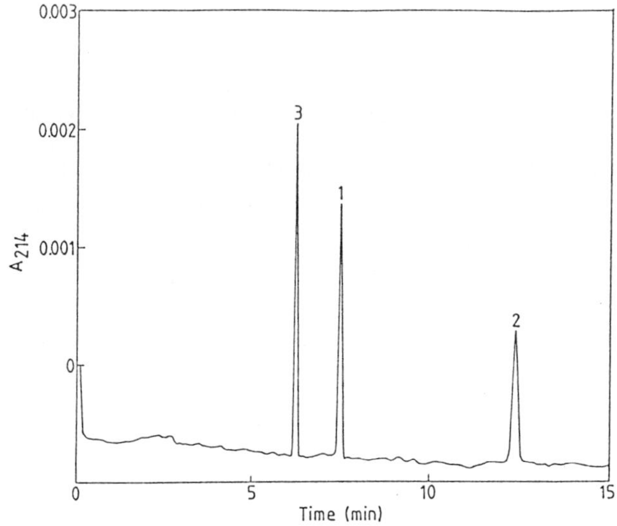

Figure 2. Separation by capillary zone electrophoresis of a Glu-C endoprotease digest of insulin B chain (200 fmol).
The peaks correspond to: 1, residues 1-13; 2, residues 14-21; 3, residues 22-30.

The separation principles of capillary zone electrophoresis, i.e. electrophoretic mobility and electroosmotic flow, can be employed to efficiently analyze the homogeneity of peptide fractions separated according to hydrophobicity by reverse phase HPLC. Capillary zone electrophoresis of such a fraction (Figure 3, trace A), resolves a single HPLC peak from a reverse phase separation of a chymotryptic digest into two components. The major component (peptide 1) was collected in five preparative runs after a 10-fold concentration to about 40 pmol/µl. It was then reseparated by capillary zone electrophoresis (Figure 3, trace B) followed by amino acid and sequence analysis (Table 2).

Table 1. N-terminal sequence analysis of peptide 3 (Figure 2) prepared by capillary zone electrophoresis.

Cycle	Residue	Amount recovered (pmol)
1	Arg	4
2	Gly	6
3	Phe	7
4	Phe	7
5	Tyr	7
6	Thr	2
7	Pro	4
8	Lys	3
9	Ala	2

Peptide 3 was collected in five runs. Before preparative runs, the digest was concentrated 24-fold by volume reduction. The total amount (24 pmol) was applied to a gas-phase sequencer. (From Bergman *et al.*, 1991.)

Table 2. N-terminal sequence analysis of peptide 1 (Figure 3) prepared by capillary zone electrophoresis.

Cycle	Residue	Amount recovered (pmol)
1	Thr	4
2	Cys	4
3	Lys	7
4	Gly	8
5	Lys	6
6	Pro	5
7	Thr	3
8	Gln	3
9	Gln	4
10	Phe	2

Peptide 1 was collected in five runs after concentration. The recovered material was divided; 75% was applied to sequence analysis and 25% to PTC-amino acid analysis. The result of the amino acid analysis reveals that 16 pmol of peptide 1 was prepared. Cys in cycle 2 was detected as the PTH derivative of carboxymethylcysteine.

Intact proteins are difficult to separate in uncoated capillaries without additives (e.g. detergents). The interaction with the negatively charged silica wall is often strong, causing severe zone broadening, and resulting in poor resolution. This effect is especially pronounced at low pH and can be diminished by a neutral coat on the glass wall or by raising the pH. On the other hand, coating also reduces the electroosmotic flow, which is one of the factors influencing separation, and at high pH the electroosmotic flow in uncoated capillaries is too rapid for sufficient resolution. To find optimal conditions for the preparation of proteins in uncoated capillaries, a wide variety of electrolytes and pH values

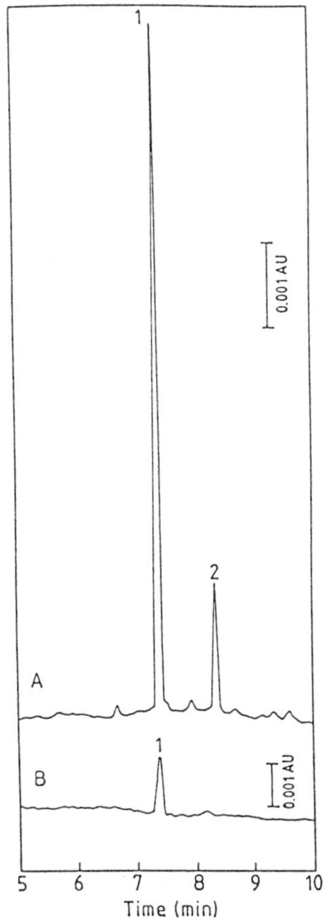

Figure 3. Capillary zone electrophoresis of an HPLC fraction from a reverse phase separation of a chymotryptic digest. The analytical separation (A) reveals two peptides previously unresolved by HPLC. Peptide 1 was collected and reseparated by capillary zone electrophoresis (B). Absorbance at 200 nm as indicated.

(from 1 to 12) were tested. The criterion was compatibility of fractions with subsequent structural analyses, i.e. absence of detergents, salts or other additives. Promising results were obtained with a 50 mM phosphate buffer at pH 7.0. Proteins in the range 14.4 - 94 kDa migrate as sharp peaks that are separated in less than 6 min (Figure 4). The sensitivity is high, about 2.5 ng of each protein (corresponding to 30 - 200 fmol) was injected, resulting in 0.003 - 0.03 absorbance units at 200 nm. Protein 1 (Figure 4) was collected in eight preparative runs and subsequently identified as α-lactalbumin by sequence analysis

(Table 3). The total amount prepared was estimated from the sequencer data to be 4 - 5 pmol and the repetitive yield in the Edman degradation was 97%. Precision and reproducibility in repetitive collections were high with a variation in migration time less than 0.3%.

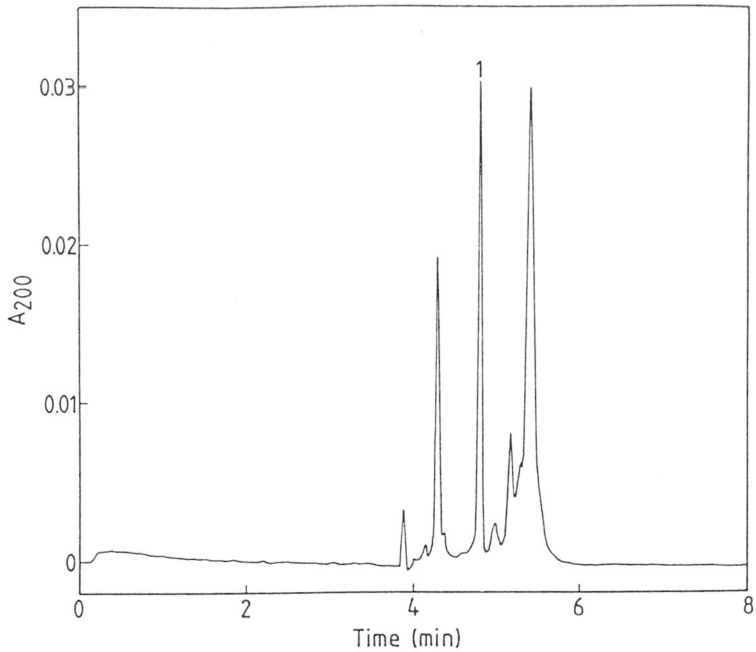

Figure 4. Capillary zone electrophoresis of proteins in the range 14.4-94 kDa (30-200 fmol) at pH 7.0.
Protein 1 was collected and identified as α-lactalbumin (below).

Table 3. N-terminal sequence analysis of α-lactalbumin (protein 1, Figure 4) prepared by capillary zone electrophoresis at pH 7.0.

Cycle	Residue	Amount recovered (pmol)
1	Glu	4.3
2	Gln	1.2
3	Leu	1.8
4	Thr	1.4
5	Lys	1.0
6	Cys[a]	-
7	Glu	1.6
8	Val[a]	-
9	Phe	2.0
10	Arg[a]	-
11	Glu	1.6
12	Leu	1.4

α-lactalbumin was collected in eight preparative runs and analyzed for 12 cycles with a gas-phase sequencer.
[a]Expected residue. Recovery insufficient to allow detection.

An interesting application is the separation and direct determination of underivatized amino acids. Ultraviolet detection at the pmol level or even in the fmol-range (less than 200 fmol for aromatic amino acids) is possible (Bergman *et al.*, 1991). A five-component amino acid mixture is completely resolved by capillary zone electrophoresis in less than 20 min (Figure 5). Structurally similar amino acids, like Ser and Thr, or amino acids with amide groups or other modifications not easily distinguished with conventional methods, are well resolved, as shown for Pro and $ProNH_2$ (Figures 5,6). This direct analysis can be used for rapid interpretation of results from carboxypeptidase digestions, as demonstrated with a 39-residue intestinal peptide (Bergman *et al.*, 1991; Agerberth *et al.*, 1991). Time aliquots were lyophilized and redissolved in electrolyte buffer, followed by capillary zone electrophoresis. This gave three residues, $ProNH_2$, Phe and Arg, of the peptide C-terminal sequence. A separation including these amino acids is shown in Figure 6.

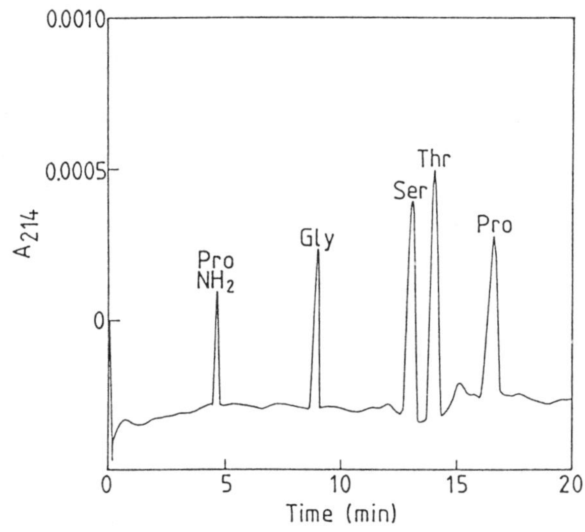

Figure 5. Separation by capillary zone electrophoresis of underivatized amino acids showing baseline resolution at high sensitivity of amidated and other similar species. Amounts: $ProNH_2$, 5 pmol; remaining amino acids, 40 pmols.

In conclusion, the results show that capillary zone electrophoresis has a high potential for peptide, protein and amino acid analyses. Like HPLC, the technique will have a strong impact on sample preparation and analysis in structural characterization of polypeptides.

Acknowledgements. This work was supported by grants from the Swedish Medical Research Council (project 03X-3532), the Swedish Cancer Society (project 1806), the Magn. Bergvall Foundation and the Swedish Society for Medical Research.

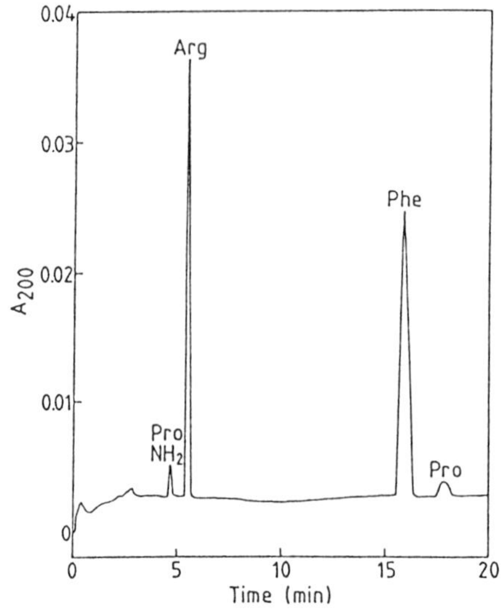

Figure 6. Capillary zone electrophoresis of underivatized amino acids showing the rapid migration of Arg, the extreme sensitivity for Phe, and a separation time exceeding 10 min for $ProNH_2$ versus Pro. Amounts: $ProNH_2$, 24 pmol; Arg, 75 pmol; Phe, 19 pmol; Pro, 120 pmol.

REFERENCES

Agerberth, B., Lee, J.-Y., Bergman, T., Carlquist, M., Boman, H. G., Mutt, V. and Jörnvall, H., 1991 Amino acid sequence of PR-39: isolation from pig intestine of a new member of the family of proline-arginine-rich antibacterial peptides, *Eur. J. Biochem.* 202:849.

Bergman, T., Carlquist, M. and Jörnvall, H., 1986 Amino acid analysis by high performance liquid chromatography of phenylthiocarbamyl derivatives, *in*: Advanced Methods in Protein Microsequence Analysis, Wittmann-Liebold, B., Salnikow, J. and Erdmann, V.A., eds., Springer, Berlin/Heidelberg, p.45.

Bergman, T., Agerberth, B. and Jörnvall, H., 1991 Direct analysis of peptides and amino acids from capillary electrophoresis, *FEBS Lett.* 283:100.

Bergman, T. and Jörnvall, H., 1992 Capillary electrophoresis for preparation of peptides and direct determination of amino acids, *in*: Techniques in Protein Chemistry III, Angeletti, R.H., ed., Academic Press, Orlando, p.129.

Bergman, T., 1992 Application of capillary electrophoresis for protein sequence analysis, *J. Protein Chem.* 11:356.

Kaiser, R., Holmquist, B., Hempel, J., Vallee, B. L. and Jörnvall, H., 1988 Class III human liver alcohol dehydrogenase: a novel structural type equidistantly related to the class I and class II enzymes, *Biochemistry* 27:1132.

Karger, B. L., Cohen, A. S. and Guttman, A., 1989 High-performance capillary electrophoresis in the biological sciences, *J. Chromatogr.* 492:585.

Kent, S. B. H., 1988 Chemical synthesis of peptides and proteins, *Annu. Rev. Biochem.* 57:957.

Large, T., 1991 Capillary electrophoresis: a new dimension in the separation sciences, *in*: Methods in Protein Sequence Analysis, Jörnvall, H., Höög, J.-O. and Gustavsson, A.-M., eds., Birkhäuser, Basel, p.79.

Omori, A., 1992 Application of microsequencing analysis on peptides and proteins from capillary zone electrophoreses, *J. Protein Chem.* 11:357.

Sillard, R., Rökaeus, Å, Xu, Y., Carlquist, M., Bergman, T., Jörnvall, H. and Mutt, V., 1992 Variant forms of galanin isolated from porcine brain, *Peptides*, in press.

Smith, A., Kenny, J. W. and Ohms, J. I., 1992 Micropreparative separation of tryptic digests by capillary electrophoresis and characterization by protein sequencing, *in*: Techniques in Protein Chemistry III, Angeletti, R.H., ed., Academic Press, Orlando, p.113.

SENSITIZATION OF GAS-PHASE PROTEIN SEQUENCER USING FLUORESCEIN ISOTHIOCYANATE (FITC)

Koji Muramoto[1], Kiyoshi Nokihara[2,3], Akira Ueda[2] and Hisao Kamiya[1]

[1]*School of Fisheries Sciences, Kitasato University, Sanriku, Iwate,*
[2]*Biotechnology Instruments Department, Shimadzu Corp., Kyoto, 604,*
[3]*Tokyo University of Agriculture and Technology, Koganei, 184, Japan*

INTRODUCTION

Since gas-phase sequencers were introduced into the market, it has become possible to analyze the amino acid sequence with 10-100 pmol of peptides or proteins (Hewick *et al.*, 1981). Further sensitization of sequence analysis has become one of the most important targets together with the development of micropurification methods and blotting technology with two dimensional gel electrophoresis. For sensitive detection of the amino acid derivatives resulting from the Edman degradation of peptides or proteins, several chromophoric or fluorescent analogs of the Edman-type reagents have been introduced. However, they are still not utilized routinely for automated sequencers. Fluorescein isothiocyanate (FITC) was used for manual sequencing to allow sensitized detection (Maeda and Kawauchi, 1968). Separation and identification of less than 50 fmol of fluorescein thiohydantoin (FTH-) amino acids using reversed-phase HPLC has been reported (Muramoto *et al.*, 1978). However, there were several difficulties in adapting the FITC-method for automated microsequencing, which included the necessity of obtaining pure FITC, of removing excess reagents as well as by-products without loss of samples, and of improving the poor coupling yield of FITC with samples. We have established the preparation of isomer free FITC and an optimized protocol for FITC reaction and the washing protocol. In the present paper we describe the adaptation of FITC-chemistry to provide a modified gas-phase sequencer with higher sensitivity. With the present instrument we could identify up to 20-25 amino acid residues from the N-terminus of 1-5 pmol of proteins.

MATERIALS AND METHODS

Heptane, 5% phenylisothiocyanate (PITC) in heptane, trifluoroacetic acid (TFA), 25%

Methods in Protein Sequence Analysis, Edited by K. Imahori
and F. Sakiyama, Plenum Press, New York, 1993

TFA, ethyl acetate and 12% trimethylamine (TMA) and sperm whale myoglobin were sequencer grade from Wako Pure Chemicals Co. Ltd. Acetone was distilled over potassium permanganate and redistilled after treatment with neutral alumina and activated charcoal. FITC and FTH-amino acids were prepared in our laboratory. Polyvinylidene difluoride (PVDF) membrane (Immobilon-P^{SQ}) and egg white lysozyme were obtained from Millipore Corp.

Sodium dodecylsulfate polyacrylamide gel electrophoresis (SDS-PAGE) and electroblotting were performed as summarized (Nokihara and Herbst, 1989).

A gas-phase protein/peptide sequencer, Shimadzu Model PSQ-1, was modified as follows: FITC-solution, which could be used for at least one month at room temperature without any problems, was delivered from an auxiliary bottle (A1), the reaction chamber was miniaturized and an on-line gradient HPLC system was combined for detection of FTH-amino acids with a fluorescence monitor (Shimadzu Model RF-550). The following reagents and solvents were used:

[Coupling]	A1: 0.1% FITC in acetone/ethyl acetate (1:9), R1: 5% PITC in heptane, R2: 12% TMA
[Extraction]	S1: Acetone, S2: Ethyl acetate, S3: Ethyl acetate
[Cleavage]	R3: TFA
[Conversion]	R4: 50% TFA, S4: 50% Acetone in water containing dithiothreitol (2 mg/100 ml)

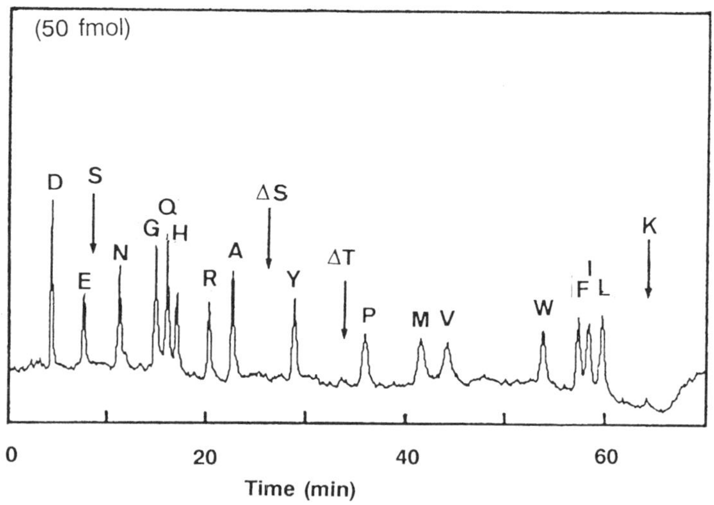

Figure 1. Chromatogram of standard FTH-amino acids (50 fmol).

Samples were applied to the reaction chamber in several ways: (1) samples were spotted onto a polybrene-coated (PB) glass filter (GF) disk (8 mm); (2) samples were spotted onto pieces of PVDF membrane (2.5 x 2.5 mm); and (3) samples were separated by polyacrylamide gel electrophoresis and electroblotted onto PVDF membrane. An excised PVDF was placed in the middle of an untreated GF disk in the reaction chamber. The reaction chamber was washed with ethyl acetate and acetone (GF), or with ethyl acetate alone (PVDF).

FTH-amino acids generated at each cycle were analyzed by on-line HPLC (Hypersil ODS column 3 μm, 4.6 x 50 mm) using a linear gradient of acetone in 10 mM sodium phosphate (pH 7.0) at 60 °C at a flow rate of 1.0 ml/min and detected by fluorescence (Ex:

490 nm, Em: 520 nm). The standard FTH-amino acids are indicated in Figure 1. The detection limit was less than 10 fmol.

RESULTS AND DISCUSSION

The replacement of PITC by FITC, which is the most sensitive and widely used fluorescent reagent, should increase the sensitivity for detection of amino acids derived from the Edman reaction. FITC shows lower reactivity with amino groups of the N-terminal residues and it is difficult to remove excess reagents as well as their by-products in the Edman degradation of liquid-phase sequencing. For the application of FITC in gas-phase sequencing, we initially evaluated the reactivity of FITC with proteins on PB-GF disks and small pieces of PVDF membranes. Figure 2 shows chromatograms for sequencing myo-

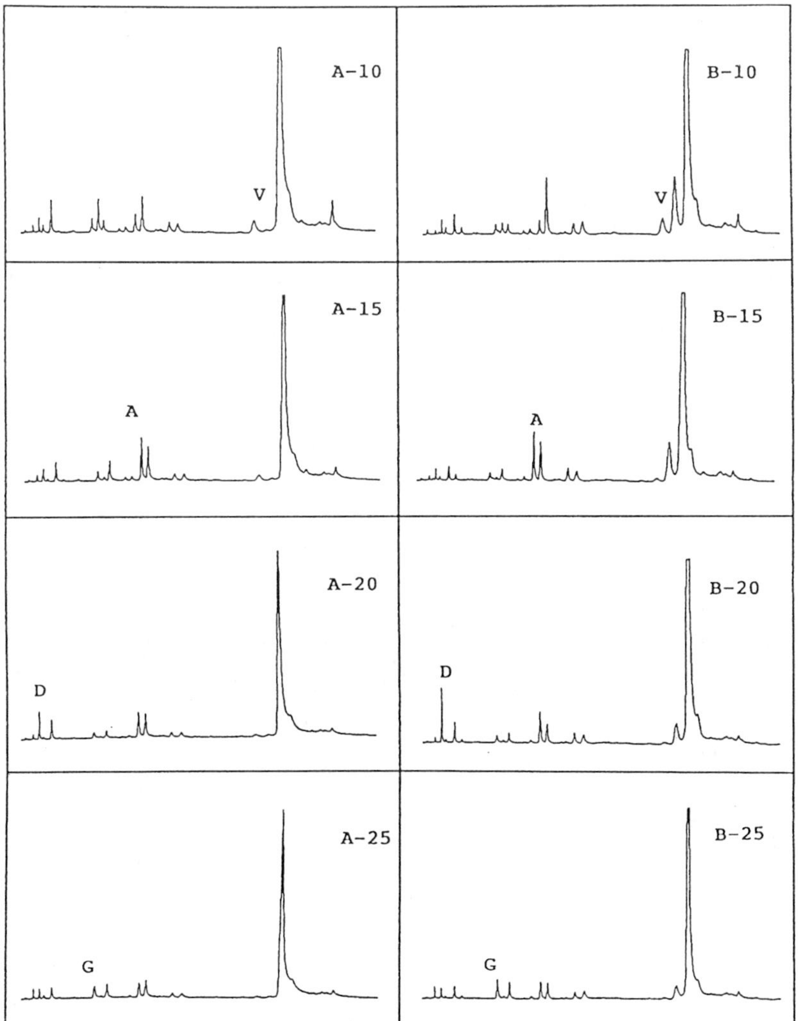

Figure 2. Sequence analysis of sperm whale myoglobin (Polybrene-coated glass fiber). Left: FITC single coupling, Right: FITC/PITC double coupling.

globin (50 pmol) on a PB-GF disk by FITC single coupling and FITC/PITC double coupling. FTH-Asp at the 20th residue could be identified even in the FITC single coupling method. The results indicate that the coupling reaction of FITC with proteins on filter disks proceeds effectively. FITC, which is a solid state reagent, was more suitable for gas-phase sequencing than for the classic liquid-phase mode. The use of the gas-phase mode can result in increased concentration of the solid state Edman-type reagent, such as FITC, on the filter disks where the coupling reaction takes place, ensure that protein samples and FITC are in

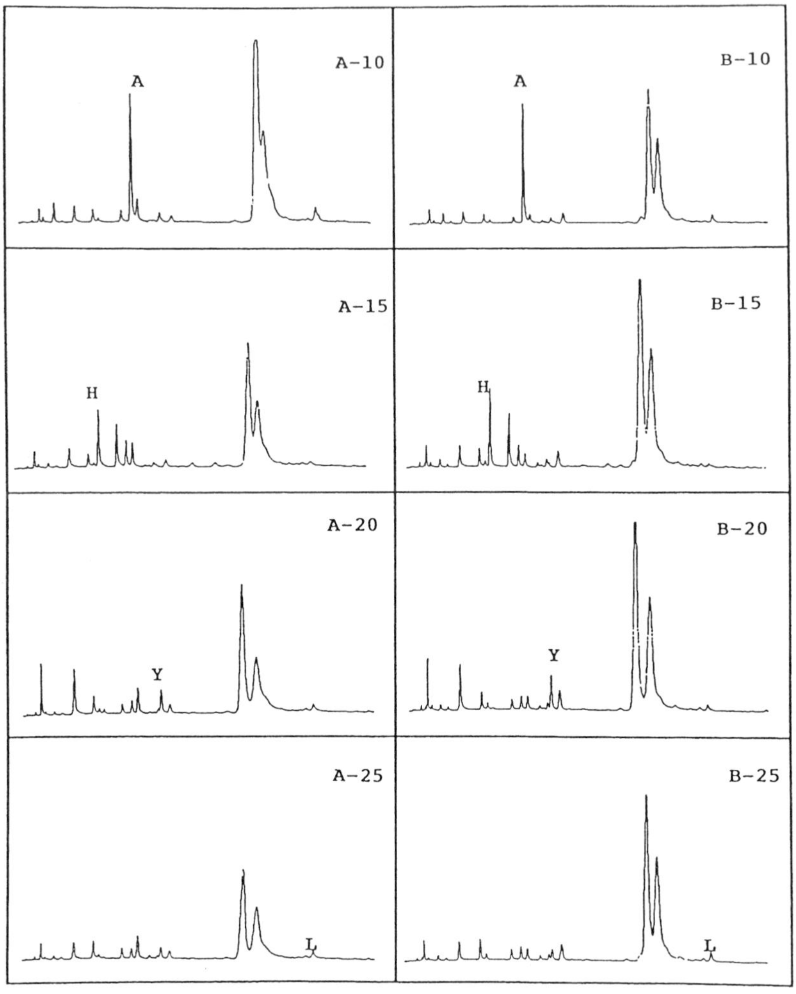

Figure 3. Sequence analysis of chicken lysozyme (Polybrene-coated glass fiber filter). Left: FITC single coupling, Right: FITC/PITC double coupling.

close proximity keeping the amount of FITC minimal. The repetitive yields by FITC single coupling was 83% (Table 1). The double coupling with FITC followed by PITC increased the repetitive yield to 89% and decreased overlapping as shown in Figure 2. FITC was supplied as a 0.1% solution as described above; however, increased concentration of FITC did not increase the yields and caused higher background on chromatograms. The Edman degradation of lysozyme (50 pmol) by the single and the double coupling method was accomplished with repetitive yields of 87% and 86%, respectively (Figure 3). Similar results were obtained with 25 pmol of myoglobin and lysozyme spotted onto a PVDF membrane (Table 1). The total cycle time of the present optimized sequencing in the modified PSQ-1 with the FITC/PITC double coupling method is 81.5 min. Extensive washing steps were required to remove excess FITC. Acetone was a good solvent to remove FITC retained on PB-GF disks. Gly and Pro derivatives were found to be resistant against cyclization under the usual conditions, hence, the conversion reaction was performed with 50% TFA at 60 °C.

Automated gas-phase sequencing using the FITC/PITC double coupling method was evaluated using myoglobin and lysozyme as standard proteins and is summarized in Table 2. Myoglobin (5 pmol) on a polybrene-coated glass filter showed the average initial yields of 55% and the average repetitive yields of 86%. Myoglobin spotted onto PVDF membranes

Table 1. Gas-phase sequencing of proteins using the FITC method.

Proteins	Amount applied (pmol)	Method	Support	Initial yield*(%)	Repetitive yield*(%)
Myoglobin	50	S	PB-GF	43	83
	50	D	PB-GF	33	89
	25	S	PVDF	21	83
	25	D	PVDF	34	89
Lysozyme	50	S	PB-GF	58	87
	50	D	PB-GF	76	86
	25	S	PVDF	25	90
	25	D	PVDF	44	91

S: FITC single coupling; D: FITC/PITC double coupling; PB-GF: polybrene-coated glass filter disk; PVDF: Immobilon-P^{SQ}. (* mean values from 2-3 experiments)

Table 2. Sequencing of protiens (5 pmol) using the FITC/PITC method.

Protein	Support	Initial yield*(%)	Repetitive yield*(%)
Myoglobin	GF	45	80
	PB-GF	55	86
	PVDF	21	90
Lysozyme	GF	24	81
	PB-GF	59	92
	PVDF	22	92

GF: glass filter disk; PB-GF: polybrene-coated glass filter disk; PVDF: Immobilon P^{SQ}. (* mean values from 2-3 experiments)

gave a comparable repetitive yield. The sequence obtained was up to 15-25 residues. The average initial yields and repetitive yields were decreased to 45% and 80%, respectively, for a load of 5 pmol of myoglobin on a glass filter without Polybrene.

Figure 4 shows the plots of the yields of the expected FTH-amino acids at each cycle for 5 pmol of myoglobin spotted onto PVDF. The initial yield varied from 21 to 37%, and the repetitive yields from 89 to 90%. Several chromatograms are shown in Figure 5. The FTH-

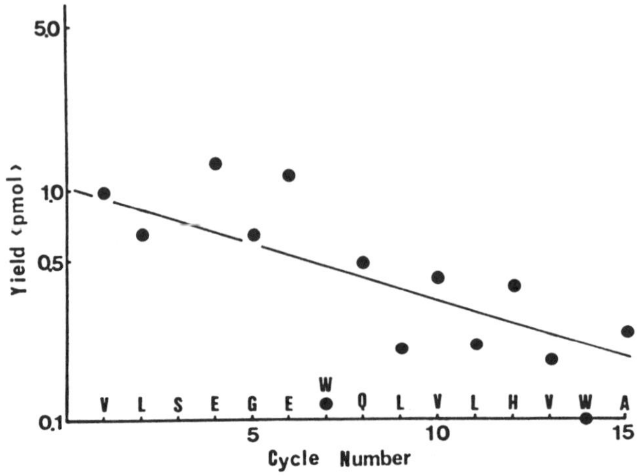

Figure 4. Sequence analysis of sperm whale myoglobin (5 pmol) spotted onto PVDF membranes.

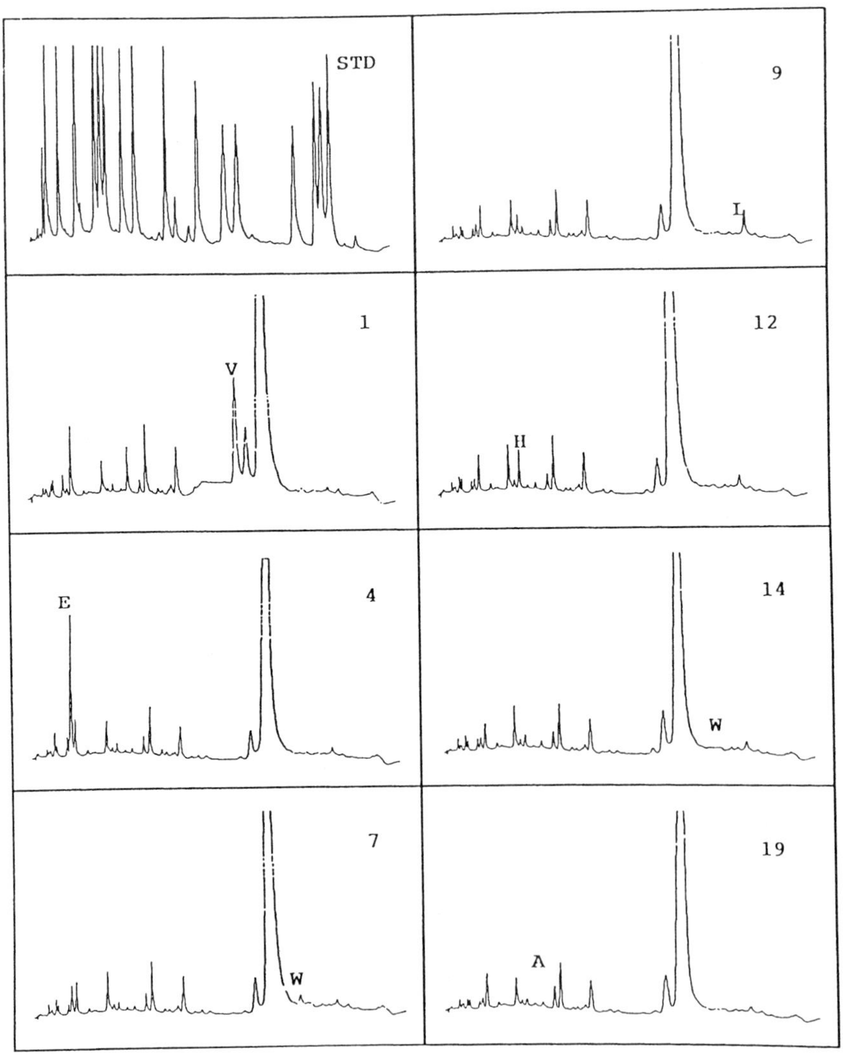

Figure 5. Sequencing of myoglobin (5 pmol) by the FITC/PITC double coupling.

Trp peak in cycle 14 was not detected. FTH-Leu and -Val gave lower yields than FTH-hydrophilic amino acids. Although repetitive yields were independent of the applied sample amounts, the initial yield was lower upon decreasing the amounts of sample. One of the most common methods for the separation of proteins on the microscale is SDS-PAGE followed by electroblotting onto PVDF membrane. Figure 6 shows sequence analysis of bovine serum albumin (BSA)(12 pmol) separated by SDS-PAGE followed by electroblotted onto a PVDF membrane. FTH-Asp (2.5 pmol)was recovered from the first cycle.

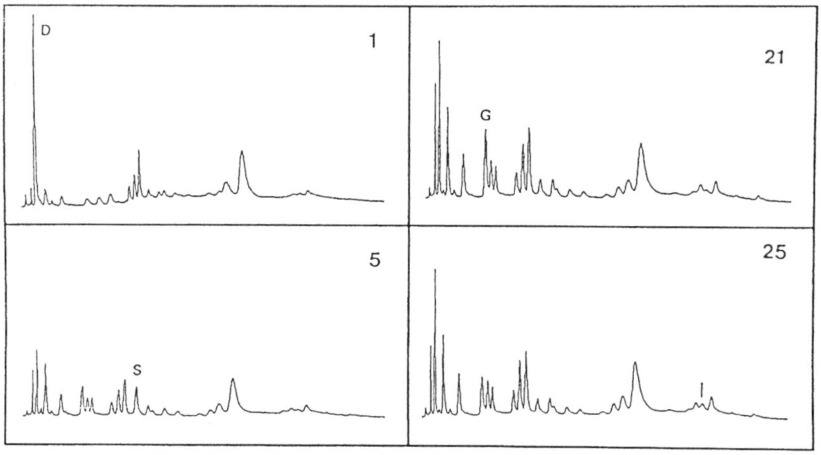

Figure 6. Sequencing of BSA (12 pmol) after SDS-PAGE followed by electroblotting onto PVDF membrane

CONCLUSION

Although FITC was proposed as an Edman-type reagent as early as 1968, it has not yet been used for sensitized sequencer, because of the low reactivity and poor extractability. As described in the present paper, the coupling efficiency of FITC with proteins in the gas-phase sequencing is high enough to allow sequencing of up to 20 residues from the N-terminus. Excess FITC could be removed by extraction with ethyl acetate and/or acetone giving a moderate repetitive yield. The modified gas-phase sequencer using the FITC/PITC double coupling method allows sequence analysis of 1-5 picomol amounts of protein to identify 15-25 residues from the N-terminus, which is enough information to proceed cDNA studies.

REFERENCES

Hewick, R. M., Hunkapiller, M. W., Hood L. E. and Dreyer, W. J., 1981 A gas-liquid solid phase peptide and protein sequenator, *J. Biol. Chem* . 256:7990.

Maeda, H. and Kawauchi, H., 1968 A new method for the determination of N-terminus of peptide chain with fluoroscein isothiocyanate, *Biochem. Biophys. Res. Commun.* 31:188.

Muramoto, K., Kawauchi, H. and Tuzimura, K., 1978 Sequence determination of peptide by the combined use fluoroscein isothiocyanate and phenylisothiocyanate, *Agric. Biol. Chem.* 42:1559.

Nokihara, K. and Herbst, F., 1989 Rapid characterization of proteins and peptides by

electrophoretic transfer and microsequencing, *in*: Proceedings of the 10th Koenigsteiner Chromatographie Tag, Millopore Corp, Eschborn, pp. 139.
Muramoto, K., Kamiya, H. and Kawauchi, H., 1984 The application of fluoroscein isothiocyanate and high-performance liquid chromatography for the microsequencing of proteins and peptides, *Anal. Biochem.* 141:446.

THIOBENZOYLATION METHOD OF PROTEIN SEQUENCING: GAS CHROMATOGRAPHY/MASS SPECTROMETRIC DETECTION OF 5-ACETOXY-2-PHENYLTHIAZOLES

Mark L. Stolowitz, Chin-Sook Kim, Steven R. Marsh and Leroy Hood

Center for Molecular Biotechnology, Division of Biology 139-74
California Institute of Technology
Pasadena, California 91125

INTRODUCTION

The thiobenzoylation method of protein sequencing was first proposed in 1967 by Barrett as an alternative to the Edman degradation (Edman, 1950). This method attracted limited interest but facilitated the subsequent development of the closely-related thioacetylation method (Doolittle *et al.*, 1977), wherein the 2-methyl-5(4*H*)-thiazolone cleavage product is identified by back-hydrolysis to the parent amino acid. The 2-methyl-5(4*H*)-thiazolone is a volatile analyte which was shown by Previero in 1977 to be potentially suitable for analysis by gas chromatography, either directly or preferably after reaction with an acetylating reagent which affords the corresponding 5-acetoxy-2-methylthiazole. This proposed use of gas chromatography preceded the development of modern capillary gas chromatography columns by several years, however, and consequently column packings which permitted recovery of basic amino acid derivatives were unavailable. We recently reexamined this approach utilizing a capillary gas chromatography/mass spectrometry system and found the method to exhibit considerable potential (Stolowitz *et al.*, 1992). Unfortunately, our efforts to further develop the thioacetylation method have been somewhat hindered by the limited reactivity of thioacetylating reagents and difficulties encountered in the synthesis of a few of the amino acid derivatives for use as analytical standards.

As thiobenzoylation reagents (Inman and Appella, 1986) exhibit substantially greater reactivity than their thioacetylation counterparts, we recently evaluated the suitability of 2-phenyl-5(4*H*)-thiazolones and 5-acetoxy-2-phenylthiazoles for detection by gas chromatography/mass spectrometry and found the later compounds to be potentially useful analytes for sub-picomole protein microsequencing. 5-Acetoxy-2-phenylthiazoles are prepared from 5-hydroxy-2-phenylthiazolium hydrochlorides which are, in turn, prepared by cyclization of *N*-thiobenzoylamino acids. The synthesis of these derivatives and their characterization by

Methods in Protein Sequence Analysis, Edited by K. Imahori and F. Sakiyama, Plenum Press, New York, 1993

matrix-assisted laser desorption time-of-flight mass spectrometry and gas chromatography/ion trap mass spectrometry is described herein.

MATERIALS AND METHODS

N-Thiobenzoylamino acids were prepared by reaction of amino acids with *S*-(thiobenzoyl)thioglycolic acid (Kurzer, 1961) (Aldrich) in alkaline aqueous solution containing 2 equiv. of NaOH as previously described (Barrett, 1965; 1966). When amino acid hydrochlorides were utilized, an additional equiv. of NaOH was added. Acidification of the reaction with 2 N H_2SO_4 and extraction into either ethyl ether or ethyl acetate provided the products in approx. 90% yield. *N*-Thiobenzoylamino acids were cyclized by dissolution in anhydrous 4 N HCl in dioxane (Aldrich). On standing overnight the crystalline 5-hydroxy-2-phenylthiazolium hydrochlorides were deposited in many instances or precipitated by addition of several volumes of cold ethyl ether. The product was collected on a glass sinter funnel, washed with cold ethyl ether and dried *in vacuo* overnight. Cyclization of *N*-thiobenzoylglycine by the aforementioned method afforded a product which proved difficult to isolate. An alternative cyclization involving the dropwise addition of 1 equiv. of PBr_3 to a solution of *N*-thiobenzoylglycine in ethyl ether yielded thiazolium hydrobromide without difficulty. 5-Hydroxy-2-phenylthiazolium hydrochlorides (or hydrobromides), prepared for all twenty amino acids, were found to be stable at room temperature for several months. The standards do not appear to be hydroscopic except in the case of the dehydro derivatives of cysteine, serine and threonine. The structures of the thiazolium salts were confirmed by matrix-assisted laser desorption time-of-flight mass spectrometry utilizing α-cyano-4-hydroxycinnamic acid as matrix. Mass spectra were recorded on a Finnigan MAT Lasermat Spectrometer in positive ion mode.

5-Acetoxy-2-phenylthiazoles were prepared on a synthetic scale by neutralization of 5-hydroxy-2-phenylthiazolium hydrochlorides with 5% $NaHCO_3$ to afford 2-phenyl-5(4*H*)thiazolones which were extracted into either ethyl ether or ethyl acetate, dried over anhydrous $MgSO_4$, and concentrated *in vacuo*. Subsequent dissolution in neat acetic anhydride containing a catalytic drop of pyridine and concentration *in vacuo* yielded the 5-acetoxy-2-phenylthiazoles, which were recrystallized from ligroine as previously reported (Barrett and Khokhar, 1969) or precipitated as the 5-acetoxy-2-phenylthiazolium hydrochlorides on addition of 1 N anhydrous HCl in ethyl ether.

Analytical scale derivatization of 5-hydroxy-2-phenylthiazolium hydrochloride (or hydrobromide) standards was achieved by dissolution of the standard in a solution containing 70% acetonitrile, 20% triethylamine and 10% acetic anhydride (7:2:1) for 2 min at room temperature. The derivatization solution must be prepared daily and discolors after standing for several hours. Concentrations were ordinarily in the range 0.5-2 mg/ml. The addition of 2% methanol (v/v) to the derivatization solution effects the concomitant esterification of the side-chain carboxylic acid moieties of the aspartic and glutamic acid thiazolium hydrochlorides during the course of the acetylation reaction. This inclusion has no adverse effect upon the acetylation of other amino acid thiazolium hydrochlorides. Samples were concentrated on a Speedvac system and dissolved in either 7:3 ethyl acetate:triethylamine or 9:1 acetonitrile:pyridine prior to analysis by gas chromatography-mass spectrometry (GC/MS) on a Finnigan MAT Magnum ion trap mass spectrometer configured for isobutane chemical ionization detection. Volumes of 1 µl were injected. An on-column injector was connected to a 0.5 m, 530 µm I.D. deactivated fused silica retention gap which was, in turn, connected to a 15 m, 250 µm I.D., DB-5ms capillary column (J and W). Attempts to utilize septum-purged and split-splitless injectors containing glass inserts proved problematic even after silanization of the insert. Similarly, the use of a glass capillary connector between the retention gap, and the capillary column proved problematic, eventually necessitating the use of a metal connector fitted with polyimide ferrules (Valco).

RESULTS AND DISCUSSION

N-Thiobenzoylamino acids were prepared and cyclized to 5-hydroxy-2-phenylthiazolium hydrochlorides (Figure 1), the structures of which were confirmed by matrix-assisted laser desorption time-of-flight mass spectrometry (Table 1). A representative mass spectrum is illustrated in Figure 2. The predicted masses were obtained for each of the amino acid derivatives if one assumes quantitative dehydration of the cysteine, serine and threonine derivatives. Dehydration is to be expected as the 4-*H* proton on the 2-phenyl-5(4*H*)-thiazolone ring is highly acidic and tends toward elimination. Further evidence of the acidity of the 4-*H* proton is participation in the tautomeric equilibrium involving the 5-hydroxy-2-phenylthiazole.

5-Acetoxy-2-phenylthiazoles were prepared from the corresponding 5-hydroxy-2-phenylthiazolium hydrochlorides on both analytical and preparative scales and analyzed by gas chromatography/mass spectrometry (Table 2). Capillary gas chromatography provided base-line resolved separation of all 5-acetoxy-2-phenylthiazoles in under 12 min with an average peak width at half height of < 3 sec. Shorter columns could be utilized to further reduce the chromatographic separation as the current resolution is more than adequate (approximately 3-fold greater resolution than a comparable HPLC separation). In principal, the chromatographic separation would be reduced to only a few minutes provided that resolution of leucine and isoleucine is retained. With the exception of these two isobaric amino acids, all other analytes can be identified by their masses alone. Electron impact

Table 1. Matrix-assisted laser desorption time-of-flight mass spectrometry of 5-hydroxy-2-phenylthiazolium hydrochlorides.[1]

Amino Acid	Formula	Predicted Mass	Expt. Mass
Gly[2]	C_9H_8NOS	178.23	178
ΔAla	$C_{10}H_8NOS$	190.24	190
Ala	$C_9H_{10}NOS$	192.25	192
ΔAbu	$C_{10}H_{11}NOS$	204.26	204
Pro	$C_{12}H_{12}NOS$	218.29	218
Val	$C_{12}H_{14}NOS$	220.30	220
Ile, Leu	$C_{13}H_{16}NOS$	234.33	234
Asn	$C_{11}H_{11}N_2O_2S$	235.28	235
Gln	$C_{12}H_{13}N_2O_2S$	249.30	249
Asp-OMe	$C_{12}H_{12}NO_3S$	250.29	250
Met	$C_{12}H_{14}NOS_2$	252.37	252
His+	$C_{13}H_{13}N_3OS$	258.32	258
Glu-OMe	$C_{13}H_{14}N_3OS$	264.31	264
Phe	$C_{16}H_{14}NOS$	268.34	268
Arg+	$C_{13}H_{18}N_4OS$	277.36	277
Tyr	$C_{16}H_{14}NO_2S$	284.34	284
Trp	$C_{18}H_{15}N_2OS$	307.38	307
Lys	$C_{19}H_{19}N_2OS$	369.51	370

[1]Mass spectra recorded utilizing α-cyano-4-hydroxycinnamic acid matrix.
[2]Determined as thiazolium hydrobromide salt.

ionization detection of 5-acetoxy-2-phenylthiazoles resulted in fragmentation of the thiazole ring which complicated identification of the analyte and substantially reduced detection limits.

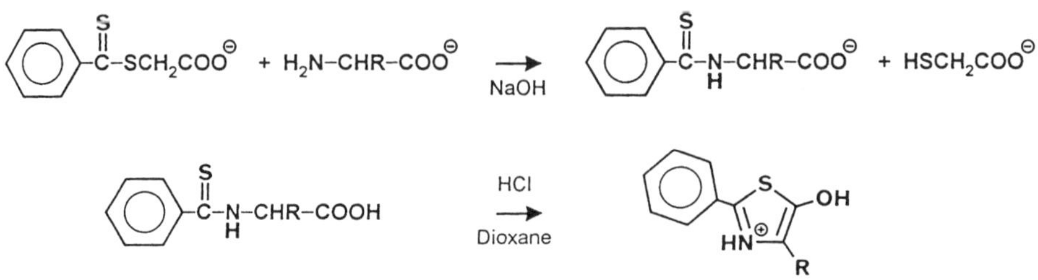

Figure 1. Synthesis of *N*-thiobenzoylamino acids by reaction with *S*-(thiobenzoyl)-thioglycolic acid in alkaline aqueous solution. 5-Hydroxy-2-phenylthiazolium hydrochlories are prepared by cyclization of *N*-thiobenzoylamino acids with anhydrous hydrochloric acid. The thiazolium hydrochlorides are stable at room temperature for several months.

Table 2. Chemical ionization mass spectrometric detection of acetylated 5-hydroxy-2-phenylthiazolium hydrochlorides.[1]

Amino Acid	Predicted Mass 5-hydroxy	Expt. Mass 5-hydroxy[2]	Predicted Mass 5-acetoxy	Expt. Mass 5-acetoxy
Gly	178.23	178	220.26	220
DAla	190.24	190	232.28	not observed[3]
Ala	192.25	192	234.29	234
DAbu	204.26	204	246.30	246
Pro	218.29	218	260.33	not observed[4]
Val	220.30	220	262.34	262
Ile, Leu	234.33	234	276.37	276
Asn	235.28	235	277.32	277
Gln	249.30	249	291.34	291
Asp-OMe	250.29	250	292.33	292
Met	252.37	252	294.41	294
His[5]	258.32	not observed	301.37	301
Glu-OMe	264.31	264	306.35	306
Phe	268.34	268	310.38	310
Arg[5]	277.36	not observed	319.40	319
Tyr[5]	284.34	284	326.38	326
Trp[5]	307.38	307	349.42	349
Lys[5,6]	369.51	370	411.55	412

[1]Reagent gases utilized for chemical ionization detection included methane and isobutane.
[2]5-Hydroxy-2-phenythiazolium ions detected only with methane chemical ionization.
[3]ΔAla derivative does not acetylate (no enol form).
[4]Pro derivative is mesoionic (thiazolium hydroxide) and does not acetylate.
[5]Side-chain acetylation of His, Arg, Tyr, Trp and Lys detected. Side-chain acetylation of Trp and Lys require heating the derivatization reaction to > 50° C.
[6]The diacyl Lys derivative results from enolization of the side-chain thiobenzamido group and acetylation of the thioenol.

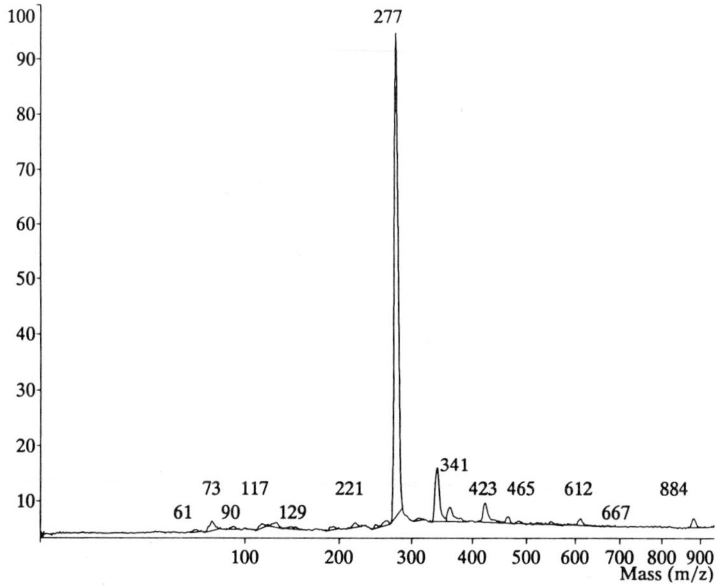

Figure 2. Matrix-assisted laser desorption time-of-flight mass spectrum of 5-hydroxy-2-phenylthiazolium hydrochloride (arginine, mass 277). α-Cyano-4-hydroxycinnamic acid matrix.

5-Acetoxy-2-phenylthiazoles were routinely detected by chemical ionization (CI) employing either methane or isobutane as reagent gas. In both instances reagent gas ions were found to act as Brønsted acids transferring a proton to the thiazole ring to produce the molecular ion $[M+H]^+$. Methane CI was found to be accompanied by partial fragmentation of the *O*-acetoxy moiety resulting in loss of an acetyl moiety (mass 42) from the molecular

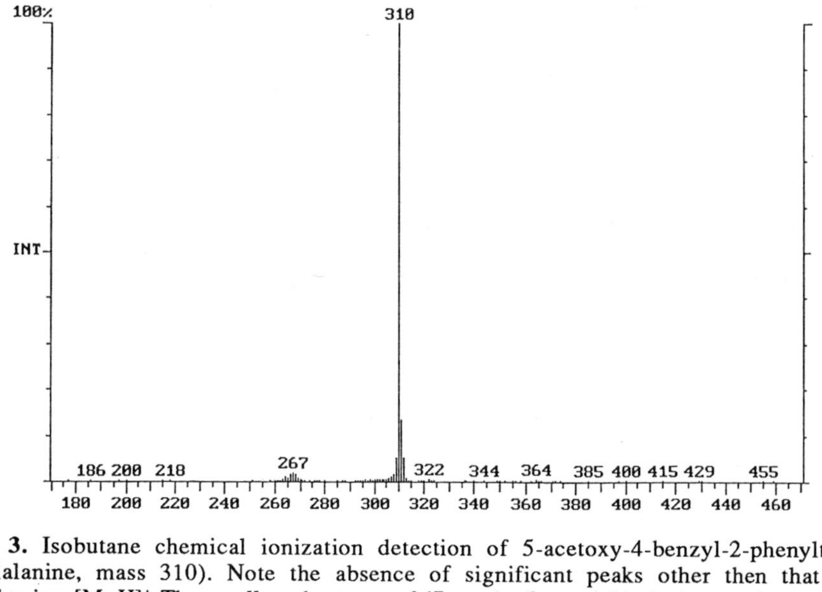

Figure 3. Isobutane chemical ionization detection of 5-acetoxy-4-benzyl-2-phenylthiazole (phenylalanine, mass 310). Note the absence of significant peaks other then that of the molecular ion $[M+H]^+$. The small peak at mass 267 results from residual electron impact related phenomena associated with the ion trap.

ion. Isobutane Cl completely eliminated fragmentation such that only the molecular ion $[M+H]^+$ is detected (Figure 3). This is advantageous with regard to detection sensitivity optimization and removes the need to interpret complex mass spectra.

Detection limits achieved employing isobutane chemical ionization are presently in the 1-10 femtomole range (Figure 4). We believe that with continued optimization, detection of 1 femtomole will be routinely obtained. A current limitation is the susceptability of 5-acetoxy-2-phenylthiazoles toward alkaline hydrolysis, which results from traces of water (parts per million) present even in gas chromatographic grade solvents. To overcome this limitation ultra-dry solvents will be required and analysis must be undertaken immediately after derivatization.

CONCLUSIONS

Isobutane chemical ionization mass spectrometric detection of 5-acetoxy-2-phenylthiazoles is compatible with the protein sequencing protocol illustrated in Figure 5. Although thiobenzoylation with *S*-(thiobenzoyl)thioglycolic acid is illustrated, alternative reagents offering improved reactivity have been described (Inman and Appella, 1986). Additionally, the experience of a decade of improvements in the field of peptide synthesis have not yet been applied to this development. Dithioesters, as a group, are only reactive in solvents of high polarity. Consequently, the pursuit of this method will require the adoption of solid-phase sequencing approaches. Appropriate instrumentation developed in our laboratory has recently been described elsewhere (Stolowitz and Hood, 1993). Cleavage of *N*-thiobenzoylpeptides occurs under conditions which are mild as compared to the Edman degradation. Although 2-phenyl-5(4*H*)-thiazolones are volatile, they elute from the gas chromatography column at temperatures above 175°C. Consequently, gas phase cleavage with trifluoroacetic acid vapor can be utilized without fear of evaporative losses. This is not the case in the related thioacetylation method (Strowitz *et al.*, 1989).

The tautomeric equilibrium involving the 2-phenyl-5(4*H*)-thiazolone and 5-hydroxy-2-

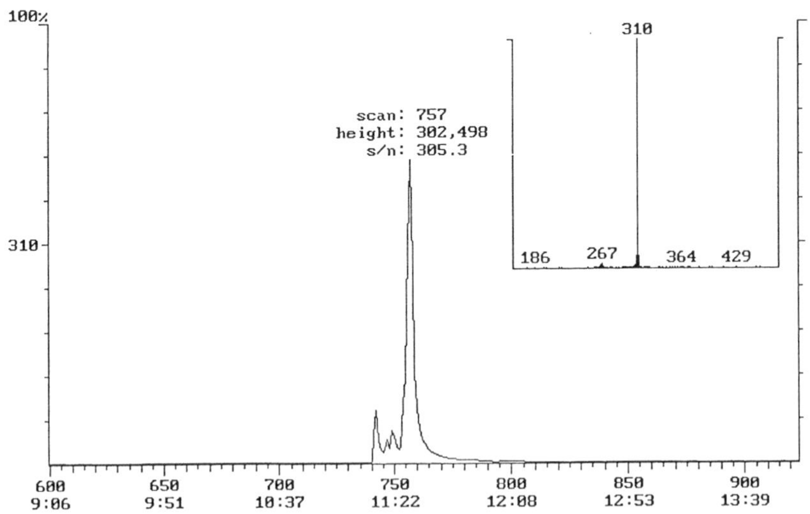

Figure 4. Isobutane chemical ionization detection of 5-acetoxy-4-benzyl-2-phenylthiazole (phenylalanine) at the 10 femtomole level. Over 300,000 area counts were obtained at a signal to noise ratio of 305 to 1.

phenylthiazole is sensitive to solvent polarity and efficient derivatization of the enol form is achieved only in polar solvents. We have found acetonitrile to be the solvent of choice in this instance. Acetylation of 5-hydroxy-2-phenylthiazoles proceeds rapidly at room temperature and affords volatile acetoxythiazoles for all twenty amino acids.

Having now prepared at least 100 mg quantities of each of the required analytical standards and properly characterized them with regard to stability, structure and detection limits, our attention is now directed toward the optimization of an interface which will allow us to place our newly developed solid-phase protein sequencer on-line to the mass spectrometer. We are hopeful that we will begin routine operation of a fully automated system in the not too distant future.

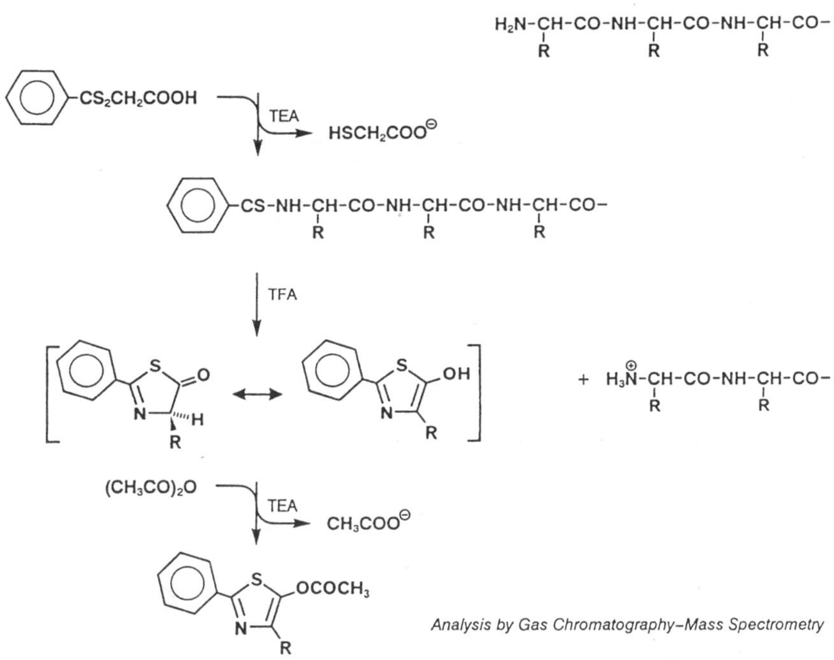

Figure 5. Thiobenzoylation method of protein microsequencing. Thiobenzoylation of the *N*-terminal amino acid is effected by reaction with *S*-(thiobenzoyl)thioglycolic acid. The reaction is catalyzed by an aqueous solution of tertiary amine. Cleavage occurs under mild condition to afford the 2-phenyl-5(4*H*)-thiazolone, which exists in a tautomeric equilibrium with the 5-hydroxy-2-phenythiazole. Acetylation of the tautomeric mixture with acetic anhydride and triethylamine affords the 5-acetoxy-2-phenylthiazole, which is suitable for analysis by GC/MS at or below the 10 femtomole level.

Acknowledgements. The authors wish to thank Dr. Graeme Currie for recording the matrix-assisted laser desorption time-of-flight mass spectra. Assistance in optimizing the performance of the gas chromatography/mass spectrometry system provided by Mr. John Ragsdale of Finnigan MAT is gratefully acknowledged.

REFERENCES

Barrett, G. C., 1965 The circular dichroism of *N*-thiobenzoyl-L-α-amino-acids in solulion in ether and in methanol, *J. Chem. Soc.* 2825.

Barrett, G. C., 1966 The circular dichroism of *N*-thiobenzoyl-L-α–amino-acids, Part II. Further examples of solvent-dependent circular dichroism, *J. Chem. Soc. (C)* 1771 .

Barrett, G. C., 1967 Cleavage of *N*-thiobenzoyl-dipeptides with trifluoroacetic acid: the basis of a new stepwise degradation of polypeptides, *Chem. Commun.* 487.

Barrett, G. C. and Khokhar, A. R., 1969 Trifluoroacetic acid as a cyclization reagent for the synthesis of thiazol-5(4*H*)-ones and 2-thiazolines, *J. Chem. Soc. (C)* 1117.

Doolittle, L. R., Mross, G. A., Fothergill, L. A. and Dooliltle, R. F., 1977 A simple solid-phase amino acid sequencer employing a thioacetylation stepwise degradation procedure, *Anal. Biochem.* 78:491.

Edman, P., 1950 Method for determination of the amino-acid sequence in peptides, *Acta Chem. Scand.* 4:283.

Inman, J. K. and Appella, E., 1986 Newer methods of solid- and liquid-phase sequence determination-personal views, *in*: Practical Protein Chemistry-A Handbook, Darbre, A. ed., Wiley.

Kurzer, F., 1961 Thioacylation with (thiobenzoylthio)-acetic acid, *Chemislty and Industry* 1333.

Previero, A., 1977 Alternative reagents in sequential degradation on solid-phase supports, *Methods Enzymol.* 47:289.

Stolowitz, M. and Hood, L., 1993 Single syringe-pump solid-phase protein sequencer, *in*: Techniques in Protein Chemistry IV, Angeletti, R.H. ed., Academic Press.

Stolowitz, M. L., Kim, C. -S. and Hood, L., 1992 Thioacelylation method of protein sequencing: gas chromatography/mass spectrometric detection of 5-acetoxy-2-methylthiazoles, *J. Protein Chem.* 11:360.

Stolowitz, M. L., Paape, B. A. and Dixit, V. M., 1989 Thioacetylation method of protein sequencing: derivatization of 2-methyl-5(4*H*)-thiazolones for high-performance liquid chromatographic detection, *Anal. Biochem.* 181:113.

DEBLOCKING AND SUBSEQUENT MICROSEQUENCE ANALYSIS OF *N*-TERMINALLY BLOCKED PROTEINS IMMOBILIZED ON PVDF MEMBRANE

Susumu Tsunasawa and Hisashi Hirano

Institute for Protein Research, Osaka University, Suita, Osaka, and National Institute of Agrobiological Resources, Tsukuba, Ibaraki, Japan

INTRODUCTION

In recent protein microsequence analyses, proteins separated by polyacrylamide gel electrophoresis (PAGE) and electroblotted onto a polyvinylidene difluoride (PVDF) membrane can be easily sequenced by a gas-phase sequencer (Matsudaira, 1987). However, even if the proteins are successfully separated by PAGE and electroblotted onto PVDF membrane, the *N*-terminal amino acid sequences of proteins with a blocking group at the *N*-terminus cannot be determined by Edman degradation. A large number of proteins are known to be *N*-terminally blocked (Aitken, 1990; Tsunasawa and Sakiyama, 1992). For example, 51 of 72 seed proteins in the rice and 14 of 36 seed proteins in the winged bean which were separated by two-dimensional gel electrophoresis were estimated to be *N*-terminally blocked (Hirano, 1989; Komatsu *et al.*, 1992). It is, therefore, important to establish simple and rapid deblocking and sequencing techniques for these proteins.

Many blocking groups have been identified so far (Table 1), with acetyl modifications being most prevalent and formyl and pyroglutamyl groups also frequently detected. Both enzymatic and chemical deblocking techniques for proteins with these blocking groups have been previously reported. Podell and Abraham (1978) could deblock the pyroglutamated proteins dissolved in a buffer solution with pyroglutamyl peptidase, Tsunasawa *et al.* (1990), Krishna *et al.* (1991) and Farries *et al.* (1991) have developed the deblocking method for acetylated proteins using acylamino acid releasing enzyme (AARE), and Sheehan and Yang (1958) dissolved the formylated proteins in HCl/methanol solution and successfully removed their *N*-terminal formyl groups.

New deblocking techniques for the proteins which were electroblotted from the polyacrylamide gel onto PVDF membrane, which were based on deblocking techniques for the proteins dissolved in the solution, have recently been developed in our and other laboratories. We have shown that PVDF membrane-bound protein blocked with the *N*-

Methods in Protein Sequence Analysis, Edited by K. Imahori and F. Sakiyama, Plenum Press, New York, 1993

terminal pyroglutamic acid on PVDF membrane could be deblocked using pyroglutamyl peptidase digestion and was subsequently sequenced (Hirano *et al.* 1991b; Moyer *et al.*, 1990). In a similar manner, an *N*-terminal formyl group was successfully removed by acid treatment from a protein on PVDF membrane (Ikeuchi and Inoue, 1988) and an *N*-acetylated protein after electroblotting onto PVDF membrane, was deblocked by both acid treatment and AARE (Hirano *et al.*, 1991a; 1992; Tsunasawa and Hirano, 1992). In this paper, we review these deblocking techniques for the proteins transferred onto PVDF membrane.

GEL ELECTROPHORESIS AND ELECTROBLOTTING

An aliquot of the *N*-terminally blocked protein was separated by SDS-PAGE, on a gel in which the ratio of acrylamide and methylenebisacrylamide concentrations was modified from 30:0.8 to 30:0.135 (final concentration of acrylamide in the gel was 15%). The proteins were electroblotted onto PVDF membranes by using a semi-dry blotting apparatus (Hirano, 1989; Hirano and Watanabe, 1990).

After electroblotting, the blots were either stained for 5-10 min with 0.1% (w/v) Coomassie Brilliant Blue R-250 in 50% (v/v) methanol containing 10% (v/v) acetic acid, followed by destaining with 60% (v/v) methanol, or were stained for 5 min with 0.1% (w/v) Ponceau S in 1% (v/v) acetic acid and destained with 1% (v/v) acetic acid.

Table 1. *N*-terminal blocking groups found in proteins.

Blocking group		Commomly affected residue
Acyl-	Formyl	Gly, Met
	Acetyl	Ser, Ala, Met, Gly, Asp, Glu, Thr, Val, Pro
	Mirystoyl	Gly
	α -Ketoacyl	Pyruvoyl[Ser] α-Ketobutyl[Thr]
	Glucuronyl	Gly
	Pyroglutamyl	Glu, Gln
	Murein	Lys
Alkyl-	Methyl (mono)	Met, Ala, Phe
	(di)	Pro
	(tri)	Ala
	Glucosyl	Val:[1-deoxy,1-(N^{α} -Val)-fructose]

DEBLOCKING OF BLOTTED PROTEINS

Removal of Acetyl Group from the *N*-Terminal Acetylserine and Acetylthreonine. The acetyl group of *N*-terminal acetylserine and acetylthreonine can be removed by exposure to trifluoroacetic acid (TFA) and an *N*→*O* acyl shift may be involved in this deblocking procedure (Wellner *et al.*, 1990).

An aliquot (1,050 pmoles) of yeast (*Saccharomyces cerevisiae*) cytochrome *c* mutant proteins, which have the *N*-terminal sequence of either AcSer-Glu-Phe-Ala-Gly-Ser-Ala-Lys- or AcThr-Glu-Phe-Leu-Ala-Gly-Ser-Ala-Lys- (Moerschell *et al.*, 1990), were separated by SDS-PAGE and electroblotted onto PVDF membrane. The pieces of PVDF membrane carrying the protein bands were excised and treated with gas-phase TFA at 60°C for 30 to 60 min, and were then directly subjected to gas-phase sequencing. Table 2 shows that these cytochrome *c* mutant proteins were deblocked and sequenced, but that the overall sequencing yields obtained by this procedure were low.

When the membrane was exposed to TFA vapor for longer than 2 h, the yields of the phenylthiohydantoin-amino acids increased. However, simultaneously, the quantities of the reaction by-products increased.

This deblocking technique has also been applied to horse heart cytochrome *c* and several cytochrome *c* mutant proteins which have acetyl group at their *N*-termini. Proteins which had either the acetylserine or acetylthreonine were successfully deblocked, while proteins which had other acetyl amino acids could not be deblocked (Table 3).

Table 2. The results of sequencing of yeast cytochrome *c* mutant proteins after deblocking on PVDF membrane.

a. AcSer-Glu-Phe-Ala-Gly-Ser-Ala-Lys-									
Residue	Ser	Glu	Phe	Ala	Gly	Ser	Ala	Lys	
Amount (pmole)	47.7	38.0	39.5	28.5	17.1	11.1	11.6	11.1	
Initial yield (%)	4.5								
b. AcThr-Glu-Phe-Leu-Ala-Gly-Ser-Ala-Lys-									
Residue	Thr	Glu	Phe	Leu	Ala	Gly	Ser	Ala	Lys
Amount (pmole)	17.0	20.4	52.8	33.9	26.1	25.8	16.8	15.6	8.4

Table 3. The results of sequencing of horse heart cytochrome *c* (a) and yeast cytochrome *c* mutant proteins (b-h) after deblocking treatment on PVDF membrane.

	Deblocking	Yield (pmole)
(a) AcGly-Asp-Val-Glu-Lys-	No	
(b) AcSer-Glu-Phe-Ser-Ala-Lys-Lys-	Yes	134.4
(c) AcSer-Glu-Phe-Gly-Ser-Ala-Lys-Lys-	Yes	273.0
(e) AcThr-Glu-Phe-Leu-Ala-Gly-Ser-Ala-Lys-Lys-	Yes	20.4
(f) AcSer-Glu-Phe-Ala-Gly-Ser-Ala-Lys-Lys-	Yes	299.9
(g) AcGly-Glu-Phe-Leu-Ala-Gly-Ser-Ala-Lys-Lys-	No	
(h) AcAla-Glu-Phe-Leu-Ala-Gly-Ser-Ala-Lys-Lys-	No	

Removal of Formyl Group. An aliquot (about 1 nmole) of *N*-formylated tryptophan synthase α subunit (MW30,000) of *Escherichia coli* produced under amplifying conditions (Sugino *et al.*, 1980) was separated by SDS-PAGE and transferred to PVDF membrane by electroblotting. The portions of PVDF membrane carrying the protein bands were cut out

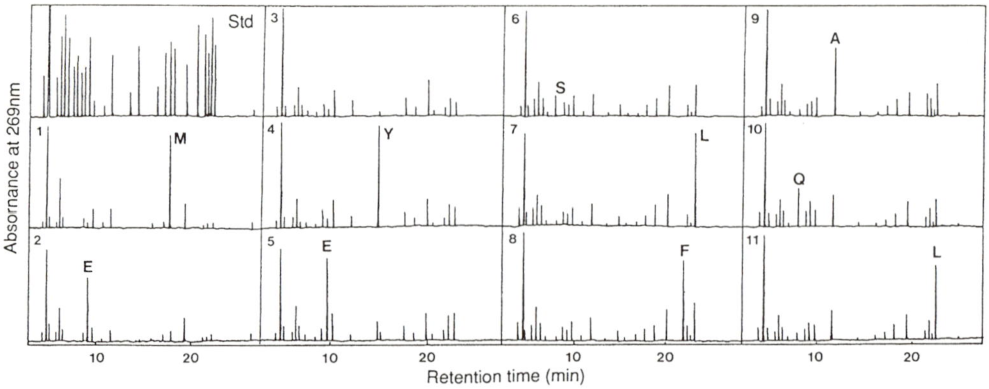

Figure 1. High performance liquid chromatography elution profile of phenylthiohydantoin-amino acids obtained by the *N*-terminal sequence analysis of *E. coli* tryptophan synthase after removal of the *N*-terminal formyl group. The protein was separated by SDS-PAGE, electroblotted onto PVDF membrane and sequenced after deblocking.

and treated with 0.6 M HCl at 25°C for 24 h as described by Ikeuchi and Inoue (1988) and was sequenced (Figure 1).

Removal of Pyroglutamic Acid. Various aliquots (500 pmoles-1 nmole) of proteins blocked with pyroglutamic acid at their *N*-termini (hen egg riboflavin-binding protein (MW34,000) and *Geotrichum candidum* lipase (MW59,000)) were separated by SDS-PAGE and electroblotted onto PVDF membrane. Each region of PVDF membrane carrying a protein band was excised and pretreated with 200 µl of 0.5% (w/v) polyvinyl-pyrrolidone (PVP)-40 in 100 mM acetic acid at 37°C for 30 min to block unbound protein-binding sites on the membrane. The membrane was washed (ten times) with 1 ml of deionized water and soaked in 100 µl of 0.1 M phosphate buffer (pH 8.0) containing 5 mM dithiothreitol and 10 mM ethylenediaminetetraacetic acid. Pyroglutamyl peptidase (5 µg) was added and the reaction solution incubated at 30°C for 24 h. The membrane was

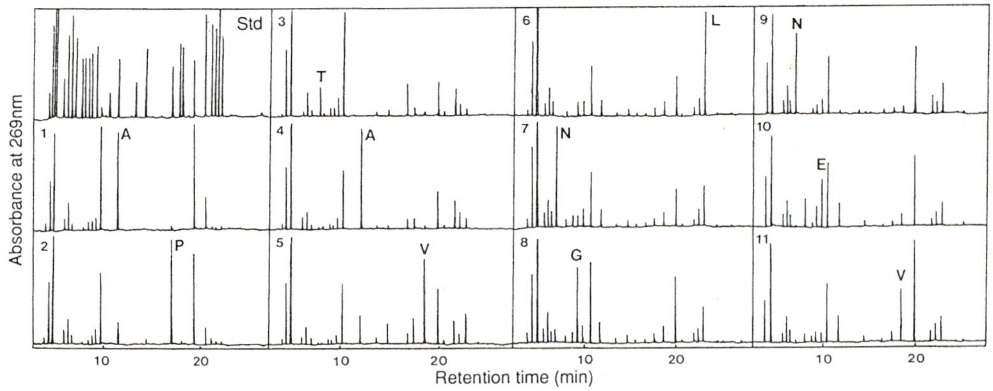

Figure 2. High performance liquid chromatography elution profile of phenylthiohydantoin-amino acids obtained by the *N*-terminal sequence analysis of *G. candidum* lipase after removal of the *N*-terminal pyroglutamic acid. The protein was separated by SDS-PAGE, electroblotted onto PVDF membrane and sequenced after deblocking.

washed with deionized water, dried *in vacuo* and applied to the gas-phase sequencer. Both proteins were shown to be deblocked and could be sequenced.

Figure 2 shows the results of the sequencing of a lipase after the removal of the *N*-terminal pyroglutamic acid. We have also applied this technique to enable the deblocking and sequencing of eight glutelin subunits and one globulin from the rice seeds (Hirano *et al.*, 1992; Komatsu and Hirano, 1992).

Removal of Acetyl Amino Acids. Increasing quantities (32, 64 and 320 pmoles) of horse heart cytochrome *c* were purified by SDS-PAGE and electroblotted onto the PVDF membrane. The regions of PVDF membrane carrying the protein bands were cut out and pretreated with 0.5% (w/v) PVP-40 in 100 mM acetic acid at 37°C for 30 min. After thorough washing with deionized water, the proteins were digested with 5-10 μg of trypsin in 100 μl of 0.1 M ammonium bicarbonate buffer (pH 8.0) containing 10% (v/v) acetonitrile, at 37°C for 24 h with shaking. The digestion buffer containing tryptic peptides was then pooled in an Eppendorf tube, the membrane was washed by vortexing with 100 μl of deionized water and the washing solution was combined with the digestion buffer. The digestion mixture was evaporated to dryness *in vacuo*, and 100 μl of 50% (v/v) pyridine and 10 μl of phenylisothiocyanate were added to bind with free, but not the blocked *N*-terminal amino acids of the tryptic peptides. Nitrogen gas was added, and the reaction solution was vortexed and centrifuged at 3,000*g* for 1 min. The resultant supernatant containing reaction by-products and excess reagents was discarded. This washing procedure was repeated three times, then sample was evaporated to dryness *in vacuo*. Performic acid was prepared by mixing 9 ml of formic acid and 1 ml of 30% hydrogen peroxide and keeping the mixture at room temparature for 1 h. Performic acid (100 μl) was added to the dried sample and following mixing the tube was placed on ice (1 h) to convert the N-terminal phenylthiocarbamyl groups of the peptides to phenylcarbamyl groups by oxidation. The sample solution was then evaporated to dryness *in vacuo*, resuspended in deionized water and again dried *in vacuo*. The sample was finally resuspended in 100 μl of 0.2 M phosphate buffer (pH 7.2) containing 1 mM dithiothreitol. Fifty mU AARE dissolved in 50 μl of the same buffer was then added and the mixture incubated (37°C for 12 h) to remove the *N*-acetylated amino acid. The sample solution was applied to a Polybrene-coated glass fiber filter mounted into the upper glass block of the reaction chamber in the gas-phase sequencer.

Figure 3 shows the results obtained following the sequencing of the deblocked cytochrome *c*. The cytochrome *c* could be deblocked and sequenced by the method described above. The overall initial yields obtained from the successive steps including electrophoresis, electroblotting, deblocking and sequencing were 23-25% (Table 4). The efficiency of deblocking and sequencing described here depended primarily on the efficiency of the tryptic digestion and the subsequent peptide elution from PVDF membrane. It was considered that the shaking of the digestion solution containing PVDF membrane during tryptic digestion was useful for the efficient elution of the tryptic peptides. The *N*-acetylated mutant proteins of yeast cytochrome *c* and the *N*-acetylated bovine erythrocyte superoxide dismutase were also deblocked and sequenced (Table 5).

Proteins, with relatively high molecular weights, with *N*-terminal pyroglutamyl groups could be deblocked on PVDF membrane by direct treatment with pyroglutamyl peptidase. In contrast, AARE could digest short peptides but could not be used to directly digest proteins with high molecular weight (Tsunasawa *et al.*, 1990). These proteins must be first digested with an appropreate protease on the membrane and the acetylated amino acid of the released *N*-terminal peptide is selectively removed using AARE.

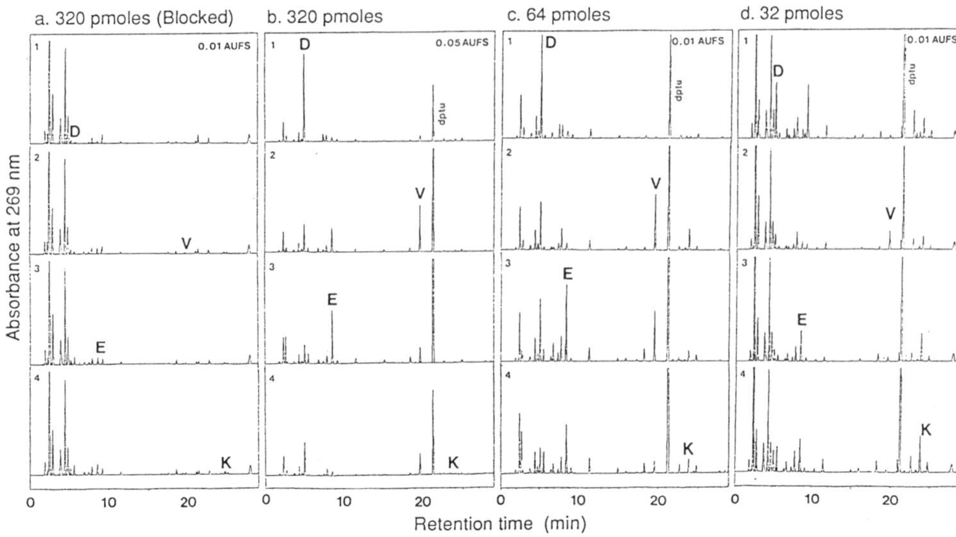

Figure 3. High performance liquid chromatography elution profiles of phenylthiohydantoin-amino acids obtained from the *N*-terminal sequence analysis of the blocked and deblocked horse heart cytochrome *c*. The cytochrome *c* was separated by SDS-PAGE, electroblotted onto PVDF membrane, and sequenced without deblocking (a) or with deblocking by AARE digestion (b-d). Amount of protein loaded onto SDS-PAGE was: a, b, 320 pmoles; c, 64 pmoles; d, 32 pmoles.

ADVANTAGE OF DEBLOCKING IMMOBILIZED PROTEINS

Deblocking of proteins electroblotted onto PVDF membrane offers several advantages: (1) proteins purified by PAGE can be efficiently deblocked and sequenced; (2) the enzymatic and chemical deblocking can be easily performed without desalting. Since after electroblotted onto PVDF membrane, the proteins become almost completely salt-free; (3) proteins can be deblocked at pmole levels; (4) sequential deblocking, as described in the next section, can be carried out.

Table 4. Results of deblocking and sequencing of five mutants of yeast cytochrome *c* (a-e) and bovine superoxide dismutase (f).

Sequence of the *N*-terminal peptide	Blockage	Initial yield (pmole)	Number of residues identified
(a) AcSer-Glu-Phe-Ser-Ala-Lys-Lys	Deblocked	172.2	4
(b) AcSer-Glu-Phe-Gly-Ser-Ala-Lys-Lys	Deblocked	785.7	4
(c) AcAla-Glu-Phe-Leu-Ala-Gly-Ser-Ala-Lys-Lys	Deblocked	28.4	6
(d) AcGly-Glu-Phe-Leu-Ala-Gly-Ser-Ala-Lys-Lys	Deblocked	7.4	4
(e) AcSer-Glu-Phe-Leu-Ala-Gly-Ser-Ala-Lys-Lys	Deblocked	35.6	3
(f) AcAla-Thr-Lys	Deblocked	128.4	2

SEQUENTIAL DEBLOCKING

These deblocking techniques may be combined to enable the sequential deblocking and sequencing of unknown proteins, which have been immobilized onto PVDF membranes (Figure 4).

(I) A protein sample is electroblotted from the PAGE gel onto PVDF membrane. Alternatively, this may be carried out by dotblotting of the protein solution directly to PVDF membrane. (II) The membrane carrying the protein is directly subjected to gas-phase sequencing. (III) If sequencing fails at step (II), the membrane is removed from the sequencer, and treated with TFA vapor to remove acetyl groups of acetylserine and acetylthreonine. If the protein is *N*-terminally blocked by acetylserine or acetylthreonine, the acetyl group is removed by this procedure and sequencing from the *N*-terminus is now possible. (IV) If sequencing fails at step (III), the membrane is incubated in 0.6 M HCl at 25°C for 24 h to remove formyl groups, and is then subjected to sequence analysis again. (V) If sequencing fails at step (IV), the sample is subjected to on-membrane pyroglutamyl peptidase digestion to remove pyroglutamic acid, and sequencing is again attempted. (VI) Finally, if sequencing fails at step (V), deblocking with AARE is performed to remove acetylamino acids that were not removed in step (III) and sequencing is then attempted again. If this attempt at sequencing fails, alternative methods are required.

Table 5. Results of deblocking and sequencing of horse heart cytochrome *c* (AcGly-Asp-Val-Glu-Lys).

Concentration (pmole/μg)	Blockage	Initial yield (pmole)	Number of residues identified
320/3.96	Blocked	0	0
320/3.96	Deblocked	87.3	3
64/0.79	Deblocked	15.0	3
32/0.40	Deblocked	8.4	3

We have applied this sequential deblocking technique to deblock a riboflavin-binding protein with the *N*-terminal pyroglutamic acid and the acetylated horse heart cytochrome *c*. It was possible to deblock and sequence these proteins at the pmole levels, although the initial yield varied depending on the experiment.

Ackonowledgements. We acknowlege the help of Dr. R. N. Waterhouse in the preparation of the manuscript and Dr. S. Komatsu for her assistance during the experiment.

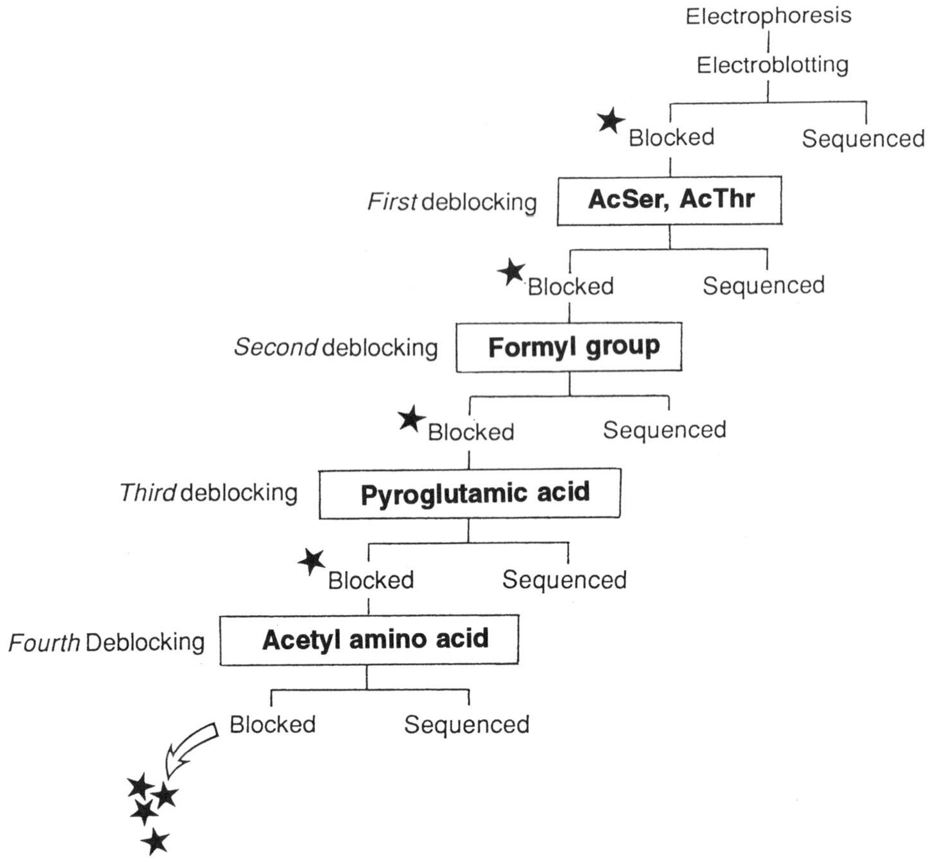

Figure 4. Sequential deblocking starategy for *N*-terminally blocked proteins blotted onto PVDF membrane.

REFERENCES

Aitken, A., 1990 Identification of Proteins Consensus Sequences, Ellis Horwood Ltd., Hemel Hempstead, UK.

Farries, T. C., Harris, A., Auffret, A. D. and Aitken, A., 1991 Removal of *N*-acetyl groups from blocked peptides with acylpeptide hydrolase, *Eur. J. Biochem.* 196:679.

Hirano, H., 1989 Microsequence analysis of winged bean seed proteins electroblotted from two-dimensional gel, *J. Protein Chem.* 8:115.

Hirano, H., Komatsu, S., Kajiwara, H., Takakura, H, Sakiyama, F. and Tsunasawa, S., 1991a Microsequence analysis of N^{α} -blocked proteins electroblotted onto an immobilizing matrix from polyacrylamide gels, *Anal. Sci.*7:suppl. 941.

Hirano, H., Komatsu, S., Nakamura, A., Kikuchi, F., Kajiwara, H., Tsunasawa, S. and Sakiyama, F., 1991b Structural homology between semidwarfism-related proteins and glutelin seed protein in rice (*Oryza sativa* L.), *Theor. Appl. Genet.* 83:153.

Hirano, H., Komatsu, S., Takakura, H., Sakiyama, F. and Tsunasawa, S., 1992 Deblocking and subsequent microsequence analysis of N^{α}-blocked proteins electroblotted onto PVDF membrane, *J. Biochem.* 111:754.

Hirano, H, and Watanabe, T., 1990 Microsequencing of proteins electroblotted onto immobilizing matrices from polyacrylamide gel electrophoresis: Application to an insoluble protein. *Electrophoresis* 11:573.

Ikeuchi, M. and Inoue, Y., 1988 A new photosystem II reaction center component (4.8 kDa protein) encoded by chloroplast genome, *FEBS Lett.* 241:99.

Komatsu, S. and Hirano, H., 1992 Rice seed globulin: A protein similar to wheat seed glutelin, *Phytochemistry* 31:3455.

Komatsu, S., Kajiwara, H. and Hirano, H., 1992 Rice protein library: a data file of rice proteins separated by two-dimensional electrophoresis, *Nippon Nogeikagaku Kaishi* 66:166.

Krishna, R. G., Chin, C. C. Q. and Wold, F., 1991 *N*-terminal sequence analysis of *N*-acetylated proteins after unblocking with *N*-acylaminoacyl-peptide hydrolase, *Anal. Biochem.* 199:45.

Matsudaira, P., 1987 Sequence from picomole quantities of proteins electroblotted onto polyvinylidene difluoride membranes, *J. Biol. Chem.* 262:10035.

Moerschell, R. P., Hosokawa, Y., Tsunasawa, S. and Sherman, F., 1990 The specificities of yeast methionine aminopeptidase and acetylation of amino-terminal methionine *in vivo.* Processing of altered iso-1-cytochrome *c* created by oligonucleotide transformation. *J. Biol. Chem.* 265:19638.

Moyer, M., Harper, A., Payne, G., Ryals, J. and Fowler, E., 1990 *In situ* digestion with pyroglutamate aminopeptidase for N-terminal sequencing electroblotted proteins. *J. Protein Chem.* 9:282.

Podell, D. N. and Abraham, G. N., 1978 A technique for the removal of pyroglutamic acid from the amino terminus of proteins using calf liver pyroglutamyl amino peptidase, *Biochem. Biophys. Res. Commun.* 81:176.

Sheehan, J. C. and Yang, D. D. H., 1958 The use of *N*-formylamino acids in peptide synthesis, *J. Am. Chem. Soc.* 80:1154.

Sugino, Y., Tsunasawa, S., Yutani, K., Ogasahara, K. and Suzuki, M., 1980 Amino-terminally formylated tryptophan synthase α-subunit produced by the *TRP* operon cloned in a plasmid vector, *J. Biochem.* 87:351.

Tsunasawa, S. and Sakiyama, F., 1992 Amino-terminal acetylation of proteins, *in*: The Post-translational Modification of Proteins, Tuboi, S., Taniguchi, N. and Katsunuma, N., eds. 113.

Tsunasawa, S. and Hirano, H., 1992 Microsequence analysis of *N*-terminally blocked proteins, *J. Protein Chem.*11:382.

Tsunasawa, S., Takakura, H. and Sakiyama, F., 1990 Microsequence analysis of *N*-acetylated proteins, *J. Protein Chem.* 9:265.

Wellner, D., Panneerselvam, C. and Horecker, B. L., 1990 Sequencing of peptides and proteins with blocked *N*-terminal amino acids: *N*-Acetylserine or *N*-acetylthreonine, *Proc. Natl. Acad. Sci. USA* 87:1947.

DEVELOPMENT OF NOVEL C-TERMINAL SEQUENCING METHODS

Akira Tsugita, Keiji Takamoto, Hiromoto Iwadate, Masaharu Kamo, Hiroyuki Yano, Norifumi Miyatake, and Kazuo Satake

Science University of Tokyo
Yamazaki, Noda 278 Japan

INTRODUCTION

Development of a reliable C-terminal sequencing method has been expected from various aspects including sequencing protein, analyzing posttranslational process, confirming recombinant proteins and cloning. Carboxypeptidase digestion has been commonly used with limitations. Since a classical isothiocyanate degradation was proposed, several modifications were reported in the past two conferences (Hawkes and Boyd, 1991; Inglis *et al.*, 1991; Miller and Shively, 1989). The present paper summarizes two related novel carboxy-terminal sequencing methods using perfluorinated carboxylic acids in monohydrate form and carboxylic acid anhydrides. These reagents produce the mixtures of C-terminal successive degraded molecules which are able to be analyzed by fast-atom-bombardment (FAB)- or electrospray ionization (ESI)-mass spectrometry (MS).

A NOVEL ORGANIC ACID SUCCESSIVE DEGRADATION (Tsugita *et al.*, 1992a)

A peptide in a solid form were exposed to the vapor of 90% pentafluoropropionic acid (PFPA) aqueous solution at 90°C for more than 2 h. The reaction mixture was dried and subjected to FAB-MS. The C-terminal successively degraded molecular ions, MH^+, including the original peptide one, were clearly observed and the differences of MH^+ corresponded to the residual mass of amino acids, which were released amino acids from the C-terminus, as shown in Figures 1a and b.

Because of the volatility and low pKa, trifluoroacetic acid (TFA), PFPA, and heptafluorobutyric acid (HFBA) were chosen as testing acid, where both PFPA and HFBA showed clear-cut C-terminal degraded molecular ion series. To avoid contamination from these acids the vapor phases of the acids were employed. We also tested acid concentration and noticed the different modes of acid hydrolysis; low acid concentrations resulted in non-specific cleavages, while high acid concentration gave the C-terminal successive degradation. Of the various acid concentrations tested (50-98%) for several peptides and proteins, around 90% were found out to be optimal for this specific C-terminal degradation.

When high concentration of acid was used, oxidation of side chains in the peptide was observed. Addition of a reducing reagent, dithiothreitol (DTT) to the acid by directly

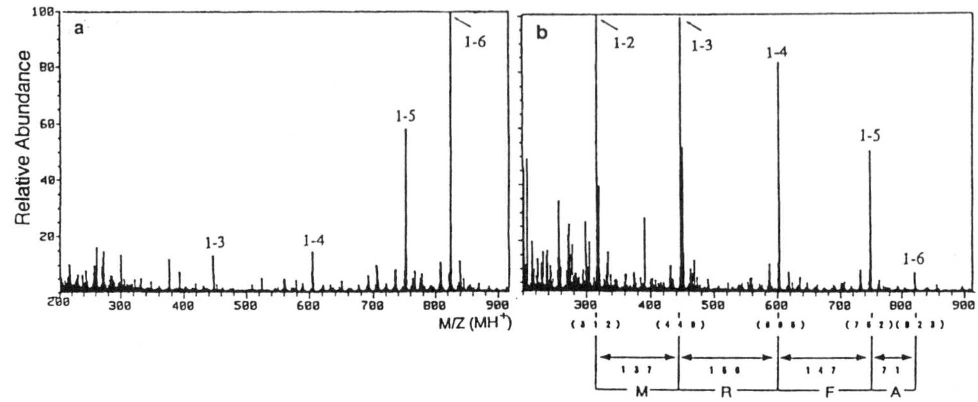

Figure 1. FAB-MS of the hydrolyzates of a hexapeptide, LWMRFA.
The peptide (2 nmol) was hydrolyzed with 90% PFPA containing 5% DTT in the vapor phase at 90°C (a) for 4 h and (b) for 24 h.

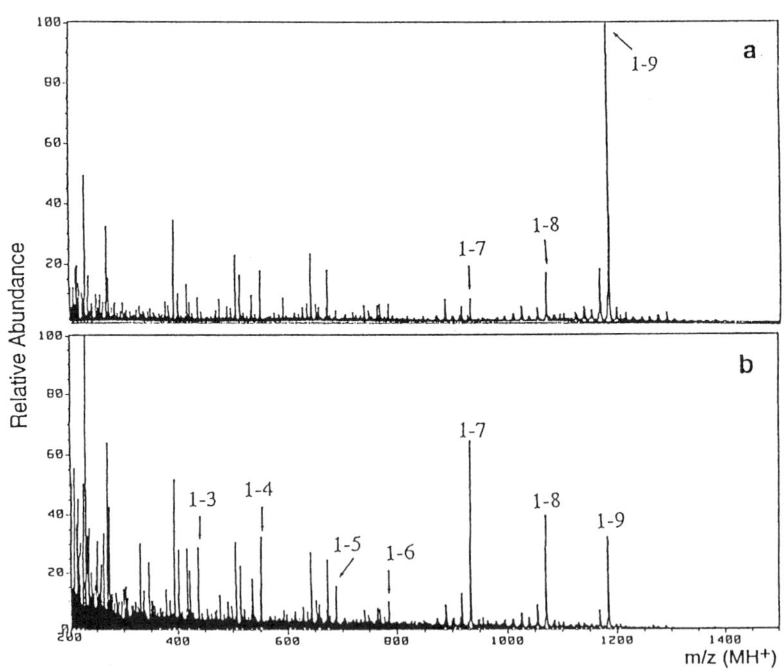

Figure 2. FAB-MS of the hydrolyzates of a nonapeptide, RVYIHPFHL.
The peptide (200 pmol) was hydrolyzed with 90% HFBA containing 5% DTT in the vapor phase at 90°C. (a) for 4 h and (b) for 24 h.

mixing (1-5%) or separately as solid was found effective to avoid the oxidative reactions of Met, Trp and Tyr residues.

The peptide bond of Pro-Xaa is partially resistant to the present successive degradation. A nonapeptide, RVYIHPFHL, was hydrolyzed for 4 h with 90% HFBA at 90°C. Analysis of the hydrolyzate indicated resistance to hydrolysis at the peptide bond between residues 6 and 7 (Figure 2a). Prolonged hydrolysis showed further successive degradation accompanying a small extent of nonspecific cleavage (Figure 2b).

The longer peptide magainin 1 having 23 amino acids was reacted with the vapor of 90% PFPA at 90°C for 2 h and analyzed by a ESI-MS (Figure 3).

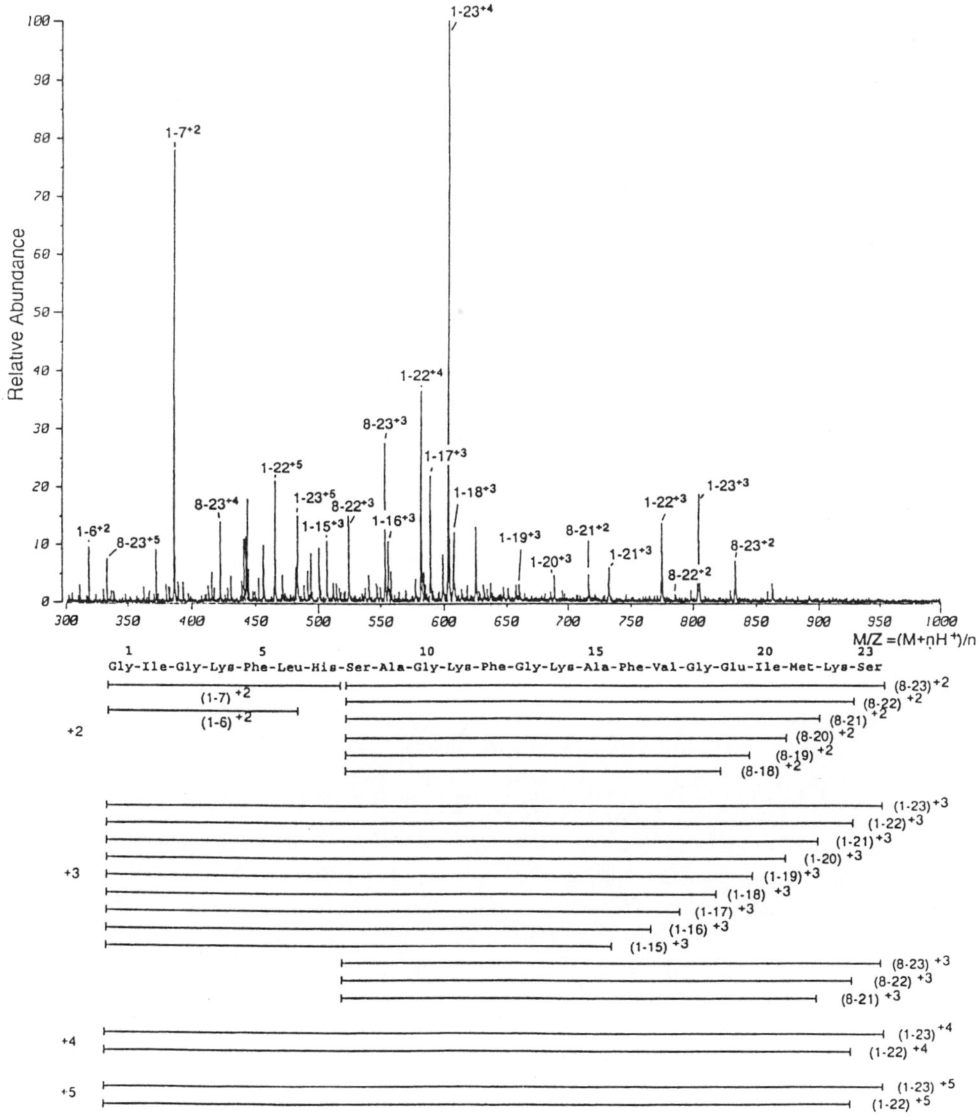

Figure 3. ESI-MS of the hydrolyzate of a peptide, GIGKFLHSAGKFGKAFVGEIMKS. The peptide (125 pmol) was hydrolyzed with 90% PFPA in the vapor phase in the presence of solid DTT (50 mg) at 90°C for 1 h. The dried hydrolyzate was dissolved in 25 μl of 2% acetic acid in 50% aqueous methanol solution.

The multicharged degraded molecular ions of both the intact peptide and a fragment specifically cleaved at the N side of the eighth Ser residue (8-23), were observed. The +5 molecular ion was the maximal chargeable ion corresponding to the N-terminal amino group and four internal Lys residues in the peptide. The absence of further degraded ions in the series of +4 and +5 molecular ions was possibly due to the charge at Lys 22. The longest successive C-terminal degradation ions, of eight residues, were detected in the +3 molecular ion series.

Both the peptide bonds at the C side of the internal Asp and the peptide bond at the N side of Ser were selectively cleaved under the present conditions. In the experiments performed on various test peptides, the N side of Thr and both sides of Gly were observed to be occasionally cleaved (data not shown).

It should be noted that under the present conditions little cleavage of the amide groups of acidic amino acids, except for the C-terminal α-carboxyl amide bond, was observed.

The method was applied to the proteins including glucagon and egg white lysozyme, where both the specific internal peptide bond cleavages and the C-terminal successive degradation of the fragments including the protein C-terminal sequences were observed (Tsugita *et al.*, 1992a).

The successive degradation molecular ions frequently accompanied with -18 molecular ions, suggesting reaction forms oxazolone ring at the C-termini as the intermediate step.

SUCCESSIVE DEGRADATION BY PERFLUOROACYL ANHYDRIDE VAPOR (Tsugita *et al.*, 1992b)

We used a vapor of 10% pentafluoropropionic anhydride (PFPAA) in acetonitrile on peptides and observed the production of the C-terminal successively degraded molecules at a very low temperature. The reaction on a dodecapeptide, ARGIKGIRGFSG, was carried out at -18°C for 10 min to 1 h (Figure 4); 10 min (Figure 4a) showed only successive degradation, and further reaction formed additional series of acylated molecular ions and the acylation almost completed after 30 min (Figure 4b).

The extent of degradation was observed to be from 10 to 20 residues. No additional nonspecific cleavage of peptide bonds has been observed as in the acid vapor degradation. Extension of reaction time (1 h) resulted in more acylation and more dehydration reactions which was shown as -18 molecular ions (Figure 4c).

FAB-MS showed two types of accompanying minor molecular ions -18 and -45 of the respective degraded fragment ions, corresponding to dehydration and decarboxylation respectively. The aqueous pyridine vapor treatment (10% pyridine at 100°C for 10 min) before applying the products to mass spectrometry lowered the -18 molecular ions and eliminate a part of acylation peaks (Figure 4d).

Acylation reaction takes place on α-amino group of the N-terminus, both O-acylation of Ser residues and oxazolone ring and ε-amino group of Lys residues in the order described. The water vapor (aqueous pyridine vapor) treatment decomposes O-acylation but not N-acylation as shown in Figure 4d. Dehydration on α-carboxyl group to oxazolone is recovered by the water vapor treatment but other dehydrations such as acidic amino acid amides to its nitrile and aspartic acid to 5 membered ring are unrecoverable by the water vapor treatment. The use of additional 20% pentafluoropropionic acid in acetonitrile solution for perfluoroacyl anhydride vapor reduces the formation of these unrecoverable dehydration.

In FAB-ionization process, peptide fragment ions were known as a, b and c series ions for the C-terminal fragments and x, y, z series ions for the N-terminal fragments. These N-terminal fragment ions were not observed in the present experiments while the similar fragments as the C-terminal fragments, a, b, and c were observed; The present FAB mass

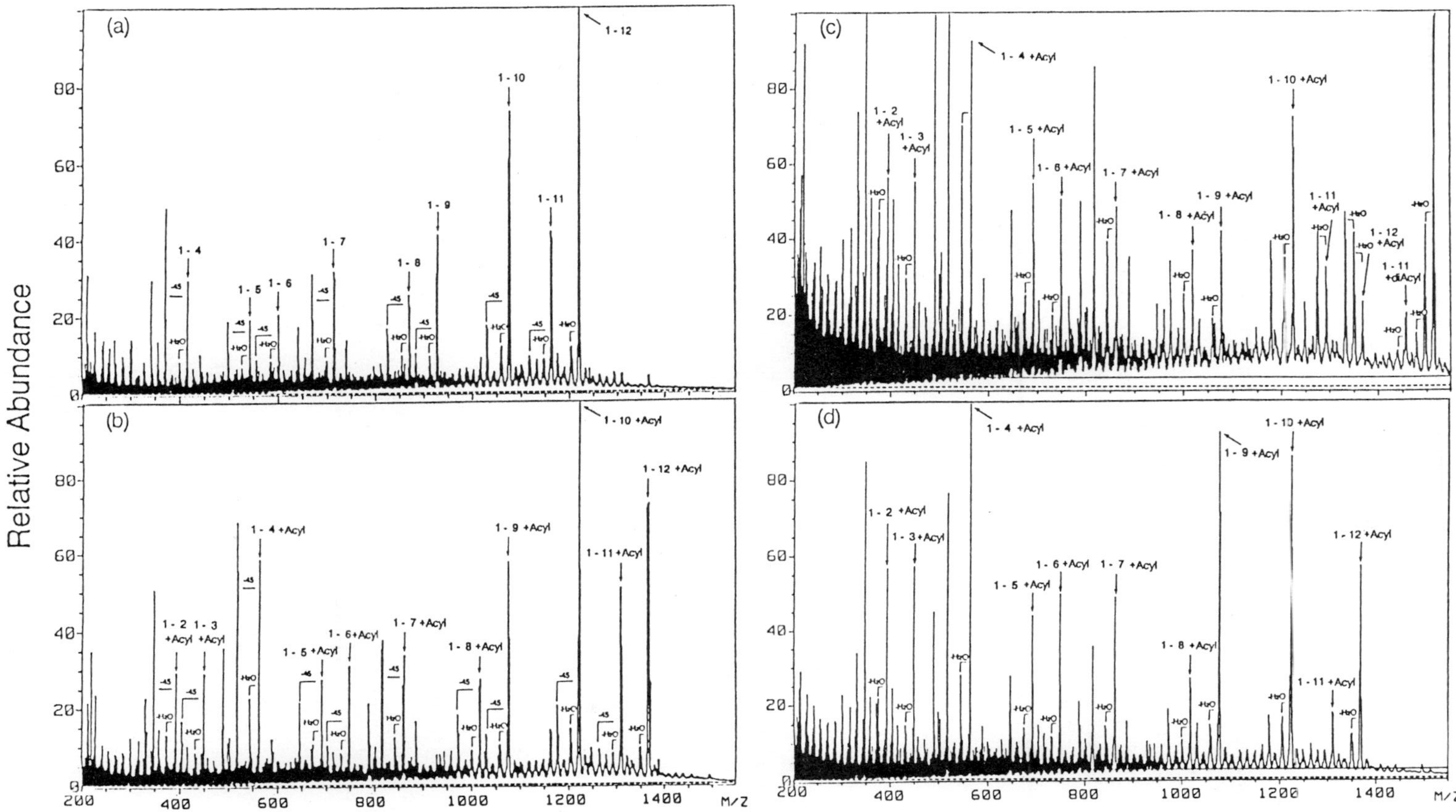

Figure 4. FAB-MS of successive degradation with PFPAA vapor on a dodecapeptide, ARGIKGIRFSG.
The reaction on a dodecapeptide (2 μg) was performed at -18°C for (a) 10 min, (b) 30 min, (c) 1 h and (d) 1 h, and treated with 10% pyridine water vapor at 100°C for 10 min.

degraded molecular ions, minus 45 mass decarboxylation ions, minus 18 mass dehydration ions and the major degraded ions are almost the same as a, b and c respectively, but a and c ion series are clearly one mass smaller than the present degraded ions.

Also both FAB MS ionization of the untreated peptide and B/E and B^2/E linked scan in FAB MS of the original peptide peak of the PFPAA products did not show any fragment ions and parent ions respectively. These observations were unfavor for the present degradation to be caused by FAB-fragmentation. Negative FAB MS spectra also resulted in reasonable degraded ions.

Release of ninhydrin positive amino groups in the PFPAA reaction products, HPLC separation and the amino acid compositions of the separated peaks confirmed the C-terminal degradation by PFPAA shown by the major degraded peaks in Figure 5. Amino acid composition of the hydrolyzates of peaks as well as FAB MS of the respective HPLC peak (data not shown) clearly demonstrated the chemical successive degradation.

Since no endo-peptide bond cleavage was observed, additional conventional enzymatic or

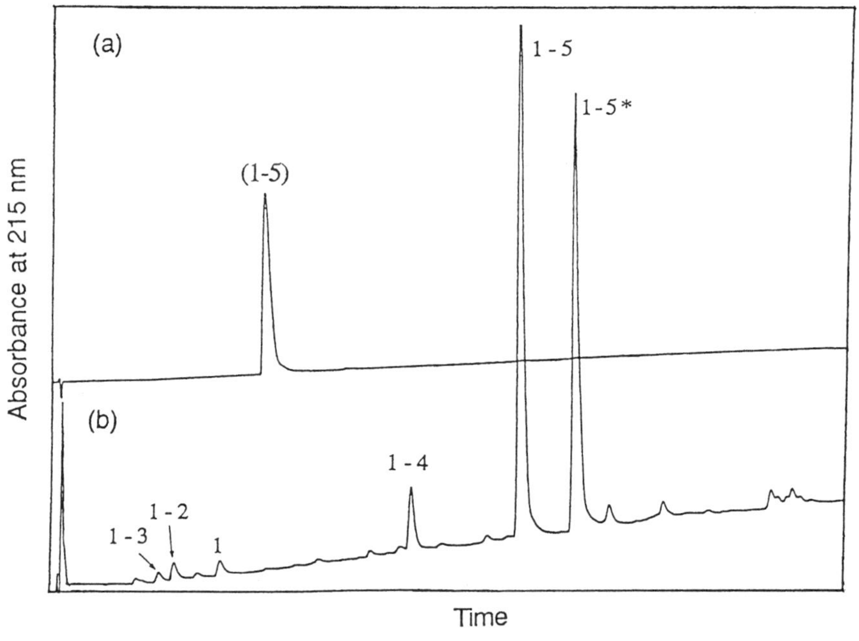

		1	1-2	1-3	1-4	1-5	1-5*
Tyr	1	1.0	1.0	1.0	0.8	1.1	1.1
Gly	2	0.1	1.3	2.3	2.1	2.3	2.3
Phe	1	0.0	0.0	0.0	1.0	1.1	1.1
Leu	1	0.0	0.0	0.0	0.0	1.0	1.0
		Y	YG	YGG	YGGF	YGGFL	YGGFL

Figure 5. HPLC elution profile of the reaction product of YGGFL.
The peptide (20 μg) was reacted with PFPAA vapor for 1 h at -18°C. The product was immediately treated with 10% pyridine water for 10 min at 100°C. Reversed-phase HPLC (AP-302, 4.6 x 150 mm) was carried out with a linear gradient of 4-48% acetonitrile in 0.1% TFA. (a) for untreated peptide (1-5) and (b) for PFPAA treated peptide, which are acylated. 1-5 and 1-5* are optical isomers.

chemical cleavages, after the successive degradation, were introduced to obtain small fragment(s). The other choice was pre-cleavage such as by CNBr and the newly introduced C-termini, homoserine lactones, were immediately blocked with dimethylhydrazine vapor or the other blocking reagents. The blockage were resistant to the successive degradation but the original C-terminus. Partial hydrazinolysis of proteins in vapor phase is also applicable for this type of cleavage.

Figure 6 shows PFPAA degradation on rice-plastocyanin and the product was cleaved by cyanogen bromide. The carboxy terminal sequence has been observed.

Glass filter membrane (Whatman GF/C) may be employed in the present degradation, which was washed by formic acid and protein samples were spotted or electroblotted. The membrane was dried and subjected to perfluoroacyl anhydride reaction. The membrane was then extracted with 70% formic acid, dried and subjected to FAB MS spectrometry. This method successfully used for the C-terminal analysis of several proteins, but still have a quite few problems to solve before generalize the method for the big proteins.

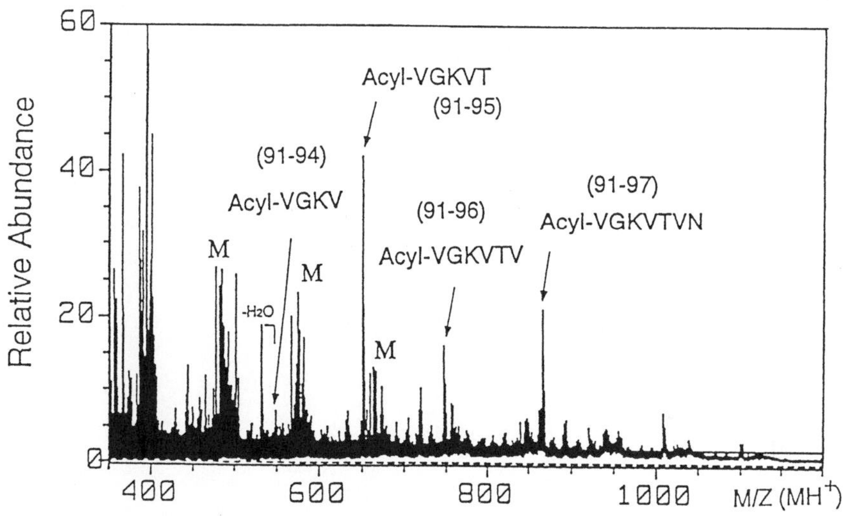

Figure 6. FAB-MS of the C-terminal degradation of CNBr fragments of plastocyanin. Rice plastocyanin was denatured by 6 M urea. The apoprotein (50 μg) was treated with the vapor of 30% (v/v) PFPAA in acetonitrile at -18°C for 2 h. After the 10% pyridine water treatment at 100°C for 10 min, the product was subjected to CNBr cleavage. M stands for matrix line.

REFERENCES

Hawke, D. H. and Boyd, V. L., 1991 Chemical C-terminal sequencing, *in*: Methods in Protein Sequence Analysis, Jörnvall, H., Höög, J. -O. and Gustavsson, A. M., eds., Birkhauser, Basel, 35.

Inglis, A. S., Moritz, R. L., Begg, G. S., Reid, G. E., Simpson, R. J. Graffunder, H., Matschull, L. and Wittmann-Liebold, B., 1991 C-terminal sequence analysis, *in*: Methods in Protein Sequence Analysis, Jörnvall, H., Höög, J. -O. and Gustavsson, A. M., eds., Birkhauser, Basel, 23.

Miller, C. G. and Shively, J. E., 1989 Carboxy-terminal sequence analysis of proteins and peptides by chemical methods, *in*: Methods in Protein Sequence Analysis, Wittmann-Liebold, B., ed., Springer-Verlag, Berlin, 145.

Tsugita, A., Takamoto, A., Kamo, M. and Iwadate, H., 1992a C-terminal sequencing of protein; a novel partial acid hydroloysis and analysis by mass spectrometry, *Eur. J. Biochem.* 206:691.

Tsugita, A., Takamoto, K. and Satake, K., 1992b Reaction of pentafluoropropionic anhydride vapor on polypeptide as revealed by mass spectrometry. A carboxypeptidase mimetic degradation, *Chem. Lett.* 235.

AUTOMATED C-TERMINAL SEQUENCING OF PEPTIDES AND PROTEINS

Jerome M. Bailey, Miro Rusnak, and John E. Shively

Division of Immunology
Beckman Research Institute of the City of Hope
Duarte, CA 91010, USA

INTRODUCTION

The development of a chemical method for the sequential degradation of a protein or peptide from the carboxy-terminus is a goal of our laboratory. Such a method, in addition to complementing existing N-terminal methods of degradation, would be invaluable for the sequence analysis of proteins with naturally occurring N-terminal blocking groups and for the detection of post-translational processing at the carboxy-terminus of expressed gene products. Although several methods for a sequential C-terminal degradation have been proposed (Ward, 1986; Rangarajan, 1988), the thiocyanate method based on the procedure originally published by Schlack and Kumpf (1926) has been the most widely studied. Recent work in our laboratory, introducing new reagents for the derivatization of the C-terminal amino acid (Hawke *et al.*, 1987) and for the specific cleavage of the derivatized amino acid (Bailey and Shively, 1991; Bailey *et al.*, 1992), has suggested that an automated chemical method for the sequential degradation of polypeptides from the carboxy-terminus, analogous to the Edman method for amino terminal degradation, may be feasible.

The development of a C-terminal method involves the combination of several diverse areas of research, all of which must be considered concurrently. These include: the design of all instrument capable of performing the sequencing chemistry, the choice of a solid support stable to the conditions used for sequencing, optimization of methodology to covalently couple polypeptide samples to the solid support in high yield or optimization of the conditions of sequencing so as not to cause washout of non-covalently coupled samples, and the development of chemistry to specifically derivatize and hydrolyze the C-terminal amino acid. This report describes the construction of a compact microsequencer specifically designed for C-terminal sequence analysis and the development of improved chemistry for the derivatization of the C-terminal amino acid to a thiohydantoin.

Methods in Protein Sequence Analysis, Edited by K. Imahori
and F. Sakiyama, Plenum Press, New York, 1993

RESULTS AND DISCUSSION

Instrumentation

A schematic of our C-terminal sequencing instrument is shown in Figure 1. The overall design is based on our new mini gas phase N-terminal sequencer described by Calaycay *et al.* (1991). The dimensions of the instrument are 11 (width) x 16.5 (height) x 23.5 (depth) inches. The hex-valve incorporated into the N-terminal machine was replaced by two square quad-valves in order to permit delivery of eight different solvents and reagents to the continuous flow reactor (CFR). One of these same quad-valves has been utilized to permit the delivery of four reagents to the conversion flask (CF) as well. The instrument is equipped with a temperature controlled CFR and conversion flask as described by Shively *et al.* (1987). Both the CFR and CF are lighted from the back with a high intensity red LED to facilitate timing of reagent and solvent deliveries. In order to minimize bubble formation during delivery of reagents to the CFR, solvent input to the CFR is from the bottom. Solvent and reagent deliveries to the CFR and CF are achieved by maintaining a low pressure on each bottle (1.5 p.s.i.) and performing timed deliveries. Argon gas is brought into the instrument gas manifold at 35 psi and distributed to five Porter (Model 8311) stainless-steel regulators. Teflon tubing was used for all plumbing. Unlike the N-terminal sequencers, the conversion flask is used only as a holding vessel for the thiohydantoin amino acid. The cleavage solution containing the C-terminal amino acid as a thiohydantoin derivative is transferred from the CFR to the conversion flask where it is then dried under a stream of argon, taken up in 2% aqueous trifluoroacetic acid, and directed to a Rheodyne 7126 pneumatically actuated injector. Liquid exiting the stainless steel loop is detected by an optical sensor which transmits a signal to the computer/controller and triggers an injection. This system has permitted the reproducible injection of up to 90% of the sample to the on-line HPLC. All the valves shown in Figure 1 are controlled by a computer and are solenoid actuated. The zero-dead volume Rusnak valve is described by Hawke *et al.* (1985) and the 3-way switching valve is described in Calaycay *et al.* (1991). Like the N-terminal instrument, the C-terminal instrument has been divided vertically by a steel bulkhead into an electronic and chemistry section.

Sequencing Chemistry

Previous work in our laboratory involved the use of acetic anhydride/trimethyl-silylisothiocyanate (TMS-ITC) for the derivatization of the C-terminal amino acid to a thiohydantoin (Bailey *et al.*, 1992). These studies revealed a number of problems with this chemistry, such as: l) precleavage, 2) the requirement of an oxazolinone for reaction of the TMS-ITC reagent to form thiohydantoins, 3) competing reactions during oxazolinone formation which formed cyclic anhydrides when C-terminal Glu or Asp was present and the formation of unsaturated oxazolinones when a C-terminal Thr or Ser was present, all of which were incapable of reacting with TMS-ITC to form the desired thiohydantoin, 4) application of this chemistry to protein samples was difficult, presumably due to the insolubility of most proteins in acetic anhydride. A major limitation of this approach was the need to activate the C-terminus with acetic anhydride. We now describe the use of a new reagent, diphenyl phosphoroisothiocyanatidate (DPP-ITC) and pyridine, which combines the activation and derivatization steps to produce peptidylthiohydantoins. Previous work by Kenner *et al.* (1953) with DPP-ITC demonstrated slow kinetics. Several days were required for complete reaction. We show here that the inclusion of pyridine was found to promote the formation of C-terminal thiohydantoins by DPP-ITC resulting in complete conversion of the C-terminal amino acid to a thiohydantoin in less than one hour. Reagents such as imidazole, triazine, and tetrazole were also found to promote the reaction with DPP-ITC as effectively as pyridine. General base catalysts, such as triethylamine, do not promote the reaction. The chemical scheme shown in Figure 2, represents the chemistry used for our automated C-terminal sequencing. It can be considered to consist of three steps: 1) activation of the C-terminal carboxylic group to form a carboxylate, 2) derivatization of the C-terminal amino acid to a thiohydantoin with DPP-ITC and pyridine, and 3) specific cleavage of the derivatized amino acid with sodium trimethylsilanolate.

Mechanism of Thiohydantoin Formation with DPP-ITC

The use of DPP-ITC for C-terminal sequencing was first described by Kenner *et al.* (1953). These authors proposed a reaction mechanism whereby isothiocyanate ion was liberated by exchange with the C-terminal carboxylate forming an acyl phosphate. The

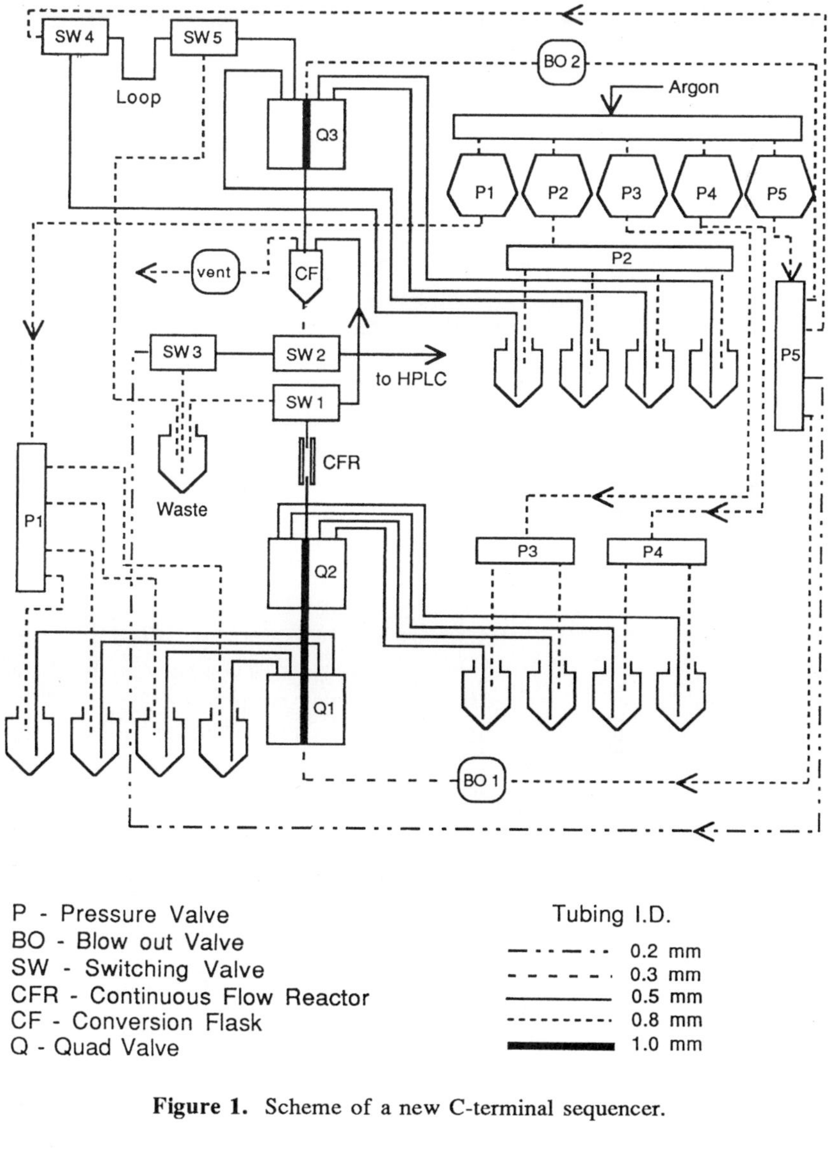

Figure 1. Scheme of a new C-terminal sequencer.

acyl phosphate was in turn postulated to be attacked by the isothiocyanate ion to form the desired acyl isothiocyanate and diphenyl phosphate ion. Although this chemistry successfully formed a C-terminal thiohydantoin it was never actively pursued, most likely since it took several days for the reaction to go to completion. We have found that inclusion of pyridine, imidazole, triazine, or tetrazole permitted the reaction of the C-terminal carboxylate with DPP-ITC, to form a thiohydantoin, to be complete in less than 30 min at 50°C. We have also found that reaction of N-acetyl-Thr-Val-Leu with DPP-ITC/pyridine formed a single peptidylthiohydantoin. This is in contrast to the reaction of this peptide with acetic anhydride/TMS-ITC for which we observed the formation of two diastereomeric peptidylthiohydantoins (Bailey and Shively, 1990). It is likely that racemization was the result of oxazolinone formation. The single peptidylthiohydantoin

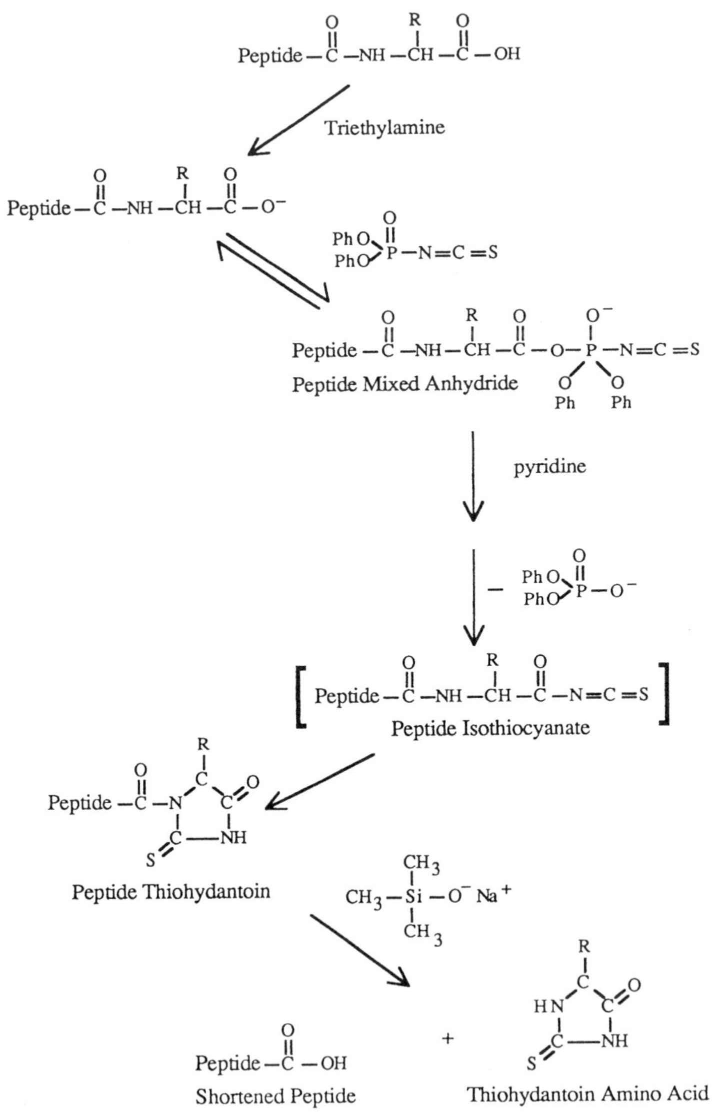

Figure 2. The formation of the C-terminal thiohydantoin.

formed on reaction with DPP-ITC/pyridine suggested that an oxazolinone was not an intermediate in the formation of the C-terminal thiohydantoin by this chemistry (Figure 2). Application of this chemistry to peptides covalently coupled to a solid support revealed that the DPP-ITC reaction could be separated from the pyridine reaction. Since excess DPP-ITC was completely removed by a solvent wash step, we concluded that pyridine does not act as a catalyst to form the acyl phosphate. Indeed this derivative must form without the elimination of -NCS as suggested by Kenner *et al.* (1953), otherwise the -NCS would be lost in the wash step preventing formation of the thiohydantoin. Since the addition of pyridine after removal of DPP-ITC is required to complete the reaction, we conclude that pyridine promotes the rearrangement of the acylphosphorylisothiocyanate to the acylisothiocyanate with concomitant release of diphenylphosphate. Pyridine has recently been shown to promote a similar type of reaction by removing carbon dioxide from the mixed anhydride formed on reaction of an amino acid carboxylate with ethyl chloroformate to form an amino acid ester (Husek, 1991). The success of this reaction was dependent on pyridine being present in a molar excess with respect to the ethyl chloroformate. Additionally, when the pyridine and DPP-ITC steps were separated, treatment of the covalently coupled peptide with a base, such as triethylamine, was found to be absolutely essential for this reaction to proceed. Presumably, this converts the C-terminal carboxylic acid group to a carboxylate, a required form for attack on DPP-ITC. When DPP-ITC and pyridine were added together, pretreatment with triethylamine was not required; thus, pyridine most likely performed this function. Nonetheless, the use of liquid pyridine contributed greatly to the background peaks on the HPLC and was not a desirable step. Thus, our optimal protocol involved first treatment of the immobilized peptide with aqueous triethylamine for a few min, removal of the triethylamine solution by drying the support, reaction of the peptide with DPP-ITC in acetonitrile, removal of the DPP-ITC

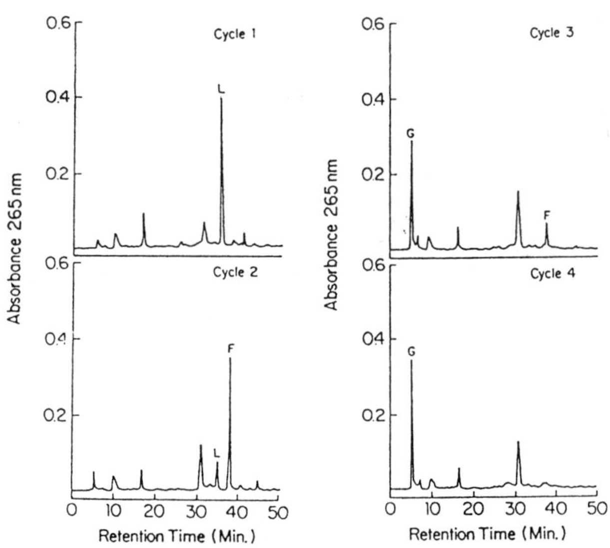

Figure 3. The result of automated C-terminal sequencing of the pentapeptide, YGGFL (5.6 nmol).

solution by washing the support with either anhydrous DMF or heptane, and then treatment of the peptide with pyridine in the gas phase. Separation of the DPP-ITC and pyridine steps was found to be essential for the successful derivatization of C-terminal aspartate to a thiohydantoin. This was most likely due to catalysis of cyclic anhydride formation by pyridine when present with an excess of DPP-ITC. Pyridine was found to catalyze cyclic anhydride formation of C-terminal glutamate when present during peptidylthiohydantoin formation by acetic anhydride/TMS-ITC (Bailey and Shively, 1990). The use of DPP-ITC obviated problems observed with the acetic anhydride/TMS-ITC reagent by minimizing the free isothiocyanate ion concentration so that precleavage was no longer apparent, and eliminating the need for oxazolinone formation. This has resulted in improved yields for Thr, Ser, Glu and Asp. Like the acetic anhydride/TMS-ITC chemistry, the DPP-ITC/pyridine chemistry is not capable of degrading C-terminal proline.

Application of the Autòmated C-Terminal Chemistry to Peptide and Protein Samples

Figure 3 shows four cycles of automated C-terminal sequencing of the pentapeptide, YGGFL (5.6 nmol), covalently coupled to carboxylic acid modified polyethylene. Procedures for the covalent coupling of peptides to this support have been described (Shenoy *et al.*, 1992). More recent work has involved application of this chemistry to the automated C-terminal sequencing of protein samples. To date, two to three cycles of sequencing has been rountinely possible on sample quantities ranging from 300 pmol to 5 nmol.

Acknowledgements. We thank Dr. Harvey Brandwein of the Pall Corp. (Long Island, NY) for providing us with the carboxylic acid modified polyethylene support, Firoozeh Nikfarjam for technical support, Dr. Narmada Shenoy for covalently coupling peptide samples, Jim Sligar for performing the amino acid compositions, and John Wieser of Column Engineering (Ontario, CA) for providing the Reliasil HPLC columns used for the thiohydantoin amino acid separation. This work was supported in part by N.I.H. Grant # GM46022 and by the Tosoh Corporation (Japan).

REFERENCES

Bailey, J. M., Shenoy, N. R., Ronk, M. and Shively, J. E., 1992 Automated carboxy-terminal sequence analysis of peptides, *Protein Science* 1:68-80.

Bailey, J. M.and Shively, J. E., 1990 Carboxy-terminal sequencing: Formation and hydrolysis of C-terminal peptidylthiohydantoins, *Biochemistry* 29:3145.

Bailey, J. M.and Shively, J. E., 1991 Carboxy terminal sequencing: Automation and application to the solid phase, *in*: Techniques in Protein Chemistry: II, Villafranca, J. J., ed., Academic Press, Inc. New York. pp. 115-129.

Calaycay, J., Rusnak, M. and Shively, J. E., 1991 Microsequencing analysis of peptides and proteins IX, An improved, compact, automated instrument, *Anal. Biochem.* 192:23.

Hawke, D. H., Lahm, H. -W., Shively, J. E. and Todd, C. W., 1987 Microsequence analysis of peptides and proteins: Trimethylsilylisothiocyanate as a reagent for COOH-terminal analysis, *Anal. Biochem.* 166:298.

Hawke, D. H., and Harris, D. C. and Shively, J. E., 1985 Microsequence analysis of peptides and proteins V. Design and performance of a novel gas-liquid phase instrument. *Anal. Biochem.* 147:315.

Husek, P., 1991 Rapid derivatization and gas chromatographic determination of amino acids, *J. Chromatogr.* 552:289.

Kenner, G. W., Khorana, H. G. and Stedman, R. J., 1953 Peptides, Part IV, Selective removal of the C-terminal residue as a thiohydantoin. The use of diphenylphosphoroisothiocyanatidate, *Chem. Soc. Jour. (London)* 673-678.

Rangarajan, M. 1988 Chemical methods of amino acid sequence analysis from carboxyterminal end, *in*: Protein/Peptide Sequence Analysis: Current Methodologies, Brown, A. S., ed., CRC Press, Bora Raton, FA., pp. 135-144.

Schlack, P. and Kumpf, W., 1926 Uber eine neue methode zur Ermittlung der konstitution von peptiden, *Z. Physiol. Chem.* 154:125.

Shenoy, N. R., Bailey, J. M. and Shively, J. E., 1992 Carboxylic acid modified polyethylene: A novel support for the covalent immobilization of polypeptides for C-terminal sequencing, *Protein Science* 1:58.

Shively, J. E., Miller, P. and Ronk, M., 1987 Microsequence analysis of peptides and proteins VI. A continuous flow reactor for sample concentration and sequence analysis, *Anal. Biochem.* 163:517.

Ward, C. W., 1986 Carboxyl terminal sequence analysis, *in*: Practical Protein Chemistry -- A Handbook, Darbre, A., ed., John Wiley and Sons, Ltd, Chichester, New York, pp. 491-525.

A NEW CHEMICAL APPROACH TO C-TERMINAL MICROSEQUENCE ANALYSIS VIA THE THIOHYDANTOIN

Adam S. Inglis and Claudio De Luca

CRIBI Biotechnology Centre, University of Padua
Via Trieste 75-35121 Padua, Italy.

INTRODUCTION

Our earlier papers reported the successful application of a thiohydantoin procedure to both manual and automatic C-terminal sequence analysis of peptides and proteins. With the exception of proline, all amino acids that might have been expected to give difficulties were shown to be degraded normally in peptides (Inglis *et al.*, 1989; Inglis *et al.*, 1990). These findings have been largely corroborated by others (Bailey *et al.*, 1992) using different reaction conditions, following some initial problems with amino acids such as aspartic acid, asparagine and glutamic acid (Bailey and Shively, 1990). Apparently dithiothreitol (DTT) in the cleavage reagent caused adverse effects; in our experience (with good quality reagents), neither the background nor the repetitive yields have been affected adversely by cleavage of the thiohydantoin with 0.5 M KOH in 33% methanol, in the presence of DTT, for just 2-3 minutes reaction at room temperature. However, the analysis of proline is now the major obstacle to a general routine method employing ammonium thiocyanate as the coupling reagent.

Stark (1968) believed that the thiohydantoin procedure would not work for proline and aspartic acid. While that has now been disproved for the latter (Inglis *et al.*, 1989), it is clear from recent work that the former is quite unreactive under some circumstances and, indeed, the premise has again been advanced that the chemical pathway is such as to preclude the reaction of thiocyanate reagents with proline (Bailey *et al.*, 1992). On the other hand, Kubo *et al.* (1971) obtained a product from acetylproline on which they did elemental, UV, IR and TLC analyses, all of which indicated that it was a pure sample of proline thiohydantoin. Our initial findings with HSCN and acetylproline (Inglis *et al.*, 1989) were completely consistent with their results in that we also obtained a major product with an absorption maximum at 267 nm; moreover, its elution time on reversed phase columns, as compared with those of other amino acid thiohydantoins, was as expected. As far as is reported, neither Stark (1968) nor Bailey *et al.* (1992) tried HSCN for this purpose. We found recently that the thiohydantoin can be made beginning with ammonium thiocyanate (Inglis

Methods in Protein Sequence Analysis, Edited by K. Imahori
and F. Sakiyama, Plenum Press, New York, 1993

et al., 1992). This approach followed our earlier work in which thiocyanic acid was prepared *in situ* by adding trifluoroacetic acid to a solution of ammonium thiocyanate, acetic acid and acetic anhydride (Inglis *et al.*, 1991) The aim in that work was to eliminate the need to chill HSCN in acetone during automatic C-terminal sequencing. Application of the ammonium thiocyanate procedure to proline also allowed us to test ideas expressed recently on the mechanism (Inglis, 1991) that suggested that the addition of a strong acid to the reaction mixture would be expected to favour the cyclization reaction. The facile preparation of proline thiohydantoin (Pro-TH) from acetylproline is consistent with this supposition.

One can conclude, therefore, that (a) acidic coupling conditions are best for the ammonium thiocyanate reaction with proline, (b) there is no impediment, a priori, to the formation of a peptidyl thiohydantoin of proline, (c) the classical reaction pathway (Inglis, 1991) is probably not applicable, (d) proline analysis may be affected by the events occurring after the coupling reaction, and (e) acid cleavage of the thiohydantoin may be preferable in principle. With respect to (c), an alternative set of reactions has been proposed (Inglis *et al.*, 1992) and those ideas are extended here; (d) and (e) have ramifications for both the analysis of proline in proteins, and for analysis using non-covalent attachment to a solid support - an advance that should greatly accelerate the routine application of the method. These aspects will be addressed in this paper.

FORMATION OF PROLINE THIOHYDANTOIN

Considering the difficulties encountered with some procedures in obtaining a sample of proline thiohydantoin (Bailey and Shively, 1990), the preparation of the compound with our procedure was remarkably easy (Inglis *et al.*, 1992). In this work we dispensed with the initial treatment of the acetylproline with an acetic acid-acetic anhydride mixture. Since ammonium thiocyanate is soluble in acetic acid we also left out the acetone that was originally required for preparation of the thiocyanic acid. It is clear from Figure 1a that even reaction at 40° for 30 min gave a reasonable yield of proline thiohydantoin (P). Although higher temperatures gave increased yields, more background products were evident.

The purified product (P, Figure 1d) was subjected to vapour phase hydrolysis at 150° and proline was liberated as established by analysis of the product after derivatization with phenylisothiocyanate. This result was unexpected in the light of the work of Cromwell and Stark (1969) who had difficulties with both acid and alkaline hydrolysis of thiohydantoins, albeit with much older methodology. The result was validated by the successful back-hydrolysis of the thiohydantoins of leucine and valine. Purified P was also analyzed by electron impact mass spectrometry (see Figure 2). The mass of the M^+ ion was found to be 156.0364 Daltons, which is the expected mass for proline thiohydantoin within 0.7 mDa. This analysis also confirmed that the acetyl group was eliminated during thiohydantoin formation. Because the major change that was made to the reaction mixture was the addition of a small amount of a strong acid (trifluoroacetic acid), we propose that an acid-catalyzed reaction mechanism operates, as depicted in Figure 3, rather than the alternatives previously postulated by Stark (1972).

RAMIFICATIONS FOR THE ANALYSIS OF PROLINE PEPTIDES

Since proline forms a thiohydantoin so readily with an amino acid, one would expect that some reaction should occur in peptides under similar reaction conditions. Hence, it follows that difficulties with this residue in peptides are more likely to be due to reactions subsequent to the coupling reaction. Now, it is also well recorded that acylamino acid

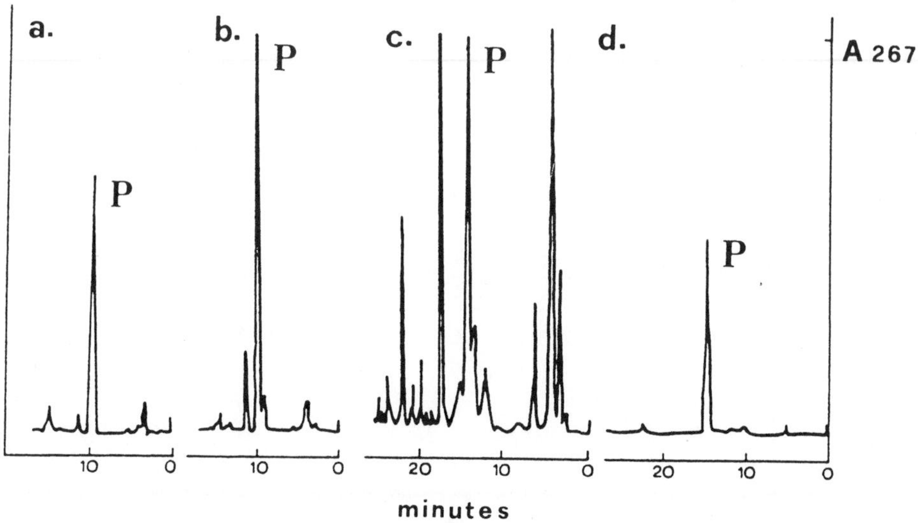

Figure 1. HPLC chromatograms of products from reaction of acetylproline and ammonium thiocyanate. a and b were obtained for reactions at 40° and 80°C respectively, c shows a trace of portion of the preparative mixture which was fractionated to give a highly enriched peak P, as indicated in d. (Inglis *et al.*, 1992)

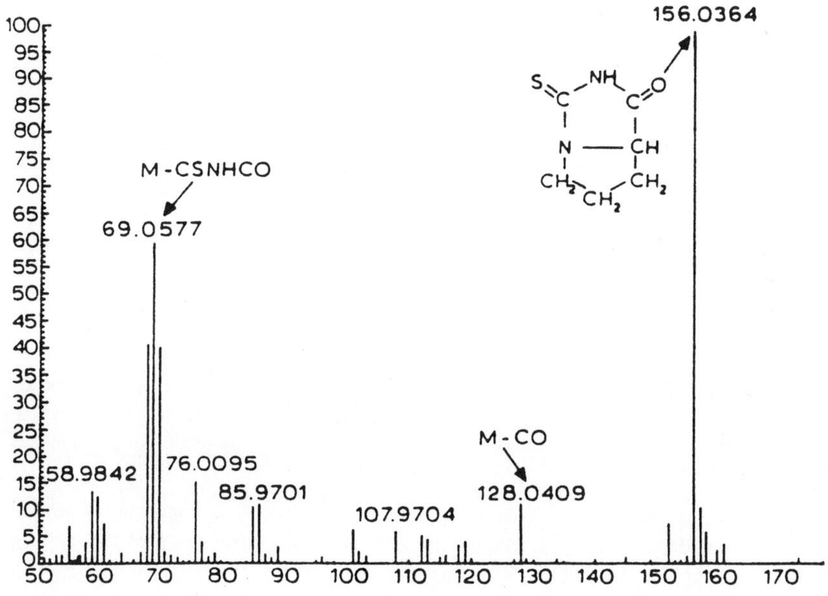

Figure 2. Mass spectrum of peak P. It shows the molecular weight for Pro-TH. ions of lower mass are consistent with loss of elements (Inglis *et al.*, 1992).

thiohydantoins can be cleaved more readily than peptide bonds (Inglis, 1991), and we therefore have the possibility that peptide bond cleavage may not occur concomitantly with cyclization (as it does for acetyl proline); if this were the case, because proline thiohydantoin contains a quaternary nitrogen atom that lends instability to the ring, addition of base may promote ring opening rather than cleavage of the peptide bond (see Figure 3). Such a thesis would imply that it would be safer to use **acidic conditions** for cleavage, at least for cleavage of proline residues.

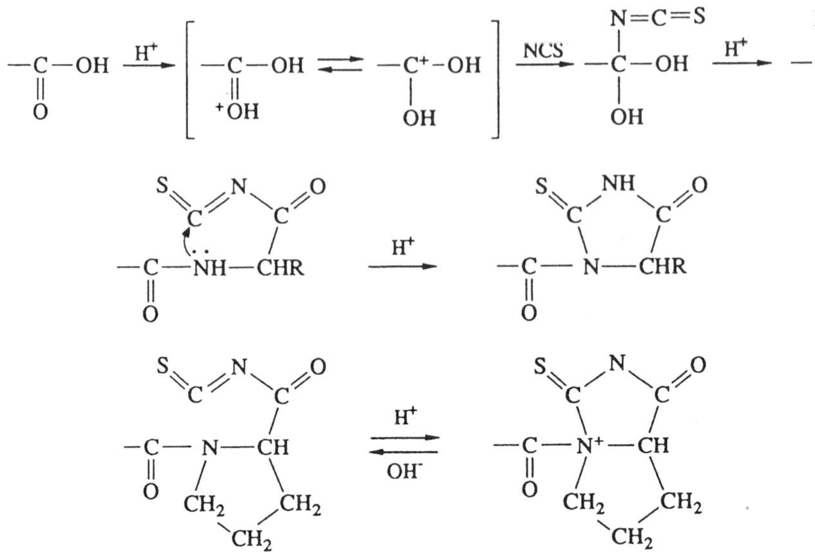

Figure 3. Reaction scheme involving an acid-catalyzed activation of the C-terminal carboxyl group for formation of a peptidyl thiohydantoin (II) via the isothiocyanate (I), and the possible reversal to the isothiocyanate for proline thiohydantoin (IIa) under basic conditions.

ACIDIC CLEAVAGE OF THE THIOHYDANTOIN

Stark (1968) tried both methanolic HCl and trifluoroacetic acid (TFA) as cleavage reagents. TFA was not effective, methanolic HCl was, but it also reacted with the residual peptide and prevented a second degradation. He then turned to 12 M HCl with some success but relinquished it in favour of acetohydroxamic acid. We obtained similar results to Stark for cleavage of acylamino acid thiohydantoins with either TFA or methanolic HCl. However, we also found that anhydrous HCl in acetic acid was superior to both of them. In these experiments (see Figure 4) the HCl was generated *in situ* from reaction of water with acetyl chloride.

The HCl cleavage procedure was then applied successfully to tyrosylphenylalanine, Met-enkephalin and neuromedin N (see Figure 5), hence establishing (a) that the cleavage also occurs with peptides, and (b) that the new carboxy terminus is free for a repeat operation. To verify the principle, the experiments were carried out very simply starting with around 0.5 μM peptide, 100 μl of coupling mixture and loading approx. 30 nm of the cleaved material. Elution was made on a C18 column using peptide elution conditions. The same tube was used throughout and just dried in the Savant between the coupling and cleavage reactions. Although background remained, or built up during successive cycles, new residues appeared at cycles 2 and 3. A small peak in the proline area also appeared in cycle 4 of neuromedin N. To confirm the assignment of these, the digests were spiked with the expected amino acid and rerun. While the conditions were not optimized, cleavage for 15 min at 60° gave results similar to those obtained using base cleavage (3 min, R.T., 0.5 M KOH in 33% methanol, plus DTT). Besides establishing the effectiveness of anhydrous acid cleavage for the thiohydantoin procedure, this work also highlights the advantage of (i) having a readily removable cleavage reagent, and (ii) not having to neutralize between the two reaction steps.

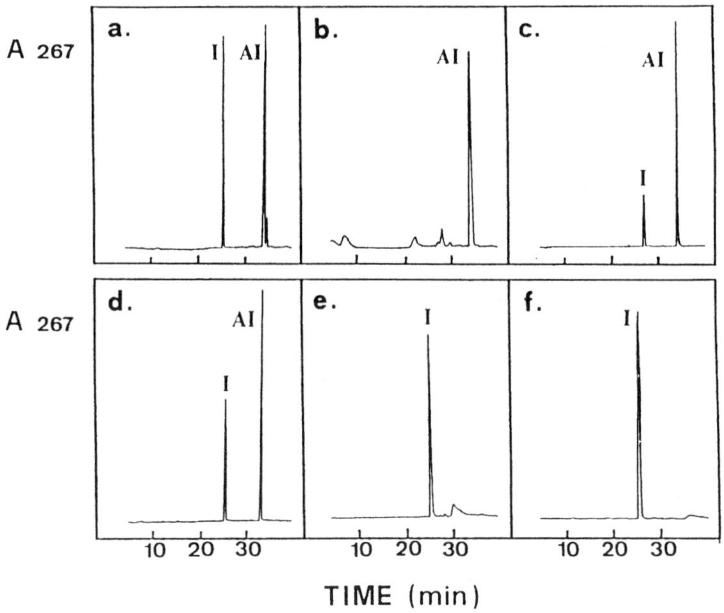

Figure 4. Cleavage of acetylisoleucine thiohydantoin (AI) under various acidic conditions. a shows the HPLC trace for pure AI and I (Ile-TH), b indicates that AI is stable in the coupling mixture, c shows partial decomposition in TFA; d, e and f show the extent of cleavage in acetyl chloride (40 µl) containing methanol (10 µl), methanol (50 µl), and water (10 µl), repectively.

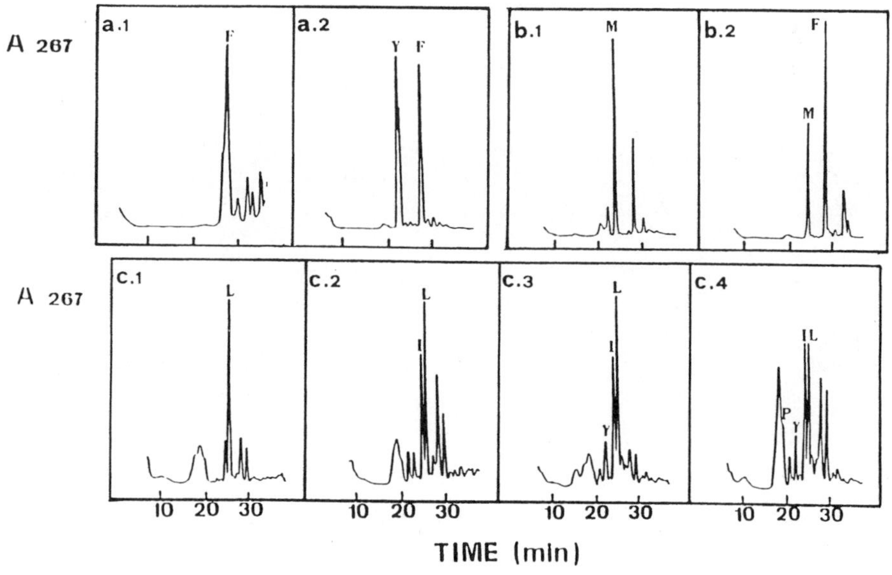

Figure 5. Results of degradation cycles on TyrPhe a and Met-enkephalin(..FM)b, and neuromedin N(..PYIL) c without extractions and with anhydrous acid cleavage.

Samples cleaved by base gave greatly reduced yields in the second cycle. We believe, therefore, that this new approach is a major step towards the development of an automated non-covalent attachment method on insoluble supports.

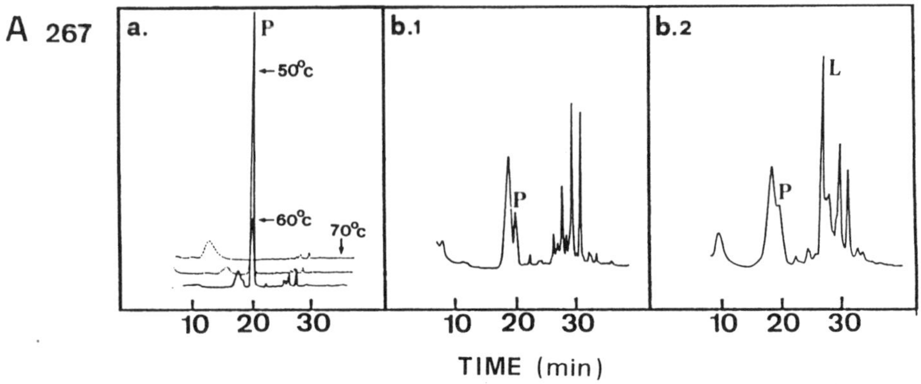

Figure 6. Effect of anhydrous cleavage on proline analysis. a shows the change in elution profile of a preparation of Pro-TH (not purified) when it degrades. b shows 2 degradation cycles for the tripeptide, PCALeu-ProNH_2 after deamidation. Pro-TH elutes with background, Leu-TH is in the second cycle, not the first.

CLEAVAGE AT PROLINE RESIDUES

The tripeptide (PCALeuProNH_2) was analyzed (see Figure 6b) after removal of the amide group with pH 2 formic acid (110°, 2 h). The results confirm that proline linked to a leucine does not cleave concomitantly (as does acetylproline), but it does react. Unequivocal assignment cannot be made at Cycle 1 because of a background peak but Leu-TH is clearly seen at Cycle 2. Decomposition of Pro-TH was found to occur in some experiments and this was more pronounced at higher temperatures (see Figure 6a). When proline cleavage occurs could well be sequence dependent; however, our work suggests that proline should no longer be a deterrent to the use of the thiohydantoin procedure. While other procedures might be applicable, the success of the simple approach employed here suggests that it could be readily refined, automated, etc.

FUTURE DIRECTIONS

In a recent paper (Inglis *et al.*, 1992) we established that proline thiohydantoin could be made easily using a method applicable to C-terminal sequencing. Here we show that it is applicable to C-terminal proline in a peptide, and that anhydrous acid removes the proline residue generating a reactive C-1 peptide. We have now shown that all of the possibly intractable residues in proteins react satisfactorily in the thiohydantoin degradation procedure. Hence, there are now no major obstacles to the development of a general procedure for C-terminal sequencing. However, we believe that the need to couple covalently to an insoluble support is a drawback to many potential users. The simple operations of coupling, drying, cleavage, drying, as used here provide evidence that covalent linking to the insoluble support should not be a pre-requisite. Of course, with some clean-up with solvents

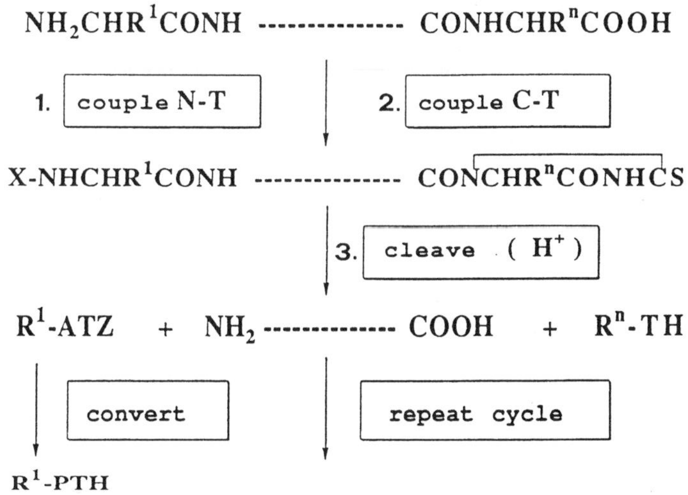

Figure 7. Scheme for dual degradation procedure of a protein.

one would expect better results; and for identification of the polar amino acids the latter is essential. This work also suggests that acid cleavage is a viable alternative to basic cleavage. The aspects of the optimal strength of the acid and the extent of non-specific cleavage need to be addressed further. However, apart from virtues that the anhydrous cleavage may have for C-terminal sequencing, it could well have opened the door to a combined N- and C-terminal sequencing method for the first time. One can now conceive of a procedure in which the N- and C-terminal amino acids are derivatized in turn using Edman and thiohydantoin chemistry and then cleaved with an anhydrous acid (scheme in Figure 7).

Acknowledgement: We wish to thank Professor A. Fontana for his encouragement of this work and our CRIBI colleagues for facilitating it.

REFERENCES

Bailey, J. M. and Shively, J. E., 1990 Carboxy terminal sequencing: formation and hydrolysis of C-terminal peptidylthiohydantoins, *Biochemistry* 29:3145.

Bailey, J. M., Shenoy, N. R., Ronk, M. and Shively, J. E., 1992 Automated carboxy-terminal sequence analysis of peptides, *J. Prot. Sci.* 1:68.

Cromwell, L. D. and Stark, G. R., 1969 Determination of the carboxyl termini of proteins with ammonium thiocyanate and acetic anhydride, with direct identification of the thiohydantoins, *Biochemistry* 8:4735.

Inglis, A. S., 1991 Review: Chemical procedures for C-terminal Sequencing of Peptides and Proteins, *Anal. Biochem.* 195:183.

Inglis, A. S., Casagranda, F., Wilshire, J. F. K. and Laslett, R. L., 1989 C-Terminal sequencing: A new look at the Schlack-Kumpf thiocyanate degradation procedure *in*: Methods in Protein Sequence Analysis, Wittmann-Liebold, B., ed., Springer-Verlag, Berlin.

Inglis, A. S., Casagranda, F. and Wittmann-Liebold, B., 1990 Progress in the development of a C-terminal sequencing method for proteins and peptides *in*: Modern Methods *in* Protein- and Nucleic Acid Research, Tschesche, H., ed., Walter de Gruyter, Berlin.

Inglis, A. S., Duncan, M. W., Adams, P. and Tseng, A., 1992 Formation of proline thiohydantoin with ammonium thiocyanate: Progress towards a viable C-terminal amino-acid-sequencing procedure, *J. Biochem. Biophys. Methods* in press.

Inglis, A. S., Moritz, R. L., Begg, G. S., Reid, G. E., Simpson, R. J., Graffunder, H., Matschull,

L. and Wittmann-Liebold, B., 1991 C-terminal sequence analysis, *in*: Methods in Protein Sequence Analysis, Jornvall, H., Höög, J.-O., Gustavsson, A. M., eds., Birkhauser-Verlag, Basel.

Kubo, H., Nakajima, T. and Tamura, Z., 1971 Formation of thiohydantoin derivative of proline from the C-terminal of peptides, *Chem. Pharm. Bull.* 19:210.

Stark, G. R., 1968 Sequential degradation of peptides from their carboxyl termini with ammonium thiocyanate and acetic anhydride, *Biochemistry* 7:1796.

Stark, G. R., 1972 Sequential degradation of peptides and proteins from their COOH termini with ammonium thiocyanate and acetic anhydride, Methods in Enzymology, Hirs, C.H.W. and Timasheff, S.N., eds.,Vol. 25, Academic Press, San Diego.

CYANOGEN BROMIDE CLEAVAGE OF PROTEINS ON BLOTS AND SUBSEQUENT SEPARATION OF THE FRAGMENTS BY POLYACRYLAMIDE GEL ELECTROPHORESIS DIRECTLY FROM THOSE BLOTS

Gerhard Frank[1], Udo Redweik[2], Monika Rümbeli[1], Gabriele Schoedon[2] and Nenad Blau[2]

[1]*Institut für Molekularbiologie und Biophysik ETH-Hönggerberg, CH-8093 Zürich, Switzerland and*
[2]*Universitätskinderklinik Zürich, Abteilung für klinische Chemie Steinwiesstr. 75, CH-8032 Zürich, Switzerland*

INTRODUCTION

Since nucleic acid sequencing is now quick and accurate, the usual goal of protein sequencing is no longer to determine an entire primary structure. A partial sequence is sufficient for the design of the oligonucleotide probes, needed to obtain starting material for sequence determinations on the nucleic acid level.

Because partial polypeptide sequences can usually be obtained from a fraction of the amount of protein necessary for a complete sequence, electro-blotting of bands from sodium dodecylsulfate polyacrylamide gel electrophoresis (SDS-PAGE) on to polyvinylidene fluoride (PVDF) has become the method of choice for the preparation of small protein amounts for sequencing. The particular advantage of this procedure is that the blotted protein can be introduced into the automated sequencers on the PVDF without further treatment (Aebersold *et al.*, 1987; Matsudaira, 1987).

There are however proteins for which this elegant procedure cannot be employed. If the protein is amino terminally blocked; or a successfully determined sequence consists of so many amino acids with multiple codons that no useful probe can be designed. Furthermore the amino terminal sequences of homologous proteins may differ too much to serve as the basis of probes, while the more conserved internal sequences cannot be determined.

In all these cases, specific cleavage of the polypeptide chain is required to obtain a useful amino acid sequence. However, all additional manipulations inevitably mean protein losses. Depending on the particular cleavage yields and the difficulties encountered with the purification of fragments, the amount of pure protein needed increases by two to three

Methods in Protein Sequence Analysis, Edited by K. Imahori and F. Sakiyama, Plenum Press, New York, 1993

orders of magnitude. The efforts to obtain the necessary amount of material can often lead to unsuccessful termination of the project.

If one can, however, cleave the protein *in situ* on the blotting membrane and subsequently electro-elute the fragments directly on to a second gel for their separation the material can be more efficiently used. Another blot yields all fragments that are not too small for the separation on SDS-PAGE in a form ready for sequencing. Since this approach involves two subsequent blots, the blotting yield and avoiding artificial amino terminal blocking of the peptides are the two most important criteria for the procedure.

EXPERIMENTAL SECTION

Standard Equipment and Chemicals

Electrophoretic protein or peptide isolations were carried out with numerous devices depending on the particular requirements for optimum separations. We used Bio-Rad and Hoefer apparatus as well as ones made in our workshop. Amino acid sequence determinations were performed on a Modular Sequencer (Dr. Ing. H. Knauer GmbH, Berlin, Germany) as described by Fischer *et al.* (1989) which was modified to allow isocratic identification of the phenylthiohydantoin amino acids (Frank, 1989).

Acrylamide, Coomassie® Brilliant Blue R 250 and sodium dodecylsulfate (SDS) were from Serva (Serva Feinbiochemica GmbH & Co. KG. Heidelberg, Germany); ammonium persulfate (APS) and N,N,N',N'-tetramethylethylenediamine (TEMED), nitrocellulose and PVDF-membranes were from Bio-Rad; Tris-(hydroxymethyl)-aminomethane (Tris) and trifluoroacetic acid (TFA) for sequence analysis were from Sigma and methanol was for HPLC gradient analysis from Fisons (Fisons Scientific Equipment, Loughborough, England). The chemicals for the sequencer (except TFA) were - with application warranty - from Knauer. All other chemicals were MicroSelect, BioChemica or purissimum quality from Fluka (Fluka Chemie AG, Buchs, Switzerland).

Blotting Apparatus and Transfer Conditions

We designed a blotting chamber to meet two main requirements: (1) that all blotting parameters, particularly those which are usually hard to obtain like the electric current actually existing in the gel, could be determined exactly, and (2) electrochemical reactions that might block the amino terminus should not occur. Thus only the gel, the membrane, glass and acrylic glass should be present in the region of the electric field. Therefore the "pads", usually employed to press but not squeeze the gel and the membrane together, were replaced by glass sinter plates, and glass fibre filters were used instead of ordinary paper filters.

The blotting sandwich is housed in a frame of acrylic glass that holds two 7.5 x 7.5 cm (8 mm thick) glass sinter plates at their edges, leaving a "window" of 6 x 6 cm for the field. The frame consists of two parts, which allow the sandwich to be assembled between them, and which are then screwed together (Figure 1). In the holding planes of one side of the frame, four expanded silicon rubber rods of 4.5 mm diameter are inserted into 4 mm grooves to provide a minimum of suspension for the sandwich. The outer sides of the frame are surrounded by tongues of 8 mm that fit into grooves in the tank. The frame fits so tightly into the tank profile that the electric flux is restricted to the area of the window.

The tank has an inner dimension of 7 x 12 x 14 cm (without the grooves). The electrodes are 8 x 8 cm and are placed at a distance of 1 cm from the walls to allow easy rinsing. The cathode is made from stainless steel while the anode consists of eight 8 cm long lines of platinum wire. These run parallel on a plate of acrylic glass and are interconnected on the rear. The tank holds a total of 650 ml of blotting buffer if filled to about 1 cm under the upper edge of the inserted frame.

The transfer buffer (7 mM Tris, 50 mM glycine, without methanol) was a more dilute version of that of Towbin *et al.* (1979). The constant transfer current was 62 mA corresponding to 1.1 mA/cm^2. The transfer time, the only variable parameter was 2 hours for polypeptides between 1.5 and 200 kDa and 3 hours for 100 to 700 kDa proteins.

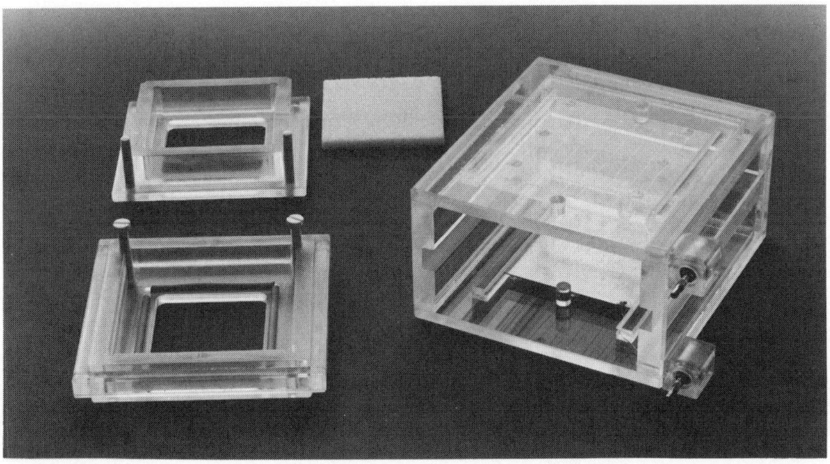

Figure 1. The parts of the blotting apparatus: The two parts of the frame are on the left. The lower part has the silicone foam rubber rods inserted into it. Next to the upper part the glass sinter plate is seen. The tank with the grooves and the platinum wire anode is shown on the right.

Before the sandwich was assembled, the glass sinter plates were immersed in transfer buffer and evacuated with a water aspirator for five minutes to remove trapped air. The transfer membranes were soaked with buffer for a few minutes, nitrocellulose directly, PVDF after it had briefly been moistened in methanol. A sinter plate was then placed in one part of the frame. One glass fibre filter was dipped in transfer buffer, put on the sinter plate and the presoaked membrane was laid on top. The gel was cut to an appropriate size and laid on the membrane carefully avoiding entrapment of air. Then the gel was covered with another glass fibre filter and the second sinter plate, and the frame was closed and screwed together. To minimise diffusion of small molecules, particularly cyanogen bromide fragments, the blotting sandwich was assembled immediately after the separation PAGE was finished, and without any treatment of the gel.

Cyanogen Bromide Cleavage on Nitrocellulose and Separation of Fragments

The high vapour pressure of cyanogen bromide favours its delivery via gas phase. The reaction can be carried out in any closed container but it is advantageous to keep the volume small. A typical device can be improvised with 5 or 10 ml autosampler vials. A holder from polypropylene, silicone rubber or teflon can then be made that stands on bottom of the vial or is inserted underneath the screw cap instead of a septum. After blotting on to nitrocellulose and staining with Amidoblack, the bands of the protein to be investigated were excised leaving some unstained area at one of their sides to serve as a handle. The strips were held at that handle with the holder in such a way that the stained parts were freely exposed to the fumes.

The strips were then moistened with 20 mM phosphoric acid, a concentration, which is easily tolerated by the nitrocellulose. The holder was placed in the reaction vessel, approximately 0.5 ml of 70% formic acid and 100 mg of cyanogen bromide were put on the bottom, and the vessel was flushed with dichlorodifluoromethane and closed. The reaction was allowed to take place in the dark at room temperature for 4-16 hours. The holder with the strips was then placed in a desiccator and evacuated for 15 minutes to remove the adsorbed cyanogen bromide. Then three 5 minute washes with water were necessary to remove the acid adequately, and the "handles" were cut off before electrophoresis.

A modified Laemmli (1970) electrophoresis system was used for the separation of the fragments. The optimum electro desorption was accomplished with glycylglycine as trailing ion. A typical gel was 12.5 x 8 x 0.1 cm, 12.5% acrylamide plus 5% bisacrylamide in 600 mM Tris/HCl of pH 8.5. This composition of the polymer was found to be suitable for small fragments down to the lowest level of detectability and must of course be adapted to the size of the fragments that are to be separated. The stacking gel contained 250 mM Tris/HCl, pH 8.5. The electrode buffer at both anode and cathode was 100 mM glycylglycine and 100 mM Tris plus 0.1% SDS. The pH was 8.4 to 8.5 and was not adjusted.

When the stacking gel was prepared, the size of the comb was selected to match the nitrocellulose strips with the CNBr-fragments. Up to 10 strips could be inserted in parallel into one gel pocket in which a small amount of sample buffer was already present. Care was taken to avoid air bubbles between the strips. The hydrophilic character of the nitrocellulose proved to be very helpful in this respect. The electrophoresis was carried out in a cold room (10 °C) typically with a constant current density of 40 mA/cm^2 at a corresponding voltage of 90 - 190 V.

RESULTS AND DISCUSSION

The Overall Procedure

The first SDS-PAGE, which is usually the last purification step for the protein, should be performed in such a way that the band to be blotted travels more than one quarter of the separation gel length (for gels that are run until the dye front just vanishes). If this is not the case a less dense gel must be prepared to facilitate good blotting results. Following this rule, hog tyroglobulin with a molecular weight of 670 kDa has been blotted so successfully that no residual material could be visualised on the gel with Coomassie stain.

If the purification is difficult and if the material that goes on to the SDS-PAGE still contains other proteins in large excess, then the special advantage of this method becomes particularly obvious: Although the band that can be obtained under such circumstances is probably quite faint and higher load is not possible because of widening of undesired bands next to the one wanted, enough material for the subsequent cleavage can be obtained simply by applying starting material to a number of slots. Up to ten bands can then be simultaneously cleaved and combined in one slot for the separation of the fragments. For those who prefer to perform such separations in just one long slot instead, a similar procedure is of course possible. The long strip of nitrocellulose is then simply cut into pieces of the right length to match the slot of the next PAGE.

The CNBr-cleavage is normally accomplished in 70% formic acid to facilitate the unfolding of the protein structure (Gross and Witkop, 1962). However, if a protein is blotted on to nitrocellulose out of an SDS-PAGE, the strong interaction between the protein and the membrane will probably hinder refolding sufficiently. Thus cleavage could take place in the presence of dilute acid as well. We have so far not experienced low cleavage yields. This is fortunate since the nitrocellulose does not withstand concentrated acids. On the other hand, PVDF does not adequately release the fragments during the electro elution procedure. The reason for this is not completely understood. Comparing the binding capacities of these two types of membrane one would expect an easier release from PVDF. It is possible that the more polar character of nitrocellulose plays an important role in this regard.

When tested with sperm whale apomyoglobin, no difference in the cleavage yield between homogeneous reaction in 70% formic acid and "gas phase" condition was seen. However in both cases the yields were far below 100%. This shows the importance of the condition of the methionine residues, which must have been partially oxidised. Thus care should be taken during the whole protein preparation procedure to minimise oxidation, which does occur rarely during electrophoresis or electro-blotting as we found.

Clearly, very small CNBr-fragments cannot be obtained with this method since they do not stain. The limit, approximately 1.5 kDa, is not well defined since small fragments have widely varying staining properties. Thus it cannot be ruled out that larger amounts in the gel or in the backup membrane escape one's attention since they are poorly stained. Even the rule that Amidoblack stains more heavily than Ponceau S may not be true. We obtained over 200 pmoles initial yield from a peptide that was hardly visible after Amidoblack staining while a strong band with Ponceau S from another peptide yielded only 80 pmoles. Nevertheless it is extremely unlikely that all CNBr-fragments have molecular weights that are too small for good staining. Furthermore, the larger fragments are much better suited for the automated sequence determination. Thus the smaller fragments are rather more a reserve for instances in which no useful results are obtained with the larger ones.

This method was first employed for the determination of internal sequences of GTP cyclohydrolase from *E. coli*. Approximately 200 pmoles of the protein were available and a fragment was obtained that could be sequenced for 30 cycles with an initial yield of 35 pmoles (two residues unidentified). With a blotting yield of 90%, and cleavage yield and initial yield both 50%, an effective initial yield of 40 pmoles would be calculated. When the genomic sequence was published (Katzenmeier *et al.*, 1991), it became obvious that the peptide comprises the carboxy terminal CNBr-fragment, which is 124 residues long. The unidentified residues were cysteine in position 12 and isoleucine in position 28 of that fragment.

The Electro Transfer

The development of the blotting system was based on the hypothesis that electroendosmosis might play an adverse role in the procedure because it would generate a liquid current directed against the migration of the protein molecules. Since the strength of electroendosmosis is directly proportional to the electric current, it was anticipated that a procedure with the minimum possible electric current would yield the best results. This would be the current that is just sufficient to maintain a homogeneous electric field. The homogeneity of the field, however, depends also on the geometry of everything that is located between the electrodes. Clearly a perforated acrylic glass plate, as a holder for the sandwich, disturbs the lines of electric flux, and the coarser the perforation the greater the disturbance. A glass sinter plate is a much better alternative in this respect and consequently the current density could be reduced to as little as 1.1 mA/cm^2.

Because of this moderate current the total energy consumed is only about 3 Wh or 2.6

kcal., divided between electrode reactions and heat generation. The observed temperature rise is only 2 to 3 degrees in the two hours, indicating that approximately 1.5 kcal are released as heat. Thus cooling is not necessary, simplifying the whole procedure.

In contrast to a recently published paper (Mozdzanowski *et al.*, 1992), we have not found it useful to add methanol to the blotting buffer, nor could we improve the results by presoaking the gels. These differences may be attributed to the fact that we use much lower ionic strength in the transfer buffer. Thus the electric resistance in the gel is initially lower than in a buffer layer of equal dimensions. Thanks to the design of our apparatus we could measure these differences. The voltage needed to produce a current of 40 mA was 75 V for the sandwich without gel but only 65 V for an identical sandwich with gel. This effect cannot be explained with an increased conductivity in the 1 mm region of the gel alone. It rather shows that the higher salt has already spread out over a considerable region of the sandwich when the voltage is applied.

If the electric field in the gel is low at the beginning and increases with the transfer time, the mobility of the proteins must change as well. Possibly, the low mobility of the proteins at the beginning of the transfer facilitates an elution of excess SDS from the gel in advance of relevant amounts of protein. Thus an effect similar to presoaking might be achieved under these conditions. While presoaking is probably not a great problem for normal proteins, smaller fragments wash out and low yields result.

Unfortunately, we cannot state exact blotting yields since it is a general problem to determine the precise amounts of structurally unknown proteins at the hundred picomol level. Even with larger amounts of well known proteins, moisture and salt content can make yield determination difficult. Furthermore, the initial yields on the sequencer vary, so that errors come from two sides. Thus we have chosen to just estimate the blotting yields. With the Bio-Rad membranes and under our blotting conditions, generally no bands visible with Coomassie stain are left in the blotted gel nor on the backup membrane. This would indicate that the blotting yields fall into the range of 90%. However, Mozdzanowski *et al.* described the loss of material through several backup membranes into the transfer buffer. They mention that this be a particular problem with low molecular weight proteins. Unfortunately they do not provide evidence for the presence of those proteins in the transfer buffer. Thus it needs to be proven that these proteins do not escape into the presoaking buffer.

CONCLUSIONS

The method of *in situ* cleavage of blotted proteins with subsequent electro elution directly on to a new separation gel considerably eases the determination of internal amino acid sequences. The amount of material needed is much smaller than that needed for homogeneous phase reaction with subsequent chromatographic separation. The method is a favourable candidate for applications with other specific cleavage reactions as long as the reaction can be varied to an on membrane version and some larger fragments can be expected.

Furthermore, the method can be used for re-electrophoresis. Minor bands that are only marginally separated from larger amounts of undesired proteins can be blotted on to nitrocellulose, excised and immediately subjected to another electrophoresis. The particular virtue of this method being, that up to 10 such blotted bands can be rerun simultaneously, thus yielding both, concentration and further purification at the same time.

Our blotting procedure appears to be a good alternative to the ones presently in use. It has the great advantage that only the transfer time needs to be adjusted to the size of the protein while buffer and current remain unchanged. Since the blotting apparatus greatly facilitates systematic work, deeper insight into the transfer physics may be gained with future experiments.

Ackowledgements. We are grateful to Werner Braun, Esslingen (Germany) for helpful discussions about the field of electrochemistry, to Hans Siegrist of the ETH workshop for his precise construction of our designs and to Catherine Schein-Braun for reading the manuscript. This work was partially supported by the Swiss National Science Foundation, Project No. 31-28797.90.

REFERENCES

Aebersold, R. H., Leavitt, J., Saavedra, R. A., Hood, L. E. and Kent, S. B. H. , 1987 Internal amino acid sequence analysis of proteins separated by one- or two-dimensional gel electrophoresis after *in situ* protease digestion on nitrocellulose, *Proc. Natl. Acad. Sci. USA* 84:6970.

Fischer, S., Reimann, F. and Wittmann-Liebold, B. , 1989 A new modular sequencer, *in*: Methods in Protein Sequence Analysis, Wittmann-Liebold, B., ed., Springer-Verlag, Berlin pp. 98-107.

Frank, G. , 1989 New aspects in isocratic HPLC separation of ionic detergents, *in*: Methods in Protein Sequence Analysis, Wittmann-Liebold, B., ed., Springer Verlag, Berlin, pp. 116-121.

Gross, E. and Witkop, B. , 1962 Nonenzymatic cleavage of peptide bonds: The methionine residues in bovine pancreatic ribonuclease. *J. Biol. Chem.* 237:1856.

Katzenmeier, G., Schmid, C., Kellermann, J., Lottspeich, F. and Bacher, A., 1991 Sequence of GTP cyclohydrolase I from *Escherichia coli, Biol. Chem. Hoppe-Seyler* 372:991.

Laemmli, U. K., 1970 Cleavage of structural proteins during the assembly of the head of bacteriophage T4, *Nature (London)* 227:680.

Matsudaira, P. , 1987 Sequence from picomole quantities of proteins electroblotted onto polyvinylidene difluoride membranes, *J. Biol. Chem..* 262:10035.

Mozdzanowski, J., Hembach, P. and Speicher, D.W., 1992 High yield electroblotting onto polyvinylidene difluoride from polyacrylamide gels, *Electrophoresis* 13:59.

Towbin, H., Staehelin, T. and Gordon, J., 1979 Electrophoretic transfer of proteins from polyacrylamide gels to nitrocellulose sheets: Procedure and some applications, *Proc. Natl. Acad. Sci. USA* 76:4350.

IDENTIFICATION OF THE DISULFIDE BONDS OF THE HUMAN COMPLEMENT COMPONENT C3

Johann Schaller, Sabine Söhndel and Egon E. Rickli

Institute of Biochemistry, University of Bern
Freiestrasse 3, CH-3012 Bern
Switzerland

INTRODUCTION

The complement component C3 plays a key role in both the classical and the alternative pathway of the human complement system. The single chain precursor comprises 1663 residues (De Bruijn and Fey, 1985) and contains 27 half-cystine residues including one thioester site. Upon activation by the C3 convertase C3 is transformed into a two-chain molecule held together by an interchain disulfide bridge. The α-chain consists of 992 residues with 24 Cys (including the C3a anaphylatoxin), the β-chain has 645 residues with 3 Cys. Each chain carries a single N-linked carbohydrate moiety. C3 shows sequence similarities with the complement components C4 and C5, and to a lesser extent with α_2-macroglobulin.

MATERIALS AND METHODS

Isolation

C3 was isolated by ion exchange chromatography on QAE-Sephadex and SP-Sephadex followed by gel filtration on Sephadex G-200 (Pharmacia, Uppsala, Sweden) according to Lundwall *et al.* (1981; 1985). The purity was checked by SDS-PAGE, amino acid and sequence analysis.

Cleavages

Native C3 was cleaved with cyanogen bromide in 70% (by vol) trifluoroacetic acid according to Schroeder *et al.*(1969). Subdigestion of Cys-containing CNBr-fragments with pepsin was carried out at an enzyme:substrate ratio of 1:50 (w/w) for 24 h at 37 °C in 0.01 M HCl, pH 2.0.

Methods in Protein Sequence Analysis, Edited by K. Imahori and F. Sakiyama, Plenum Press, New York, 1993

Isolation of Fragments

The generated fragments were isolated by RP-HPLC on a Bakerbond Butyl column (wide pore, 33 nm; 5 μm; 4.6 x 250 mm; Baker, Deventer, The Netherlands) and rechromatographed on Aquapore Butyl and Phenyl columns (wide pore, 30 nm; 7 μm; 2.1 x 100 mm; Applied Biosystems, Foster City, USA).

Identification of Disulfide Containing Peptides

Aliquots of the HPLC fractions were reduced with tri-n-butylphosphine and the generated thiol groups specifically labeled with ammonium 7-fluorobenz-2-oxa-1,3-diazole-4-sulfonate (SBD-F) according to Sueyoshi *et al.* (1985). The resulting fluorescence intensities were measured with excitation at 385 nm and emission at 515 nm.

Amino Acid and Sequence Analysis

To obtain maximum yield of cystine, samples were hydrolyzed in the gas phase with 6 M HCl containing 0.1 % (v/v) phenol for 24 h at 115 °C under vacuum according to Chang and Knecht (1991). The liberated amino acids were reacted with PITC and the resulting PTC-amino acids analyzed by RP-HPLC on a Nova Pak C18 column (4 μm; 3.9 x 150 mm; Waters, Milford, USA) according to Bidlingmeyer *et al.* (1984).

Sequence analysis was carried out in a pulsed-liquid-phase sequenator 477A from Applied Biosystems. PTH-amino acids were analyzed on-line according to instructions of the manufacturer. It is essential to use only DTT-free reagents and solvents. Di-PTH-Cys was detected as double peak in the vicinity of PTH-Tyr.

RESULTS AND DISCUSSION

Isolation of Fragments

Native C3 (4 mg) was cleaved with cyanogen bromide and the fragments separated by RP-HPLC (Figure 1). An aliquot (10 %) of each fraction was reacted with SBD-F to locate the Cys-containing fragments. Cys-containing peptides were predominantly clustered in three pools: CB7, CB9 and CB11 (Figure 1).

Pools CB7, CB9 and CB11 were further cleaved with pepsin and the fragments separated by RP-HPLC (chromatograms not shown). An aliquot (20 %) of each fraction was treated with SBD-F and the Cys-containing fragments identified. Rechromatography of Cys-containing peptides was achieved on Aquapore columns (chromatograms not shown).

Amino Acid and Sequence Analysis

The Cys-containing peptides were characterized by amino acid analysis, enabling a preliminary localization of the fragments in the known polypeptide chain of the C3 molecule.

The disulfide bonds were identified by amino acid sequence analysis. The results are summerized in Table 1.

Assignment of the Disulfide Bonds of C3

The assignment of the disulfide bonds and the positioning within the polypeptide chain is given in a schematic model of the C3 molecule (Figure 2): Cys 537 in the β-chain forms the single interchain disulfide bridge with Cys 794 of the α-chain. The remaining

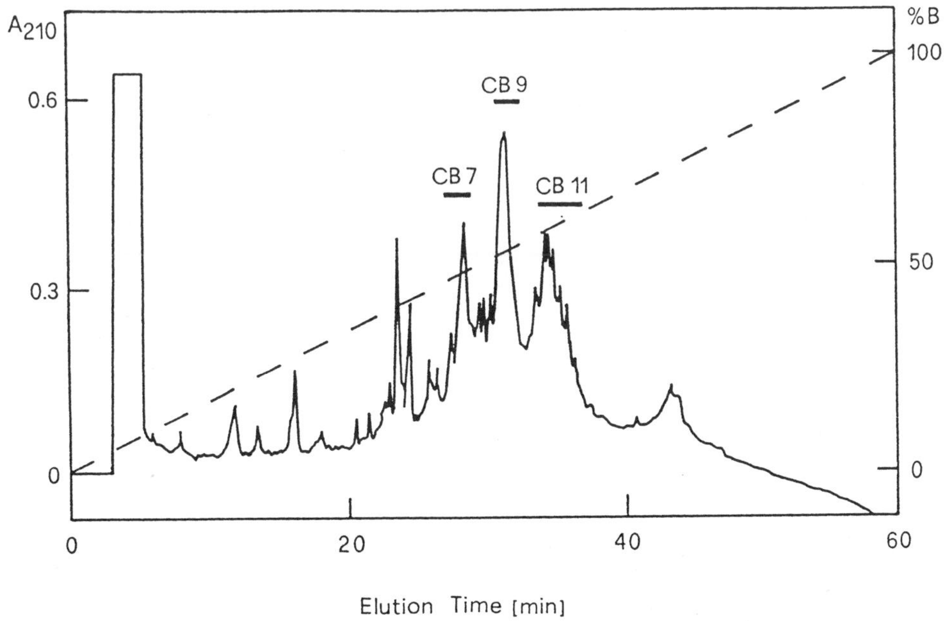

Figure 1. Separation of CNBr-fragments of native C3. RP-HPLC on Bakerbond Butyl (wide pore, 33 nm; 5 µm; 4.6 x 250 mm) using a linear acetonitrile gradient (0 - 100 % B/60 min). Solution A: 0.1 % (by vol) TFA in H_2O; solution B: 0.1 % (by vol) TFA in 80 % (by vol) acetonitrile. Flow rate: 1 ml/min; detection: 210 nm. Pools corresponding to CB7, CB9 and CB11 are indicated by horizontal bars.

two Cys residues in the α-chain (Cys 605 - Cys 640) form the single intrachain bridge of the β-chain. These data confirm a preliminary localization by Matsuda *et al.* (1985) and Janatova (1986).

In the course of the activation process the C3 convertase releases the C3a anaphylatoxin from the N-terminal region of the α-chain. The arrangement of the three disulfide bonds in C3a (Cys 671- Cys 698, Cys 672 - Cys 705, Cys 685 - Cys 706) is in agreement with crystallographic data from Huber *et al.* (1980).

Three disulfide bridges in the C-terminal part of the α-chain, Cys 1336 - Cys 1467, Cys 1515 - Cys 1639 and Cys 1615 - Cys 1624 could be definitively assigned.

The remaining five disulfide bonds were tentatively assigned: Cys 851 is connected in a cluster with either Cys 1484, 1489 and/or 1491 (the remaining two Cys of this cluster form the other disulfide bond), and Cys 1496 - Cys 1568 (see Table 1). Proteolytic cleavage experiments of native C3 by Janatova (1986) gave evidence for the disulfide bond Cys 1079 - Cys 1136. Cys 1367 - Cys 1436 form the last remaining disulfide bridge.

Table 1. Edman degradation data and identification of disulfide bonds in peptic fragments of C3.

Peptide	Structural data	Disulfide bond
CB9P23-4	531 V D V K D S C V G S L 788 S D K K G I C V A D P F E	537-794
CB11P13-4	600 K A D I G C T P G S G K D 635 R A E L Q C P Q P A A	605-640
CB7P19-7/ CB7P20-10	659 D K V G K Y P K E L R K X X E D G . . . 692 I S L G E A X K K V F L D X X N Y I . . . 682 R F S X Q R R T R F	671-698 672-705 685-706
CB9P21-2/4	1326 Y H A K A K D Q L T X N K F 1464 E E S X T R F	1336-1467
CB9P15-6/ CB9P17-3	1512 D K A X E P G V D 1635 V V F G C P.N	1515-1639
CB7P9-6/ CB9P7-4/ CB11P8-6	1613 D E X Q D E E 1629 A G L D Q C Q K Q N	1615-1624
CB9P22-9	845 L H N P A F X S L 1481 N K L X R D E L X R X A E E	851 1484, 1489, 1491
CB9P27-6	845 L H N P A F X S L 1481 N K L X R D E L X R X A E E N X F 1562 F I S P I K X R E A L	851 1484, 1489, 1491, 1496 1568

X: Half-cystine residues according to amino acid sequence, but not detected during Edman degradation; C: Detection of di-PTH-Cys; ---: Tentative assignment of disulfide bridges. The numbering of the cysteine residues is according to the mature protein (De Bruijn and Fey,1985)

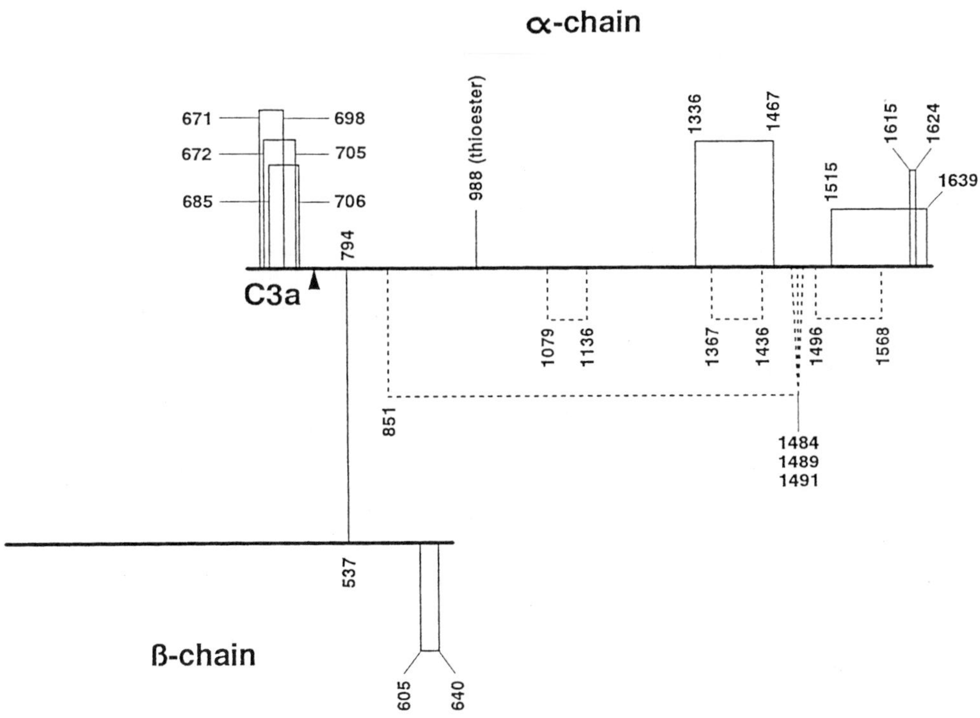

Figure 2. Schematic model of the C3 molecule: Assignment of the disulfide bonds. Disulfide bridges with solid lines are definitively, those with dashed lines tentatively assigned. The numbering of the Cys residues is as in Table 1.

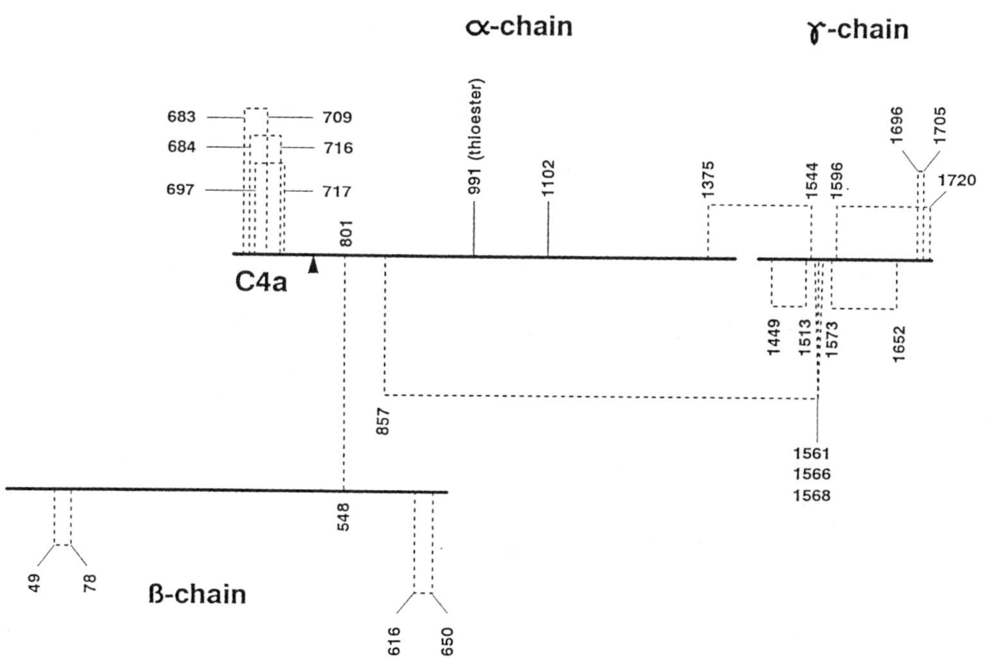

Figure 3. Schematic representation of the C4 component of human complement. The tentative assignment of the disulfide bonds was obtained by sequence comparison with C3 using the multiple sequence alignment program CLUSTAL (Higgins and Sharp,1988). The numbering of the Cys residues is according to the mature protein (Belt *et al.*,1984). Position 1102: Variant Ser -- Cys.

Possible Disulfide Bridge Pattern of Complement Components C4 and C5

Sequence comparison of C3 with components C4 and C5, calculated with the multiple sequence alignment program CLUSTAL (PC/GENE) according to Higgins and Sharp (1988), and comparison of the location of the Cys residues in the polypeptide chain revealed:

1) Twelve disulfide bridges in C4 are at identical or similar positions as in C3 including the single interchain bridge between the α- and the β-chain and the two interchain bridges between the α- and the γ-chain of C4. One bridge is missing, corresponding to Cys 1079 - Cys 1136 in the α-chain of C3. Cys 49 and Cys 78 are unique in the β-chain of C4 and probably form an additional bridge (Figure 3).

2) Thirteen disulfide bridges in C5 are at identical or similar positions as in C3 including the single interchain bridge. Cys 445 and Cys 472 in the α-chain are only present in component C5 and might form an extra disulfide bridge (Figure 4).

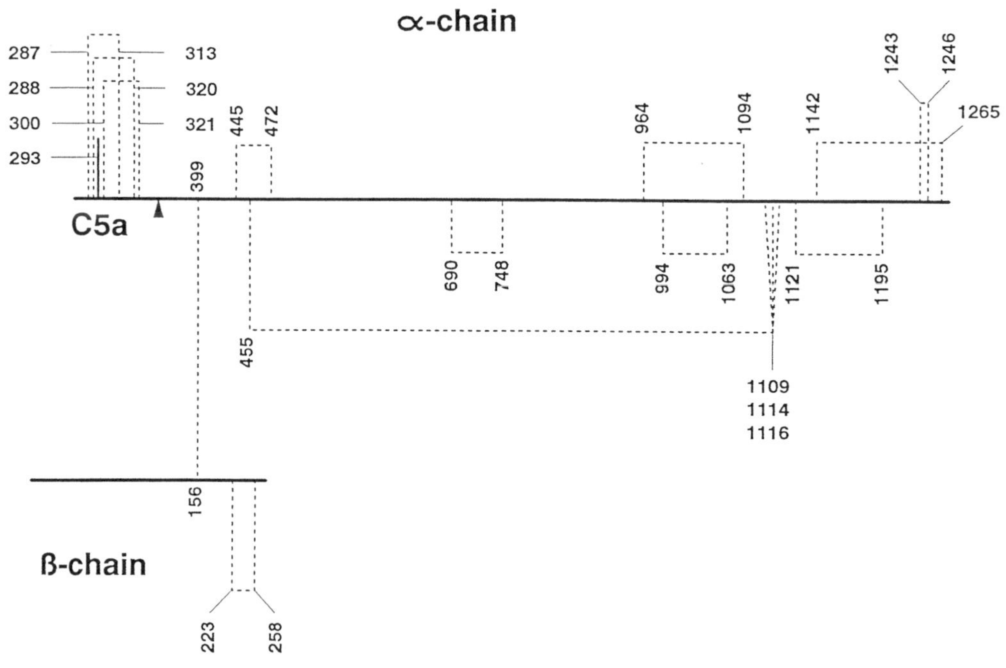

Figure 4. Schematic representation of the C5 component of human complement. The tentative assignment of the disulfide bonds was obtained in the same way as in Figure 3. The numbering of the Cys residues is according to the mature protein (Wetsel *et al.*, 1988). Position 293 (in human C5a): Unpaired Cys.

Acknowledgements. We thank Urs Kämpfer and Claudia Straub for skillful technical assistance. We are grateful to the Central Laboratory of the Blood Transfusion Service, Swiss Red Cross, for the generous supply of human plasma.

REFERENCES

Belt, K. T., Carroll, M. C. and Porter, R. R., 1984 The structural basis of the multiple forms of human complement component C4, *Cell* 36:907.

Bidlingmeyer, B. A., Cohen, S. A. and Tarvin, T. L., 1984 Rapid analysis of amino acids using pre-column derivatization, *J. Chromatogr.* 336:93.

Chang, J.-Y. and Knecht, R., 1991 Direct analysis of the disulfide content of proteins: Methods for monitoring the stability and refolding process of cystine-containing proteins, *Anal. Biochem.* 197:52.

De Bruijn, M. L. H. and Fey, G. H., 1985 Human complement component C3: cDNA coding sequence and derived primary structure, *Proc. Natl. Acad. Sci. USA* 82:708.

Higgins, D. G. and Sharp, P. M., 1988 CLUSTAL: a package for performing multiple sequence alignment on a microcomputer, *Gene* 73:237.

Huber, R., Scholze, H.,Pâques, E. P. and Deisenhofer, J., 1980 Crystal structure analysis and molecular model of human C3a anaphylatoxin, *Hoppe-Seyler's Z. Physiol. Chem.* 361:1389.

Janatova, J. 1986 Detection of disulfide bonds and localization of interchain linkages in the third (C3) and fourth (C4) components of human complement, *Biochem. J.* 233:819.

Lundwall, A. and Eggertsen, G., 1985 Isolation of human complement factors C3, C5 and H, *J. Immunol. Methods* 81:147.

Lundwall, A., Malmheden, I., Stålenheim, G. and Sjöquist, J., 1981 Isolation of component C4 of human complement and its polypeptide chains, *Eur. J. Biochem.* 117: 141.

Matsuda, T.,Seya, T. and Nagasawa, S., 1985 Location of the inter-chain disulfide bonds of the third component of human complement, *Biochem. Biophys. Res. Commun.* 127:264.

Schroeder, W. A., Shelton, J. B. and Shelton, J. R., 1969 An examination of conditions for the cleavage of polypeptide chains with cyanogen bromide: Application to catalase, *Arch. Biochim. Biophys.* 130:551.

Sueyoshi, T.,Miyata, T.,Iwanaga, S.,Toyo'oka, T. and Imai, K., 1985 Application of a fluorogenic reagent, ammonium 7-fluorobenzo-2-oxa-1,3-diazole-4-sulfonate for detection of cystine-containing peptides, *J. Biochem.* 97:1811.

Wetsel, R. A., Lemons, R. S., Le Beau, M. M., Barnum, S. R., Noack, D. and Tack, B. F., 1988 Molecular analysis of human complement component C5: Localization of the structural gene to chromosome 9, *Biochemistry* 27:1474.

A NOVEL PROTEASE FROM JACK-BEAN SEEDS: ASPARAGINYL ENDOPEPTIDASE

Shin-ichi Ishii[1], Yukichi Abe[1], Masanori Mitta[2], Hideyuki Matsushita[2] and Ikunoshin Kato[2]

[1]*Department of Biochemistry, Faculty of Pharmaceutical Sciences, Hokkaido University, Kita-ku, Sapporo 060 and*
[2]*Biotechnology Research Laboratories, Takara Shuzo Co., Ltd Otsu, Shiga 520-21, Japan*

INTRODUCTION

Enzymatical and chemical reactions that induce the peptide-bond cleavage at definite amino acid residues and so produce a limited number of fragments from proteins are essential tools in protein sequence analysis. Although several kinds of residue-specific proteases are now widely used for this purpose, the asparaginyl-bond specific protease has not been available yet . Carrington *et al.* (1985) and Bowles *et al.* (1986) have reported that concanavalin A, a lectin of jack bean (*Canavalia ensiformis*), is produced by post-translational proteolysis and transpeptidation (or ligation) at the carboxyl side of Asn-residues of its precursor. Their results suggest that the bean at least in its premature state contains an asparaginyl endopeptidase which may have a dual function, proteolysis and transpeptidation. We confirmed this suggestion by isolating it from commercially available jack-bean meal and showed its high utility in protein sequence analysis. Purification and characterization of this novel protease are described in the following.

PURIFICATION OF THE ENZYME

Our study started to prove the presence of asparaginyl endopeptidase in premature jack-bean seeds and to isolate it. For the enzyme assay, we adopted DNP-Pro-Glu-Ala-Asn-NH_2 as a substrate. This peptide was synthesized by imitating one of the target sequences for the limited proteolysis in pro-concanavalin A (Bowles *et al.*, 1986) The substrate (20 μM) was treated with the enzyme at 35°C for 10 min in a buffer, pH 5.0, and analyzed by reversed-phase HPLC under isocratic conditions (45 mM ammonium acetate, pH 4.5, containing 22.5% acetonitrile at a flow rate of 1 ml/min). Absorption at 370 nm of the DNP group

Methods in Protein Sequence Analysis, Edited by K. Imahori and F. Sakiyama, Plenum Press, New York, 1993

allowed easy monitoring of clear separation among the substrate, the reaction product DNP-Pro-Glu-Ala-Asn-OH, and an internal standard DNP-Ser. Accurate determination of asparaginyl-bond hydrolyzing activity was achieved within 6 min in this way. Such an enzyme activity was detected by this method also in the extract of various other legume seeds including soybean.

After our purification work of the enzyme from an extract of premature jack-bean seeds proceeded for several steps, we noticed that mature jack-bean meal purchased from Sigma Chemicals also contained a protease that behaved in almost the same manner as the premature-seed enzyme did. Therefore, we chose the meal as a constantly available enzyme source for a larger-scale work and succeeded in obtaining a highly purified preparation of the enzyme. Table 1 gives an overview of the purification steps. The enzyme seemed to be a cysteine protease, because it was inhibited by *p*-chloromercuribenzoic acid and this inhibitory effect was canceled by 2-mercaptoethanol or dithiothreitol. Actually, chromatography on organomercurial-Sepharose (immobilized *p*-chloromercuribenzoic acid (Yokosawa *et al.*, 1977)) offered the most efficient step for the purification. In each of the purification steps, protein concentration of a preparation was determined by measuring absorbance at 280 nm, with the exception of the initial meal extract which was deeply colored. The datum of specific activity for the extract was, therefore, missing. Even when compared with the preparation after the second step (Phenyl-Toyopearl 650M), the final enzyme specimen was found to have been purified about 5,500-fold.

This enzyme specimen gave a single protein peak in HPLC on a reversed-phase column (TSKgel Octadecyl-NPR, 0.46 x 3.5 cm) and a single protein band with the mobility corresponding to a molecular weight of about 37,000 on SDS-polyacrylamide gel electrophoresis under the reductive conditions. The material in the HPLC peak-fraction gave a single N-terminal sequence on the Edman reaction for 25 cycles. The sequence is shown later (see Figure 4).

Table 1. Purification of asparaginyl endopeptidase from jack-bean meal.

Step	Protein mg[a]	Activity milliunit[b]	Specific activity unit/g	Yield %
Extract	-	6800	-	100
Phenyl-Toyopearl 650M	6380	5680	0.890	84
Organomercurial-Sepharose	138	2330	16.9	34
Phenyl-Toyopearl 650S	84.2	1640	19.5	24
Ion Exchanger, SP-Toyopearl	11.8	1900	161	28
TSKgel Phenyl-5PW	3.68	1590	432	23
G3000SW Gel-Permeation (1st)	0.105	443	4220	7
G3000SW Gel-Permeation (2nd)	0.072	352	4900	5

a: One mg/ml was estimated to give the absorbance of 1 at 280 nm.
b: One unit was defined as the enzyme activity yielding 1 μmol of the product per min from DNP-Pro-Glu-Ala-Asn-NH_2.

SITES OF CLEAVAGE IN POLYPEPTIDE SUBSTRATES

Strict specificity toward asparaginyl bonds of this enzyme was verified by using 14 kinds of polypeptide substrates including reduced *S*-alkylated derivatives of hen-egg lysozyme and bovine ribonucleases A and B. Each substrate (10 nmol) was incubated with the enzyme

(0.2 or 0.1 milliunit) at 37°C for 15 h in 100 μl of 20 mM sodium acetate buffer, pH 5.0, containing 1 mM EDTA and 10 mM dithiothreitol. The peptide fragments produced were separated from each other by reversed phase HPLC and identified by analyzing their amino acid sequences. The sites of enzymatic cleavage in the substrates were assigned in this way.

Figure 1 shows the result with reduced *S*-alkylated lysozyme. All but one of the peptide bonds with Asn residues at the P_1 position (according to the nomenclature by Schechter and Berger, 1967) were susceptible to the enzyme. These were indicated by arrows. Only the exception was the bond between Asn77 and Ile78. Peptide bonds having any other amino acid at the P_1 position were not cleaved. Figure 2 explains the result on reduced *S*-alkylated ribonuclease A. All the asparaginyl bonds were susceptible in this case. No exception was found; even the bond between Asn113 and Pro114 was split. Any other bonds were not cleavable again. In the case of S-alkylated ribonuclease B, the linkage between Asn34 and Leu35 was not split. The reason may be that this Asn is glycosylated in ribonuclease B.

```
1                 10                     20                   30
K V F G R C E L A A A M K R H G L D N Y R G Y S L G N W V C
                                    ↑                 ↑

                  40                     50                   60
A A K F E S N F N T Q A T N R N T D G S T D Y G I L Q I N S
            ↑               ↑   ↑                         ↑

                  70                     80                   90
R W W C N D G R T P G S R N L C N I P C S A L L S S D I T A
        ↑                 ↑     ⇑

                  100                    110                  120
S V N C A K K I V S D G N G M N A W V A W R N R C K G T D V
    ↑                   ↑     ↑                 ↑

                129
Q A W I R G C R L
                                                  ⇑, not cleaved
```

Figure 1. Peptide bonds susceptible to jack-bean asparaginyl endopeptidase in reduced *S*-3-(trimethylamino)propylated lysozyme (↑).

```
1                 10                     20                   30
K E T A A A K F E R Q H M D S S T S A A S S S N Y C N Q M M
                                              ↑     ↑

                  40                     50                   60
K S R N L T K D R C K P V N T F V H E S L A D V Q A V C S Q
      ⇑                   ↑

                  70                     80                   90
K N V A C K N G Q T N C Y Q S Y S T M S I T D C R E T G S S
  ↑         ↑       ↑

                  100                    110                  120
K Y P N C A Y K T T Q A N K H I I V A C E G N P Y Y V P V H F
      ↑                 ↑                   ↑

    124
D A S V     The bond with ⇑ is not cleaved in ribonuclease B
            which has glycosylated Asn at No. 34.
```

Figure 2. Peptide bonds susceptible to jack-bean asparaginyl endopeptidase in reduced *S*-carboxymethylated ribonuclease A (↑ and ⇑).

Although the Asn-Ile bond was not cleaved in lysozyme as mentioned above, the same bond in gastric inhibitory polypeptide was split. The Asn-Pro bond in neuropeptide Y was not cleaved, in opposition to the results with ribonucleases A and B. The reason of these discrepanicies is not clear. Properties of the residues surrounding P₁ and P₁' positions might modify the susceptibility of these peptide bonds to the enzyme.

```
Vasoactive intestinal peptide (VIP)
    HSDAVFTDNYTRLRKQMAVKKYLNSILN-NH2
            ↑              ↑   ↑

Parathyroid hormone [1-24] (PTH)
    SVSEIQLMHNLGKHLNSMERVEWLRKKLQDVHNF
             ↑     ↑                ↑

Reduced S-pyridylethylated erabutoxin c
 RICFNHQSSQPQTTKTCSPGESSCYHKQWSDFRGTIIERGCGCPTVKPGINLSCCESEVCNN
     ↑                                             ↑          ↑

[Asn1,Val5]-Angiotensin I               Neuromedin C
    NRVYVHPFHL                              GNHWAVGHLM-NH2
    ✕          not cleaved                   ✕          not cleaved

Mastoparan
    INLKALAALAKKIL-NH2
     ✕         not cleaved
```

Figure 3. Peptide bonds susceptible to jack-bean asparaginyl endopeptidase in some peptide substrates (↑).

As shown in Figure 3, angiotensin I, neuromedin C, and mastoparan are not digested with the enzyme even though they contain one Asn. This suggests that the enzyme is not accessible to the Asn residue that occupies the N-terminus or the second position from the N-terminus of a peptide. On the other hand, the linkage with Asn at the penultimate position was cleavable as seen in the cases of vasoactive intestinal peptide, parathyroid hormone, and reduced S-pyridylethylated erabutoxin c. The other substrates we examined in this study were: gastrin releasing peptide, pancreatic polylpeptide, RCM- insulin B-chain, tyrosine kinase substrate. The P_1 positions of the peptide bonds cleaved by the enzyme were occupied exclusively with Asn in all the substrates. On the contrary, there were 20 varieties of the P_1' residues including S-alkylated cysteines. We have not used the peptide substrates containing intact cysteine or cystine residues yet.

These results promise the high utility of this enzyme as a tool for specific cleavage at asparaginyl bonds in protein sequence analysis. In fact, we have already had successful results in the application of this enzyme to the primary structure determination of canine haptoglobin (Kumazaki *et al.*, 1992) and galactose-specific lectin of solitary ascidian. Strict specificity toward asparaginyl bonds of the enzyme was confirmed also in these experiments.

VARIOUS OTHER PROPERTIES

The purified enzyme shows the maximal activity in a range of pH 5.0-6.5 and requires the presence of 5 or 10 mM 2-mercaptoethanol or dithiothreitol as a stabilizer. The addition of non-ionic detergent Brij 35 in about 0.01 % and EDTA in 1 mM are also helpful to stabilize it. The enzyme was found to have ability to catalyze transpeptidation reactions at pH 7.7, as we had expected at the beginning of this study. An example of the reactions is shown below.

His-Tyr-Ile-Asn-Leu-Ile-Thr-Arg-Gln-Arg-Tyr-NH_2 + Gly-Gly-His $\rightarrow$
His-Tyr-Ile-Asn-Gly-Gly-His + Leu-Ile-Thr-Arg-Gln-Arg-Tyr-NH_2

Either *p*-chloromercuribenzene sulfonic acid (0.1 mM) or *N*-ethylmaleimide (0.1 mM) suppressed the activity of jack-bean asparaginyl endopeptidase almost completely. Peptide aldehydes such as leupeptin showed the weak inhibitory effect. Serine-protease inhibitors, diisopropyl fluorophosphate and phenylmethanesulfonyl fluoride, an aspartylprotease inhibitor, pepstatin, and a metallo-protease inhibitor, phosphoramidon, had no effect. EDTA is a stabilizer for the enzyme as mentioned before. The effects of protein protease inhibitors are shown in Table 2. Substantial inhibition was observed only with cystatin egg white and high-molecular-weight kininogen. *Streptomyces* subtilisin inhibitor, SSI, is known as a potent inhibitor for serine proteases and some metallo-proteases of microbial origins. Even recombinant SSI (Kojima *et al.*, 1990), whose reactive site Met73 had been replaced with Asn, did not inhibit this enzyme.

Table 2. Effects of various protein protease inhibitors on the activity of jack-bean asparaginyl endopeptidase.

Inhibitors	Conc.(μM)	Inhibition(%)
High-molecular-weight kininogen (human)	5	32
Cystatin EW (chicken egg white)	10	47
Cystatin α (rat)	5	5
Cystatin A (human)	40	2
Calpastatin domain I (human)	5	0
Aprotinin	5	0
Lima bean trypsin inhibitor	5	0
Bowman-Birk inhibitor	5	0
Chicken ovomucoid	5	0
Turkey ovomucoid	5	0
Streptomyces subtilisin inhibitor (Met73Asn)	5	0

COMPARISON WITH TYPICAL CYSTEINE PROTEASES

The properties described above suggest that jack-bean asparaginyl endopeptidase belongs to the cysteine protease. Some differences are observed, however, between this enzyme and typical cysteine proteases. For example, susceptibility of the jack-bean enzyme toward cystatin EW and high-molecular-weight kininogen is relatively low. Cystatin α and cystatin A show almost no inhibitory effect on it. Furthermore, the N-terminal sequence of the enzyme determined for 25 amino acid residues was unable to find any counterpart among the primary structures of proteins that had been reported to be typical cysteine proteases. The only exception was hemoglobinase produced by a kind of schistosome, *Schistosoma mansoni*. The primary structures of its precursor had been recently deduced by the analysis of a cDNA encoding this protein (Meanawy *et al.*, 1990). The precursor consisted of 429 amino acid residues, and the mature enzyme was assumed to be composed of 260 residues. As shown in Figure 4, 19 identical residues were detectable between the N-terminal sequences of hemoglobinase and jack-bean asparaginyl endopeptidase. Recently we succeeded to isolate a cDNA clone encoding a precursor of the asparaginyl endopeptidase by using a

DNA mixed probe synthesized to represent all possible base sequences predicted from the N-terminus of this enzyme and an mRNA mixture obtained from premature seeds of jack bean. The primary structure of the precursor deduced by the analysis of this cDNA consisted of 475 amino acid residues and contained this N-terminal sequence at the region from the residue No.36 to No.60. About 34% sequence homology was found between pro-asparaginyl endopeptidase and pro-hemoglobinase.

In 1984, Csoma and Polgár reported that kidney bean contains an enzyme that catalyzes the preferential hydrolysis of *t*-Boc-asparagine *p*-nitrophenyl ester, and this enzyme has a thiol group essential to the catalytic activity. Although they have not used any peptide

```
    1                                           25
  (36)                                         (60)
    E V G T R W A V L V A G S N G Y G N Y R H Q A D V –
              | | | | | | | | | | |   | | | | | | | |
V S D N N K W A V L V A G S N G Y P N Y R H Q A D V –
1                                                25
(32)                                            (57)
```

Residue numbers are for mature enzymes and the parenthesized numbers are for their precursors.

Figure 4. Comparison of N-terminal amino acid sequences between jack-bean asparaginyl endopeptidase (1st line) and *Schistosoma mansoni* hemoglobinase (2nd line).

substrates for the examination of its substrate specificity, this enzyme seems to be analogous to asparaginyl endopeptidase that we isolated from jack bean. Their paper further stated that considerable differences were found between the kidney-bean enzyme and papain, particularly in their reactivity to iodoacetic acid and iodoacetamide, and so the former enzyme may represent a new variant in the family of cysteine proteases. I would like to present an idea, therefore, that legume asparaginyl endopeptidase and schistosome hemoglobinase might constitute a new subfamily in the cysteine protease family. To prove this hypothesis we have to determine the active-site structure of these proteases .

REFERENCES

Bowles, D. J., Marcus, S. E., Pappin, D. J. C., Findlay, J. B. C., Eliopoulus, E., Maycox, P. R. and Burgess, J., 1986 Posttranslational processing of concanavalin A precursor in jackbean cotyledons, *J. Cell Biol.* 102:1284.

Carrington, D. M., Auffret, A. and Hanke, D. E., 1985 Polypeptide ligation occurs during post-translational modification of concanavalin A, *Nature* 313:64.

Csoma, C. and Polgár, L., 1984 Proteinase from germination bean cotyledons, *Biochem. J.* 222:769.

Kojima, S., Obata, S., Kumagai, I. and Miura, K., 1990 Alteration of the specificity of the *Streptomyces* subtilisin inhibitor by gene engineering, *Bio/Technology* 8:449.

Kumazaki, T., Urushibara, N. and Ishii, S., 1992 Amino acid sequence and disulfide-bridge location of canine haptoglobin, *J. Biochem.* 112:11.

Meanawy, M. A. E., Aji, T., Phillips, N. F. B., Davis, R. E., Salata, R. A., Malhotra, I., McClain, D., Aikawa, M. and Davis, A. H., 1990 Definition of the complete *Schistosoma mansoni* hemoglobinase mRNA sequence and gene expression in developing parasites, *Am. J. Trop. Med. Hyg.* 43:67.

Schechter, I. and Berger, A., 1967 On the size of the active site in protease. I. Papain, *Biochem. Biophys. Res. Commun.* 27:157.

Yokosawa, H., Ojima, S. and Ishii, S., 1977 Chemical transformation of the active-site serine residue of *Streptomyces griseus* trypsin to a cysteine residue, *J. Biochem.* 82:869.

LYSINE-SPECIFIC SERINE PROTEASE FROM *ACHROMOBACTER LYTICUS*: ITS SUBSTRATE SPECIFICITY AND COMPARISON WITH TRYPSIN

Shigemi Norioka and Fumio Sakiyama

Institute for Protein Research, Osaka University
3-2 Yamadaoka, Suita, Osaka 565, Japan

Achromobacter protease I (API) is an extracellular protease isolated from *Achromobacter lyticus* M497-1 (Masaki *et al.*, 1978) and is specific to lysyl bonds including Lys-Pro bond. In addition, API has several attractive properties (Masaki *et al.*, 1981a; Masaki *et al.*, 1981b) : (i) Its proteolytic activity is an order of magnitude higher than that of bovine trypsin, (ii) It has a broad pH optimum ranging from 8.5 - 10.7, (iii) It exhibits full activity in the presence of 5 M urea or 0.1% SDS. Accordingly, API has been widely used as a tool for the primary fragmentation of polypeptide in protein sequence analysis. In this paper, we describe the structural basis of the specificity for lysine and compare it with that of bovine trypsin.

The primary structure of API has already been elucidated (Tsunasawa *et al.*, 1989). API consists of a single polypeptide chain of 268 residues with three disulfide bonds (Figure l). API is a member of trypsin family. Our site-specific mutagenesis and subsite mapping experiments (Sakiyama *et al.*, 1990) revealed that His57, Aspl13 and Serl94 are the three residues composing the catalytic triad and His210, Gly211 and Gly212 are residues composing substrate binding subsites S_1, S_2 and S_3, respectively.

The most interesting enzymatic property of API is the specificity for lysyl bonds. X-Ray crystallographic analysis of bovine trypsin complexed with pancreatic trypsin inhibitor (Huber *et al.*, 1974) or benzamidine (Bode and Schwager, 1975; Bode *et al.*, 1984) has shown that Aspl89 locating at the bottom of the S_1 binding pocket of trypsin interacts specifically with the side chain of arginine and lysine. The positive charge of the arginyl side chain interacts directly with the negative charge of Aspl89 and the lysyl side chain interacts with the negative charge through two bound water molecules. This is the reason why trypsin is specific to these two basic amino acids. Aspl89 is called the substrate specificity determinant of trypsin.

Figure 2 shows the comparison of the amino acid sequence around the active site of API with those of bovine trypsin and related proteases. In API, the position aligned to the Aspl89 of trypsin is occupied by valine of which side chain has no negative charge. This fact suggests that a different substrate specificity determinant exists in place of the Aspl89.

Methods in Protein Sequence Analysis, Edited by K. Imahori
and F. Sakiyama, Plenum Press, New York, 1993

Two acidic amino acids, Glu190 and Asp225, were considered as a candidate for the specificity determinant, because no other acidic amino acids exist among the boxed residues (Figure 2). To elucidate the structural basis of the lysine-specificity of API, we initiated to explore a substrate specificity determinant by a use of protein engineering technique.

```
1                                                                              25
GlyValSerGlySerCysAsnIleAspValValCysProGluGlyAspGlyArgArgAspIleIleArgAlaVal
                                                                               50
GlyAlaTyrSerLysSerGlyThrLeuAlaCysThrGlySerLeuValAsnAsnThrAlaAsnAspArgLysMet
                                                                               75
TyrPheLeuThrAlaHisHisCysGlyMetGlyThrAlaSerThrAlaAlaSerIleValValTyrTrpAsnTyr
                                                                              100
GlnAsnSerThrCysArgAlaProAsnThrProAlaSerGlyAlaAsnGlyAspGlySerMetSerGlnThrGln
                                                                              125
SerGlySerThrValLysAlaThrTyrAlaThrSerAspPheThrLeuLeuGluLeuAsnAsnAlaAlaAsnPro
                                                                              150
AlaPheAsnLeuPheTrpAlaGlyTrpAspArgArgAspGlnAsnTyrProGlyAlaIleAlaIleHisHisPro
                                                                              175
AsnValAlaGluLysArgIleSerAsnSerThrSerProThrSerPheValAlaTrpGlyGlyGlyAlaGlyThr
                                                                              200
ThrHisLeuAsnValGlnTrpGlnProSerGlyGlyValThrGluProGlySerSerGlySerProIleTyrSer
                                    *     *     *                             225
ProGluLysArgValLeuGlyGlnLeuHisGlyGlyProSerSerCysSerAlaThrGlyThrAsnArgSerAsp
                                                                              250
GlnTyrGlyArgValPheThrSerTrpThrGlyGlyGlyAlaAlaAlaSerArgLeuSerAspTrpLeuAspPro
                                                          268
AlaSerThrGlyAlaGlnPheIleAspGlyLeuAspSerGlyGlyGlyThrPro
```

Figure 1. Primary structure of *Achromobacter* protease I. Amino acid residues composing a catalytic triad are boxed. Asterisks indicate the residues corresponding to subsites S_1, S_2 and S_3, respectively.

```
             190        200        210        220        230
BT      EGGKDSCQGDSGGPVVCS------GKLQGIVSWGS-GCAQK--NKPGVYTKVCN
SGT     TGGVDTCQGDSGGPMFRKDNADEWIQVGIVSWGY-GCARP--GYPGVYTEVST
SET     EGGVDTCQGDSGGPMVVN------NKLIGVTSWGE-GCARP--GKPGVYARVGA
PK      PGGKDTCMGDSGGPLICN------GMWQGITSWGHTPCGSA--NKPSIYTKLIF
API     SGG-VTEPGSSGSPIYSP----EKRVLGQLHGGPSSCSATGTNRSDQYGRVFT
              190        200            210        220        230
```

Figure 2. Alignment of the amino acid sequences around the active sites of trypsin and related proteases. The residues composing the S_1 binding pocket of bovine trypsin are boxed. BT, Bovine trypsin; SGT, *Streptomyces griseus* trypsin; SET, *Streptomyces erythraeus* trypsin; PK, porcine kallikrein; API, *Achromobacter* protease I.

GENE STRUCTURE AND EXPRESSION

The gene encoding API has been cloned from *Achromobacter lyticus* and the expression system of the cloned API gene in *E. coli* has been established (Ohara *et al.*, 1989). Figure 3 shows the structure of the gene encoding API. API is synthesized *in vivo* as a precursor protein comprising 653 amino acids. The region corresponding to the 268-residue mature API is located at the central portion. The N-terminal 205 amino acid prepro-region and the C-terminal 180 amino acid threonine/serine-rich region are flanking the mature protein. In *Achromobacter lyticus*, the mature protein is secreted to culture media by releasing both the long N- and C-terminal extensions. However, when the entire API gene was inserted downstream *trc* promoter and expressed in *E. coli*, an active 448 amino acid protease, named API', was accumulated in the periplasm and was not secreted to culture media. The protease secreted to the periplasm is composed of mature API and the C-terminal 180-residue extension, suggesting that the C-terminal extension may be associated with secretion across outer membrane in gram-negative *Achromobacter lyticus* cells. API′ retained the same catalytic activity and the same substrate specificity as those of API, but it was sensitive to autolysis. To express mature API in *E. coli*, a stop codon was inserted to the position encoding Asn269. Consequently, mature API, which possesses the same enzymatic property as that of API from *Achromobacter lyticus*, was secreted to the periplasm. We used this expression system for protein engineering experiments, as described below.

For the inactive precursor to become active API, the cleavage of the Lys-Gly bond is essential, suggesting that the maturation of the precursor is autocatalytic. This was clearly demonstrated by an experiment for the preparation of an API mutant in which the essential Ser194 is replaced by alanine. This mutant was secreted as pro-API to the periplasm but could not be activated to mature API.

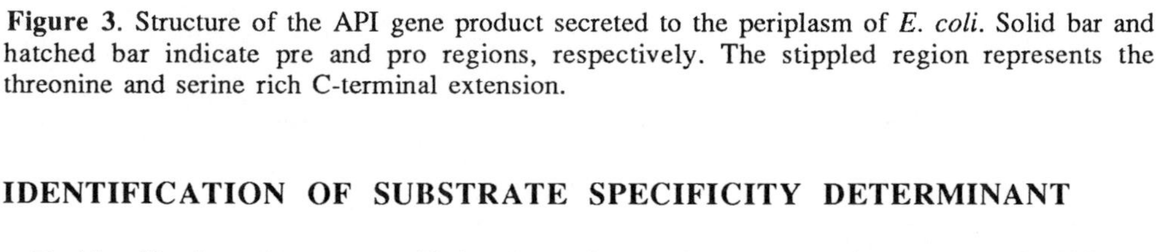

Figure 3. Structure of the API gene product secreted to the periplasm of *E. coli*. Solid bar and hatched bar indicate pre and pro regions, respectively. The stippled region represents the threonine and serine rich C-terminal extension.

IDENTIFICATION OF SUBSTRATE SPECIFICITY DETERMINANT

To identify the substrate specificity determinant of API, several mutants at Glu190 and Asp225 were prepared. Glu190 was replaced by Gln to eliminate the negative charge of the side chain, by Asp to shorten the side chain length, and by Leu to increase the hydrophobicity of the side chain. Similarly, Asp225 was replaced by Asn, Glu and Leu. These mutants are noted in the order of parental amino acid, residue number, mutated amino acid. For example, E190D is a mutant in which Glu190 is replaced by Asp.

E. coli was transformed by a expression vector containing a mutated API gene and cultured at 25°C overnight. After collecting the cells, periplasmic proteins were extracted by osmotic shock (Kreil, 1981), separated by SDS-PAGE and stained by Western blotting with anti-API rabbit sera (Figure 5). A thick band corresponding to a 31 kD protein seen in every lane was nonspecific. In lanes 3 to 5, a band corresponding to mature API was clearly detected, suggesting that the precursors of all three mutants at Glu190 were perfectly processed to mature API. However, the mature API was not detected in all Asp225 mutants (lanes 6 and 7). Instead of the mature protein, a new band corresponding to a 45 kD protein was seen in the case of D225N. The 45 kD protein was inactive and is thought to be a pro-API, since no amidase activity for Boc-Val-Leu-Lys-MCA was detected in the periplasmic fraction. The formation of pro-API was also seen for the D225L mutant, and any API activity was not detected as well as for the D225N mutant (data not shown). The failure of maturation of pro-API indicates that the catalytic activity for lysyl bonds was drastically decreased upon substitution of Asp225 to Asn or Leu. Nevertheless, when a larger amount of the periplasmic fraction of D225E was loaded on a gel, a faint band of mature API was observed in addition to the band of pro-API, the retention of a weak activity being indicated. In order to examine the significance of the Glu190 and Asp225 in detail, we have purified and characterized the four mutants secreted as mature proteins, E190Q, E190D, E190L and D225E.

API mutants were easily purified by affinity chromatography with ovomucoid. A periplasmic fraction prepared by the osmotic shock method (Kreil, 1981) was passed through a DEAE-cellulofine column, and the unadsorbed proteins were loaded on an ovomucoid-Sepharose 4B column. After the column had been washed with 50 mM Tris-HCl (pH 8.0) and subsequently distilled water, the adsorbed API mutant was eluted with 10 mM ammonium acetate (pH 3.0). All of the four mutants thus obtained gave single bands on SDS-PAGE.

To examine the substrate specificity of each purified mutant, *Vibrio parahaemolyticus* hemolysin was digested with the mutant and analyzed by reversed phase HPLC. Wild type API cleaved the lysyl bonds in hemolysin specifically to produce 9 peptides. The elution patterns of the digests obtained with the four mutants were the same as that obtained with the wild type enzyme, showing that the four mutants are active and specific to lysyl bonds.

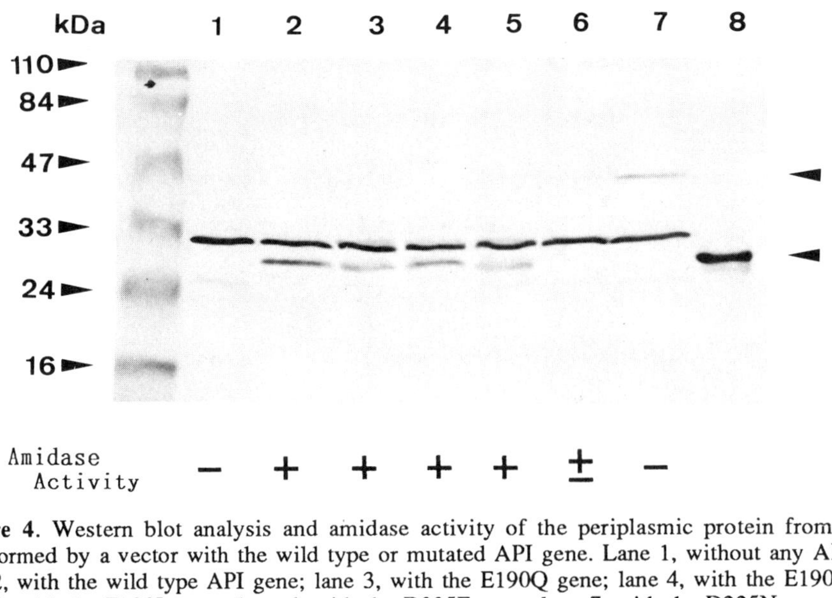

Figure 4. Western blot analysis and amidase activity of the periplasmic protein from *E. coli* transformed by a vector with the wild type or mutated API gene. Lane 1, without any API gene; lane 2, with the wild type API gene; lane 3, with the E190Q gene; lane 4, with the E190D gene; lane 5, with the E190L gene; lane 6, with the D225E gene; lane 7, with the D225N gene; lane 8, API from *Achromobacter lyticus*. Amidase activities were assayed with Boc-Val-Leu-Lys-MCA.

Since the yield of each peptide obtained with D225E was much smaller than those recorded with the Glu190 mutants, a decrease in the proteolytic activity of D225E was suggested.

Kinetic constants estimated with Boc-Val-Leu-Lys-MCA are summarized in Table 1. It is clear that Km, kcat and kcat/Km of the three Glu190 mutants are the same as or very similar to that of wild type API. However, the enzyme activity (kcat/Km) of D225E was significantly decreased (about 1/400) mainly due to the 140-fold increase of Km. Both D225N and D225 L mutants could not be maturated as described above.

These results suggest that Asp225 is essential to the proteolytic activity of API and that a negative charge at Asp225 plays a key role in the binding of the typical substrate, lysine residue. We conclude that Asp225 is the most potent candidate as a substrate specificity determinant of API.

STRUCTURAL BASIS OF LYSINE-SPECIFICITY

Recently the three dimensional structure of API has been elucidated by X-ray crystallography (Kitagawa *et al.*, in preparation). The tertiary fold and the structure of the active site are quite similar between API and bovine trypsin. Each protein is composed of two β-barrels, and the active site is constructed between these two β-barrels. There is a unique negative charge, Asp225, in the S_1 binding pocket. The side chain of Glu190 orients outside from the pocket. All these structural features are consistent with the results of our mutagenesis experiments.

Table 1. Kinetic constants of wild type API and mutants at Glu190 and Asp225 for the hydrolysis of Boc-Val-Leu-Lys-MCA,

Mutant	Gene Product	Km (μM)	kcat (sec^{-1})	kcat/Km ($\mu M^{-1}sec^{-1}$)
Wild Type	Mature	2.1	93	44
E190Q	Mature	3.8	84	22
E190D	Mature	7.4	88	12
E190L	Mature	2.1	74	35
D225E	Mature	295	31	0.11
D225N	Pro-API	-	-	-
D225L	Pro-API	-	-	-

Enzyme activity was assayed in 0.2 M Tris-HCl at 9.0 and 37°C.

It is interesting to consider the structural basis of lysine-specificity in API in conjunction with that of the substrate specificity of bovine trypsin. Trypsin cleaves both arginyl and lysyl bonds, but API cleaves only lysyl bond. The critical structural difference of these enzymes exists in the S_1 binding pocket. API has a shallow binding pocket, whereas trypsin has a deep pocket. Asp225 in API is located on the wall of the pocket and its side chain sticks out inside the pocket, making the bottom of the pocket about 2 Å shallower than that of trypsin. Trypsin has Gly at the same position as Asp225 in API. In API, the side chain of lysine in the bound substrate can well fit to the pocket and interact with Asp225 by electrostatic force. Since the side chain of arginine is longer than that of lysine, arginine cannot reside in the shallow binding pocket of API. Namely, the strict specificity of API for lysyl bonds is primarily based on the presence of Asp225.

CONCLUDING REMARKS

Asp225 is the substrate specificity determinant of API. Lys-C is another lysine-specific protease isolated from *Lysobacter enzymogenes*. However, since the amino acid sequence of Lys-C was found to be identical with that of API, the presence of a negative charge at the positon of 225 is considered to be one of the mechanisms that determine the lysine-specificity of the mammalian-type serine protease. In addition to API, murine granzyme B, a cytotoxic lymphocyte aspartic acid-specific serine protease, has a putative specificity determinant, arginine, at the same position as Asp225 in API (Brunet *et al.*, 1986; Lobe *et al.*, 1986; Odake *et al.*, 1991). These ideas suggest that an amino acid at the position of 225 plays a key role in determining the specificity of a single amino acid-specific serine protease. Accordingly, it is interesting whether V8 protease and Blase, glutamic acid-specific serine proteases, have their specificity determinants at the position of 225 or not.

REFERENCES

Bode, W. and Schwager, P., 1975 The refined crystal structure of bovine β-trypsin at 1.8 Å resolution, *J. Mol. Biol.* 98: 693.

Bode, W., Walter, J., Huber, R., Wenzel, H. R. and Tshesche, H., 1984 The refined 2. 2 Å (0. 22 nm) X ray crystal structure of ternary complex formed by bovine trypsinogen, valine-valine and Argl5 analogue of bovine pancreatic trypsin inhibitor, *Eur. J. Biochem.* 144: 185.

Brunet, J. F., Dosseto, M., Denizot, F., Mattei, M. G., Clark, W. R., Haqqi, T. M., Ferrier, P., Nabholz, M., Schmitt-Verhulst, A. M., Luciani, M. F. and Golstein, P., 1986 The inducible cytotoxic T-lymphocyte-associated gene transcript CTLA-1 sequence and gene location to mouse chromosome 14, *Nature* 322: 268.

Huber, R., Kukla, D., Bode, W., Schwager, P., Bartels, K., Deisenhofer, J. and Steigemann. W., 1974 Structure of the complex formed by bovine trypsin and bovine pancreatic trypsin inhibitor, *J. Mol. Biol.* 89: 73.

Kitagawa, Y., Sasi, K., Matsuura, Y., Katsube, Y., Masaki, T., Soejima, M., Norioka, S. and Sakiyama, F., in preparation.

Kreil, G., 1981 Transfer of proteins across membranes, *Annu. Rev. Biochem.* 50: 317.

Lobe, C. G., Finlay, B. B., Paranchych, W., Paetkau, V. H. and Bleackley, R. C., 1986 Novel serine proteases encoded by two cytotoxic T lymphocyte-specific genes, *Science* 232: 858.

Masaki, T., Nakamura, K., Isono. M. and Soejima, M., 1978 A new proteolytic enzyme from *Achromobacter lyticus* M497-1, *Agric. Biol. Chem.* 42:1443.

Masaki, T., Tanabe, M., Nakamura. K. and Soejima, M., 1981 Studies on a new proteolytic enzyme from *Achromobacter lyticus* M497-1.I. Purification and some enzymatic properties, *Biochim. Biophys. Acta* 660: 44.

Masaki, T., Fujihashi, T., Nakamura, K. and Soejima, M., 1981 Studies on a new proteolytic enzyme from *Achromo-bacter lyticus* M497-1. II. Specificity and inhibition studies of Achromobacter protease I, *Biochim. Biophys. Acta* 660: 51.

Odake, S., Kam, C.-M., Narasimhan, L., Poe, M., Blake, J. T., Krahenbuhl, O., Tschopp, J. and Powers, J. C., 1991 Human and murine cytotoxic T lymphocyte serine protease: subsite mapping with peptide thioester substrates and inhibition of enzyme activity and cytolysis by isocoumarins, *Biochemistry* 30: 2217.

Ohara, T., Makino, K., Shinagawa, H., Nakata, A., Norioka, S. and Sakiyama, F., 1989 Cloning, nucleotide sequence, and expression of *Achromobacter* protease gene, *J. Biol. Chem.* 264: 20625.

Sakiyama, F., Suzuki, M., Yamamoto, A., Aimoto, S., Norioka, S., Masaki, T. and Soejima, M., 1990 Mapping of substrate binding subsites of a lysine-specific serine protease, *Achromobacter* protease I, *J. Protein Chem.* 9:297.

Tsunasawa, S., Masaki, T. , Hirose, M., Soejima, M. and Sakiyama, F., 1989 The primary structure and structural characteristic of *Achromobacter lyticus* protease I, a lysine-specific serine protease, *J. Biol. Chem.* 264: 3832.

STRUCTURE AND FUNCTION OF SNAKE VENOM METALLOPROTEINASE FAMILY

Sadaaki Iwanaga and Hiroyuki Takeya

Department of Biology, Faculty of Science
Kyushu University
Fukuoka 812, Japan

INTRODUCTION

Venoms of snakes belonging to families *Viperidae* (viper) and *Crotalidae* (pit viper) produce striking local effects, consisting of hemorrhage, necrosis, and edema, and often induce marked alterations of blood coagulation system as well (Iwanaga and Suzuki, 1979). Among these pathological effects, hemorrhage is a most common occurrence in a victim bitten by crotalid and viperid snakes and various components, such as hemorrhagic factors and metalloproteinases, involved in these venoms have been isolated and characterized (Osaka, 1979). The hemorrhagic metalloproteinases cause localized hemorrhage by direct action on the blood vessel wall. Electron microscopic studies indicate that erythrocytes are leaked in a one-by-one fashion through widened inter-endothelial gaps when capillaries are exposed to these hemorrhagic proteins. Hemorrhagic metalloproteinases degrade basement membrane preparation, in addition to the isolated components including type IV collagen, laminin, nidogen, and fibronectin (Baramova *et al.*, 1989). The enzymes may disrupt the pericellular basement membrane through a proteolytic activity and with subsequent damage to the integrity of the vessel wall after which hemorrhage occurs. On the other hand, crotalid and viperid venoms contain an abundance of metalloproteinases which are completely free from hemorrhagic activity (Iwanaga *et al.*, 1976). Some of them have originally been reported as anticoagulant proteases with fibrinogenolytic or fibrinolytic activity (Markland, 1991). Additional group of venom metalloproteinases are procoagulants having very strict substrate specificities. Two well known examples of this type are Russell's viper venom factor X activator (RVV-X) and *Echis carinatus* venom prothrombin activator (Morita and Iwanaga, 1981; Furie and Furie, 1976).

The hemorrhagic and non-hemorrhagic metalloproteinases isolated from *Trimeresurus (T.) flavoviridis* (Habu snake) and *Crotalus (C.) atrox* (Western diamondback rattlesnake) and RVV-X have been well characterized and their amino acid sequences have been established recently. This chapter describes the structure and function relationships of these well studied venom metalloproteinases.

Methods in Protein Sequence Analysis, Edited by K. Imahori and F. Sakiyama, Plenum Press, New York, 1993

HEMORRHAGIC METALLOPROTEINASES

In the 1970 era, four hemorrhagic metalloproteinases HRlA, HRlB, HR2a, and HR2b, and one non-hemorrhagic H_2-proteinase have been purified from the venom of *T. flavoviridis*. High molecular mass hemorrhagic factors (Mr 60,000) HRlA and HRlB express 10-25 times higher hemorrhagic activities than do low molecular mass hemorrhagic factors (Mr 25,000) HR2a and HR2b, thereby indicating major lethal factors in the venom. For example, minimum hemorrhagic doses (which is expressed as ng of protein required to produce a hemorrhagic spot of 10 mm diameter on the back skin of rabbit, when injected intracutaneously) of HRlB and HR2a are 11 ng and 66 ng, respectively, and the LD_{50} values of HRlB and HR2a are 4,9 and 144 μg/mouse, respectively. In addition to these pathological functions, HRl (mixture of HRlA and HRlB) as well as the crude venom has been shown to inhibit ADP-stimulated platelet aggregation. HRl has the potential to inhibit 60% of the platelet aggregation at a concentration of 6 ng/ml in platelet rich plasma. *In situ* microscopic observation also revealed that hemorrhagic factors present in venom induce hemorrhage with little or no formation of white thrombi at the site of the hemorrhage.

No significant difference in the biochemical property has heen observed between hemorrhagic and non-hemorrhagic metalloproteinases (Takeya *et al.*, 1989). The increased proteolytic activity is generally not evidenced by increased hemorrhagic activity. The difference of biological activity might possibly be related to the difference of proteolytic specificity of these metalloproteinases. Although there are few reports about the proteolytic

Table 1. Values of kcat/Km for hydrolysis of Abz-peptide-Dna substrates by snake venom metalloproteinases and thermolysin.

Peptides	HR1A	HR2a	H_2[1]	RVV-X	Thermolysin
P3 P2 P1 P1' P2' P3'	kcat/Km ($M^{-1}s^{-1}$)				
Abz-Ser-Pro-Met-Leu-Dna	180,000	340,000	20,000	N.H.	10,000
Abz-Thr-Glu-Lys-Leu-Val-Dna	130,000	120,000	32,000	N.H.	770,000
Abz-Phe-Ser-Pro-Met-Leu-Dna	92,000	-[2]	-	N.H.	8,400
Abz-Gly-Phe-Arg-Leu-Leu-Dna	18,000	12,000	7,800	76,000	1,000,000
Abz-Ala-Gly-Leu-Ala-Dna	2,800	-	-	N.H.	1,100,000
Abz-Asn-Ala-Pro-Leu-Ala-Dna	1,500	-	-	N.H.	310,000
Abz-Asn-Ala-Pro-Ser-Ala-Dna	N.H.[3]	N.H.	N.H.	N.H.	N.H.
Abz-Ala-Thr-Asp-Ile-Val-Dna	N.H.	N.H.	N.H.	N.H.	N.H.
Abz-Gly-Phe-Arg-Leu-Dna	N.H.	N.H.	N.H.	N.H.	3,800
Abz-Gly-Pro-Leu-Gly-Pro-Dna	N.H.	N.H.	N.H.	N.H.	N.H.

[1]H_2, H_2-proteinase. [2]-, not determined. [3]N.H., not hydrolyzed.

specificity between hemorrhagic and non-hemorrhagic metalloproteinases, no clear difference of proteolytic specificity, however, has been observed as far as using fibrinogen as a substrate (Takeya *et al.*, 1992b; Retzios and Markland, 1992). For instance, a non-hemorrhagic metalloproteinase Cbfib2 isolated from *Crotalus basiliscus basiliscus* and a hemorrhagic metalloproteinase HR2a both cleave mainly at Pro-516-Met-517 bond in the Aα-chain of fibrinogen.

An abundance of studies done on the substrate specificity of venom metalloproteinases using oxidized insulin B chain have shown that these enzymes generally have a primary specificity toward a hydrophobic and bulky Pl' residue, such as Leu and Met, and that a peptide longer than a pentapeptide is needed for substrate hydrolysis (Iwanaga and Suzuki, 1979). Despite numerous studies on the substrate specificity of venom metalloproteinases, using insulin, glucagon, ribonuclease A and lysozyme in addition to insulin B chain (Iwanaga *et al.*, 1976), there is no available substrate for a simple, sensitive and accurate assay of metalloproteinases. Intermolecularly quenched fluorogenic peptide substrates have recently been developed, based on data with sequences in close proximity to the autoproteolytic cleavage sites in high molecular mass metalloproteinases (described later) and the sites of fibrinogen cleaved by HR2a and H_2-proteinase (Takeya *et al.*, 1992b). The substrates contain 2-aminobenzoyl group (Abz) as fluorophore and 2, 4-dinitroanilinoethylamide (Dna) as quenching group. These aromatic groups linked via amide

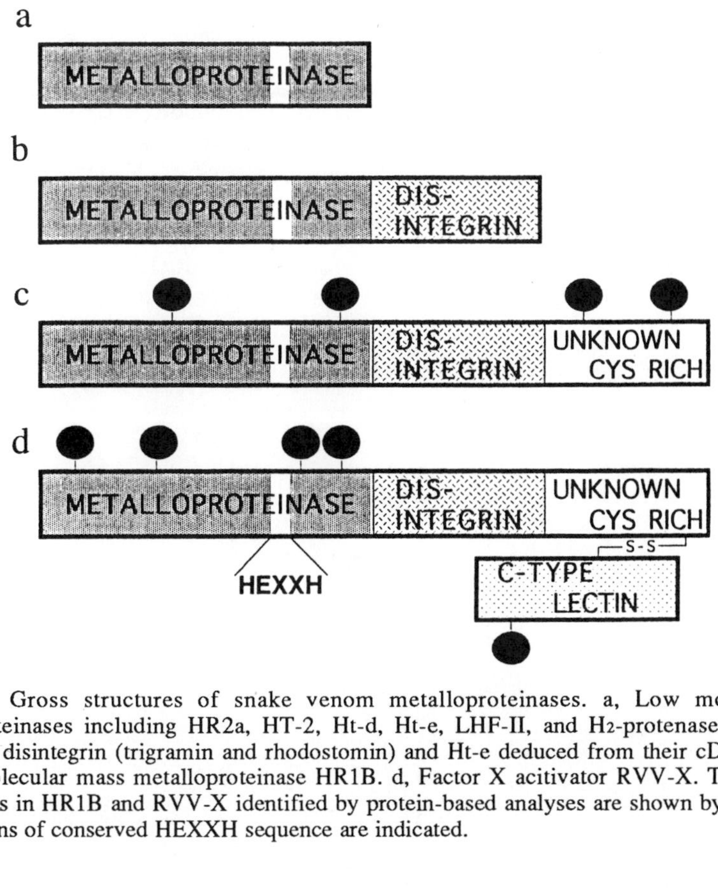

Figure 1. Gross structures of snake venom metalloproteinases. a, Low molecular mass metalloproteinases including HR2a, HT-2, Ht-d, Ht-e, LHF-II, and H_2-protenase. b, Precursor proteins of disintegrin (trigramin and rhodostomin) and Ht-e deduced from their cDNA sequence. c, High molecular mass metalloproteinase HR1B. d, Factor X acitivator RVV-X. The Asn-linked sugar chains in HR1B and RVV-X identified by protein-based analyses are shown by *closed circle*. The locations of conserved HEXXH sequence are indicated.

bonds to the N- and C-terminal ends of the synthetic peptides are separated upon enzymatic hydrolysis with a resultant increase in fluorescence that is utilized to measure hydrolysis rates. Among the ten fluorogenic substrates synthesized (Table 1), Abz-Ser-Pro-Met-Leu-Dna and Abz-Thr-Glu-Lys-Leu-Val-Dna, which are mimiced with the sequence surrounding cleavage sites of fibrinogen Aα-chain by HR2a and H_2-proteinase, proved to be the best substrates for these enzymes, while the bacterial metalloproteinase thermolysin is inert toward this substrate. It should be noted in Table 1 that Russell's viper venom factor X activator (RVV-X) effectively hydrolyzes Abz-Gly-Phe-Arg-Leu-Leu-Dna at the Pl site of the Arg residue but not the other synthetic substrates. The sequence of this substrate is similar to those with the cleavage sites of factors X (-Leu-Thr-Arg-Ile-Val-) and IX (Phe-Thr-Arg-Val-Val-) in zymogen activations catalyzed by RVV-X. Thus, differing from other venom metalloproteinases, RVV-X seems to require the Arg residue at thc Pl site. All the venoms belonging to *Crotalidae* and *Viperidae* families which contain several types of hemorrhagic and non-hemorrhagic metalloproteinases, effectively hydrolyze the synthetic substrates (Takeya *et al.*, 1992b), thus indicating that these substrates are useful for simple and sensitive assays of venom metalloproteinases.

METALLOPROTEINASE DOMAIN

Figure 1 shows the domain structures of the venom metalloproteinases so far determined. Low molecular mass hemorrhagic and non-hemorrhagic metalloproteinases, HR2a, HR2b and H_2-proteinase from *T. flavoviridis*, HT-2 from *C. ruber ruber*, Ht-d and Ht-e from *C. atrox*, and LHF-II from *Lachesis muta muta* have a very similar structure consisting of 201-204 amino acid residues (Takeya *et al.*, 1989; 1990b; 1992c; Miyata *et al.*, 1989; Shannon *et al.*, 1989; Hite *et al.*, 1992; Sanchez *et al.*, 1991). The high molecular mass hemorrhagic protein, HR1B (Takeya, *et al.*, 1990a), is a mosaic protein consisting of three distinct domains, an amino-terminal metalloproteinase domain (residues 1-203) similar to those of HR2a, HT-2, Ht-d, Ht-e, LHF-II, and H_2-proteinase, a disintegrin (platelet aggregation inhibitor)-like structure (residues 204-300) and unique carboxyl-terminal domain (residues 301-416). RVV-X has an extra light chain (123 residues) structurally resembling Ca^{2+} dependent (C-type) animal lectins, in addition to a heavy chain (427 residues) similar in overall sequence to that of HR1B (Takeya *et al.*, 1992a). The cDNAs for two disintegrins, trigramin and rhodostomin and for low molecular mass metalloproteinase Ht-e have recenty

Astacin numbering					92	93			96			99			102		
H2-Proteinase numbering					142	143			146			149			152		
Astacin	:	T	I	I	H	E	L	M	H	A	I	G	F	Y	H	E	H
H2-pro.	:	T	M	T	H	E	L	G	H	N	L	G	M	E	H	D	D
RVV-X-H	:	I	M	A	H	E	L	S	H	N	L	G	M	Y	H	D	G
HR1B	:	I	M	T	H	E	M	G	H	N	L	G	I	P	H	D	G
HR2a	:	T	M	T	H	E	I	G	H	N	L	G	M	E	H	D	D
HT-2	:	T	M	A	H	E	L	G	H	N	L	G	M	E	H	D	G
Ht-d	:	T	M	A	H	E	L	G	H	N	L	G	M	E	H	D	G
Ht-e	:	T	M	T	H	E	L	G	H	N	L	G	I	H	H	D	T
LHF-II	:	T	M	A	H	E	L	G	H	N	L	G	M	K	H	D	E
Trig.Pre.	:	T	M	T	H	E	M	G	H	N	L	G	M	H	H	D	E
Rhod.Pre.	:	I	M	A	H	E	M	G	H	N	L	G	V	R	H	D	G
Thermolysin	:	V	V	G	H	E	L	T	H	A	V	T	D	Y	T	A	G

Figure 2. Alignment of the sequence around the consensus sequence of astacin and venom metalloproteinases. The important residues for zinc-chelating and the catalytic residue conserved in astasin and verom metalloproteinases are *boxed* (see text). His and Glu residues in the His-Glu-X-X-His consensus sequence of thermolysin are also *boxed*.

been cloned and sequenced (Hite *et al.*, 1992; Neeper and Jacobson, 1990; Au *et al.*, 1991). All the results suggest that these disintegrins and metalloproteinases might be translalted as similar protein precursors (Figure 1b) to be processed posttranslationally. Since the protein features shown in Figure 1 indicate a structural and evolutional relationship among these proteins, the venom metalloproteinases belong to a new gene family.

The venom metalloproteinases share a highly conservative region corresponding to the residues 138-149, which includes the His-Glu-X-X-His sequence (residues 142-146, H_2-

proteinase numbering) similar to the zinc-chelating region of thermolysin. In thermolysin, this sequence occurs in a helical region, as demonstrated by X-ray analysis. As expected, the predicted secondary structures of snake venom metalloproteinases show that the His-Glu-X-X-His sequence also occurs in a helical region (Takeya *et al.*, 1990b). Thus, it seems likely that His-145 and His-149 contribute to a zinc ligand and the corresponding Glu-146 is one of the catalytic residues, based on the data of sequence comparison. However, thc third zinc ligand Glu-166 (thermolysin numbering) and other functional residues closely associated with a catalytic site, such as Asn-112, Ala-113, Trp-115, Tyr-157, and His-231, all of which have been identified for thermolysin, could not be found in any of the venom metalloproteinases. Therefore, the catalytic residues of the venom metalloproteinase may differ greatly from those of thermolysin. Recently, Bode *et al.* have reported the X-ray crystal structure of astacin, a digestive metalloproteinase from the crayfish *Astacus astacus* L (Bode *et al.*, 1992). His-102 in astacin appears the third zinc-ligand replacing Glu-166 in thermolysin (Figure 2). Bode *et al.* have also stated that the main-chain turning at Gly-99 is important to allow His-92, His-96 and His-102 to ligand zinc. As shown in Figure 2, these His and Gly residues are conserved in the venom metalloproteinases, thereby suggesting that the third zinc ligand in the venom metalloproteinases corresponding to His-102 in astacin is His-155.

As to the sequence comparison between hemorrhagic and non-hemorrhagic metalloproteinases, it is now possible to compare the primary structures of six hemorrhagic proteinases with non-hemorrhagic RVV-X and H_2-proteinase (Takeya *et al.*, 1992a). Although we could not find no clear structural differences between the hemorrhagic and non-hemorrhagic metalloproteinases, there are four amino acid residues shared by hemorrhagic proteinases but not by non-hemorrhagic ones (Takeya *et al.*, 1992a). Thus, these conservative residues may be related to the structural element with the hemorrhagic activity and also responsible for binding with the basement membrane, which constitute a solid barrier against erythrocyte escape. Furthermore, the conserved residues among all proteins may be candidates for catalytic residues and/or important residues to form fundamental conformation of the venom metalloproteinases (see Takeya *et al.*, 1992a).

```
                 210       220       230       240       250       260       270    ***280       290       300
RVV-X-HC   : LRKDIVSPPVCGNEIWEEGEECDCGSPANCQNPCCDAATCKLKPGAECGNGLCCYQCKIKTAGTVCRRARDECDVPEHCTGQSAECPRDQLQQNGKP
HR1B       : SKTDIVSPPVCGNELLEAGEECDCGSPENCQYQCCDAASCKLHSWVKCESGECCDQCRFRTAGTECRAAESECDIPESCTGQSADCPTDRFHRNGQP
HR1A-32K   :     IVSPPVCGNELLEVGEECDCGSPATCRYPCCDAATCKLHSWVECESGECCEQCRFRTAGTECRARRSECDIAESCTGHSADCPTDRFHRNGQP
HT-1-31K   : LGTDIISPPVCGNELLEVGEECDCGFPRNCRDPCCDATTCKLHSWVECESGECCGQCKFTSAGNECRPARSECDIAESCTGQSADCPMDDFHRNGQP
PH-30 β    :   PVYRSNPVCGNNRVEQGEDCDCGSQEECQDTCCDAATCRLKSTSRCAQGPCCNQCEFKTKGEVCRESTDECDLPEYCNGSSGACQEDLYVINGHR
PH-30 α    :                                                                             YCTGQSGKCPLDTYKQDGTP
Ht-e Pre.  : LRTDTVSTPVSGNELLEAGIECDCGSLEN---PCCYATTCKMRPGSQCAEGLCCDQCRFMKKGTVCRVSMVD-RNDDTCTGQSADCPRNGLYG
Bitistatin :       SPPVCGNKILEQGEDCDCGSPANCQDQCCNAATCKLTPGSQCNHGECCDQCKFKKARTVCRIARGD-WNDDYCTGKSSNCPWNH
Trigramin  :                 EAGEDCDCGSPAN---PCCDAATCKLIPGAQCGEGLCCDQCSFIEEGTVCRIARGD-DLDDYCNGRSAGCPRNPFH
Applaggin  :                 EAGEECDCGSPEN---PCCDAATCKLRPGAQCAEGLCCDQCKFMKEGTVCR-ARGD-DVNDYCNGISAGCPRNPFFH
Echistatin :                                               ECESGPCCRNCKFLKEGTICKRARGD-DMDDYCNGKTCDCPRNPHKGPAT

                   310       320       330       340       350       360       370       380
RVV-X-HC   : CQNNRGYCYNGDCPIMRNQCISLFGSRANVAKDSCFQE-NLKGSYYGYCRKE-N-GRKIPCAPQDVKCGRLFCLNNSPRNKNP-------------
HR1B       : CLYNHGYCYNGKCPIMFYQCYFLFGSNATVAEDDCFNN-NKKGDKYFYCRKE-N-EKYIPCAQEDVKCGRLFCDNKKYP-----------------
HR1A-32K   : CLHNFGYCYNGNCPIMYHQCYALWGANATVAKDSCFED-NQKGNDYGYCRKE-N-GRKIPCEPQDVKCGRLYCSLGNQLP----------------
HT-1-31K   : CLNNFGYCYNGNCPILYHQCYALFGSNVYEAEDSCFER-NQKGDDDGYCRKE-N-GEKIPCAPEDVKCGRLYCKDNSPGPNDS-------------
PH-30 β    : CANEEWICMNGRCLSGKAQCQETFGTEMEMGSVDCFEQLNTKNDITGNCG-ILSPGNYKACGASNWKCGKLICSYDKSEILRNKEGMTIYANISGHI
PH-30 α    : C-NEGFFCVSKGCTDPGIQCATYFGHGARSAPDACYTTLNSIGNIFGNCGQSGNPTTYVGCSGDSTKCGKLICTGISSIPPIRALFAAIQIPHGDDW

                            390       400       410       420
RVV-X-HC   : CNMH-Y---SCMDQHKGMVDPGTKCEDGKVCNNKRQCVDVNTAYQSTTG
HR1B       : CHYN-Y---S-EDLDFGMVDHGTKCADGKVCSN-RQCVDVNEAYKS
HR1A-32K   : CRFF-Y---TPTDENIGMVDTGTKCGDKKVCSN-RQCVDVNTAY
HT-1-31K   : CKTF-N---SNEDDHKEMVLPGTKCADGKVCSN-GHCVDVASAY
PH-30 β    : CVSIEYPPGHAKSAL-MWVRDGTVCGPSEVCRQ-QQCVSSSYLGYDCTP
PH-30 α    : CWSISNFGDPASSPTEGAVSAGTSCASGKACVN-AQCSTFTLDTANCSA
```

Figure 3. Alignment of the amino acid sequences of disintegrin-like and Cys-rich domains with disintegrins. Disintegrin polypeptides are aligned with the carboxyl-terminal regions of RVV-X-heavy chain (RVV-X-HC), HRlB (*HR1B*) and Ht-e precursor *(HT-1 Pre.)*, 32 kDa and 31 kDa fragments derived from HRlA (*HR1A-32* kDa) and HT-1 (*Ht-1-31* kDa), and the amino-terminal regions of α- and β- chains of PH-30. Residue numbering is referred to that of RVV-X. *Asterisks* indicate the Arg-Gly-Asp-sequence. *Dashes* are inserted to maximize similarity. All cysteine residues are boxed.

DISINTEGRIN AND CYS-RICH DOMAINS

Platelet aggregation inhibitory peptides, termed trigramin, echistatin, applaggin, and bitistatin have been isolated from various crotalid and viperid snake venoms, and the amino acid sequences were determined (Gould *et al.*, 1990). These peptides are homologous cysteine rich peptides consisting of 49-83 amino acid residues. A highly conserved structural feature in these peptides is the tripeptide unit Arg-Gly-Asp, a sequence that appears in a variety of adhesive protein ligands, such as fibronectin, vitronectin, fibrinogen, and von Willebrand factor, and contributes to their interaction with specific membrane receptors, called integrins. These peptides, now known as the disintegrin family, are competitive inhibitors of fibrinogen and von Willebrand factor binding to the activated platelet membrane integrin, glycoprotein IIb/IIIa (GPIIb/IIIa) complex. This inhibitory activity of disintegrin is about 1,000-30,000 times more potent than that of the tetrapeptide Arg-Gly-Asp-Ser, perhaps owing to the favorable conformational restraints of the Arg-Gly-Asp-sequence (Adler *et al.*, 1991).

As shown in Figures 1 and 3, the cysteine-rich middle portions (residues 204-300) following the metalloproteinase domains of HR1B and RVV-X have a close similarity to disintegrins. However, the Arg-Gly-Asp-sequence conserved among disintegrins is absent in HRlB and RVV-X. The Arg-Gly-Asp-sequence is substituted by Glu-Ser-Glu-Cys in HRlB and Arg-Asp-Glu in RVV-X (Figure 3). We tested the platelet aggregation inhibitory activity of RVV-X, and found that RVV-X dose-dependently inhibited collagen- and ADP-induced platelet aggregations (Takeya *et al.*, 1992a), as reported for HRlB (Ohsaka, 1979). Several lines of evidence suggest that platelet aggregation inhibitory activity resides within the disintegrin-like domains of HRlB and RVV-X, despite the absence of the Arg-Gly-Asp-sequence. First, it is suggested that a certain conformation of disintegrin is a prerequisite for its binding activity to the GPIIb/IIIa complex, since the chemically reduced sample does not inhibit fibrinogen and von Willebrand factor binding to activated platelets. Secondly, tyrosylpentadecapeptide (Tyr-Gly-Gln-His-His-Leu-Gly-Gly-Ala-Lys-Gln-Ala-Gly-Asp-Val) corresponding to the carboxyl-terminal sequence of the fibrinogen γ chain blocks fibrinogen binding and trigramin binding to platelets, thereby suggesting that the Arg-Gly-Asp-sequence present in the snake venom disintegrin family is not only sequence binding to platelets. It remains to be determined which structural segment of HRlB and RVV-X is essential for the inhibition of platelet aggregation. Finally, HRlB and RVV-X might recognize other integrins but not GPIIb/IIIa. All of twenty-five disintegrins so far isolated contain the RGD-sequence with one exception, barbourin containing Lys-Gly-Asp instead (Scarborough *et al.*, 1991). The substitution in barbourin is suggested to impart integrin specificity to barbourin, since it has been isolated as a specific antagonist for glycoprotein IIb/IIIa, while other RGD-containing disintegrins interact with various integrins of β_1 and β_3 families, such as the vitronectin receptor(s) and fibronectin receptor. The integrin specificities of RVV-X and HR1B which do not contain Arg-Gly-Asp sequence remain to be determined.

Trigramin was noted to inhibit the adhesion of melanoma cells to fibronectin (Gould *et al.*, 1990). In this regard, the disintegrin-like domain of high molecular mass metalloproteinases may be important to localize these molecules to the peripheral blood vessel, especially to integrins located in the basement membrane attachment site of the endothelial cells, when the amino-terminal metalloproteinase domain degrades the basement membrane. This localization of high molecular mass metalloproteinase to the subendothelium, in addition to its platelet aggregation inhibitory activity, may be one reason why it expresses 10-25 times higher hemorrhagic activity than does that of low molecular mass ones.

Blobel *et al.* (1992) have just reported that PH-30, a sperm surface protein involved in sperm-egg fusion, has a sequence similarity to those of disintegrin-like and Cys-rich domains originally found in HRlB. Surprisingly, they share a significant amino acid identity and all the cysteine residues are aligned with identical positions in the sequences (Figure 3). It is also of interest that the disintegrin domain present in PH-30 does not contain the Arg-Gly-Asp sequence, and the corresponding region contains the sequence of Thr-Asp-Glu-Cys similar to that of RVV-X. These significant sequence similarities indicate a common ancestry.

PRECURSOR PROTEINS

As shown in Figure 1, we and others found structural similarity between HR1B and the protein precursors of disintegrins, all of which contain metalloproteinase- and the disintegrin-domains (Takeya *et al.*, 1991; 1992a; Au *et al.*, 1991). This finding led to the notion that all these disintegrins are released through an autoproteolysis from metalloproteinases abundantly present in crotalid and viperid venoms. Thus, on incubation of the purified high molecular mass hemorrhagic metalloproteinases, HR1A and HR1B from *T. flavoviridis*, and HT-l from *C. ruber ruber*, at 37°C for several hours in the absence of Ca^{2+}, they released, respectively, a large fragment which apparently derives from the carboxyl-terminal portions of these proteins (Takeya, 1992b). The autoproteolyses did not occur when 5 mM Ca^{2+} was present in the incubation mixtures, hence, Ca^{2+} may stabilize some conformation of the metalloproteinases susceptible to autoproteolysis. The entire amino acid sequences for the two fragments from HRlA and HT-l revealed that each contains disintegrin-like and Cys-rich domains which correspond to those detected in HRlB (Figure 3). However, the putative precursors of trigramin and rhodostomin and high molecular mass metalloproteinases are essentially different in the following aspects. First, the cDNA for both trigramin and rhodostomin have a stop codon located soon after the coding regions for mature disintegrins, whereas HRlA, HRlB, HT-l and RVV-X all have an additional Cys-rich domain following the disintegrin domains. Second, the Arg-Gly-Asp sequence extensively conserved in the venom disintegrin family is substituted by Glu-Ser-Glu-Cys in HRlB, Arg-Ser-Glu-Cys in HRlA and HT-l, and Arg-Asp-Glu-Cys in RVV-X (Figure 3). Therefore, as suggested by our autoproteolysis experiments, it seems very likely that the disintegrin polypeptides free from the metalloproteinases may be autoproteolytically generated not from high molecular mass metalloproteinases but from other sources, such as the disintegrin precursor proteins shown Figure lb.

Recently, the precursor protein of Ht-e, the low molecular mass hemorrhagic metalloproteinase isolated from the venom of *C. atrox,* has also been deduced from its cDNA sequence (Hite *et al.*, 1992). The structure of this precursor protein is very similar to those of disintegrins (Figure lb), all of which contain metalloproteinase- and disintegrin-domains but not Cys-rich domain. Therefore, a low molecular mass metalloproteinase and a disintegrin polypeptide both originate from this type precursor protein, although this precursor protein has never been identified in any snake venoms. It is noteworthy, however, that the disintegrin domain of Ht-e precursor does not contain the Arg-Gly-Asp sequence (Figure 3).

RUSSELL'S VIPER VENOM FACTOR X ACTIVATOR (RVV-X)

RVV-X (M_r 79,000) consists of a disulfide-bonded two-chain glycoprotein with a heavy chain of M_r 59,000 and a light chain of M_r 18,000. We have revealed that RVV-X is a new member of the venom metalloproteinase family (Takeya *et al.*, 1992a). The heavy chain

(427 residues) is similar in entire sequence to HRIB (Figure 1). On the other hand, the light chain of RVV-X shows a striking similarity to the C-type lectin-like structure (Ca^{2+}-dependent carbohydrate recognition domain). Factor IX/factor X-binding protein (IX/X-bp) is the most interesting among the proteins having the C-type lectin structure (Atoda *et al.*, 1991), since IX/X-bp and RVV-X share several key characteristics, in addition to sequence similarity. IX/X-bp has been purified as an anticoagulant protein and it binds specifically to factors IX and X, but not to other vitamin K-dependent coagulation factors. This functional activity of IX/X-bp as well as RVV-X is dependent on Ca^{2+} and the amino-terminal Gla-domains of factors IX and X. IX/X-bp is able to inhibit the RVV-X-catalyzed factor X activation, in a dose dependent manner, thereby indicating that IX/X-bp occupies a RVV-X-recognized site on factor X (Takeya *et al.*, 1992a). Taken together, it is suggested that the light chain of RVV-X might play an important role in the binding of RVV-X with factor X, and that it recognizes the Gla-domain of factor X. There are two potential roles of the exogeneous Ca^{2+}, which is required for the factor X activating activity but not for the proteolytic activity of RVV-X: First, the C-type lectin-like light chain of RVV-X requires Ca^{2+} for its function, and second, Ca^{2+} induces conformational change of the Gla-domain, an event which might be a prerequisite for recognition by RVV-X and/or for concomitant conformational change of factor X susceptible to RVV-X.

REFERENCES

Adler, M., Lazarus, R. A., Dennis, M. S. and Wagner, G., 1991 Solution structure of kistrin, a potent platelet aggregation inhibitor and GP IIb-IIIa antagonist, *Science* 253:445.

Atoda, H., Hyuga, M. and Morita, T., 1991 The primary structure of coagulation factor IX/factor X-binding protein isolated from the venom of *Trimeresurus flavoviridis, J. Biol. Chem.* 266:4903.

Au, L.-C., Huang, Y.-B., Huang, T.-F., Teh, G.-W. Liu, H.-H. and Choo, K.-B., 1991 A common precursor of a putative hemorrhagic protein and rhodostomin, a platelet inhibitor of the venom of *Calloselasma rhodostoma, Biochem. Biophys. Res. Commun.* 181:585.

Baramova, E. N., Shannon, J. D., Bjarnason, J. B. and Fox, J. W., 1989 Degradation of extracellular matrix proteins by hemorrhagic metalloproteinases, *Arch. Biophys. Biochem.* 275:63.

Blobel, C. P., Wolfsberg, T. G., Turck, C. W., Myles, D. G., Primakoff, P. and White, J. M., 1992 A potential fusion peptide and an integrin ligand domain in a protein active in sperm-egg fusion, *Nature* 356:248.

Bode, W., Gomis-Rutll, F. X., Huber, R., Zwilling, R. and Stocker, W., 1992 Structure of astacin and implications for activation of astacins and zinc-ligation of collagenases, *Nature* 358:164.

Furie, B. C. and Furie, B., 1976 Coagulant protein of Russell's viper venom, *Methods Enzymol.* 45:191.

Gould, R. J., Polokoff, M. A., Friedman, P. A., Huang, T.-F., Holt, J. C., Cook, J. J. and Niewiarowski, S., 1990 Disintegrins: a family of integrin inhibitory proteins from viper venoms, *Proc. Soc. Exp. Biol. Med.* 195:168.

Hite, L. A., Shannon, J. D., Bjarnason, J. B. and Fox, J. W., 1992 Sequence of a cDNA clone encoding the zinc metalloproteinase hemorrhagic toxin e from *Crotalus atrox, Biochemistry* 31:6203.

Iwanaga, S., Oshima, G.,and Suzuki, T., 1976 Proteinases from the venom of *Agkistrodon halys blomhoffii, Methods Enzymol.* 45:459.

Iwanaga, S. and Suzuki, T., 1979 Enzymes in snake venom, *in*: Handbook of Experimental Pharmacology Vol. 52, Snake Venoms, Lee, C.-Y., ed, pp. 61-158, Springer Verlag, New York.

Markland, F.S., Jr., 1991 Inventry of, α- and β-fibrinogenases from snake venoms, *Thromb. Haemostas.* 65:438.

Miyata, T., Takeya, H., Ozeki, Y., Arakawa, M., Tokunaga, F., Iwanaga, S. and Omori-Satoh, T., 1989 Primary structure of hemorrhagic protein, HR2a, isolated from the venom of *Trimeresurus flavoviridis, J. Biochem. (Tokyo)* 105:847.

Morita, T. and Iwanaga, S., 1981 Prothrombin activator from *Echis carinatus* venom, *Methods Enzymol.* 80:303.
Neeper, M. P. and Jacobson, M. A., 1990 Sequence of a cDNA clone encoding the platelet aggregation inhibitor trigramin, *Nucleic Acids Res.* 18:4255.
Ohsaka, A., 1979 Hemorrhagic, necrotizing and edema-forming effects of snake venoms, *in*: Handbook of Experimental Pharmacology Vol. 52, Snake Venoms, Lee, C.-Y., ed, pp. 480-546, Springer Verlag, New York.
Retzios, A. D. and Markland, F. S. Jr., 1992 Purification, characterization, and fibrinogen cleavage sites of three fibrinolytic enzymes from the venom of *Crotalus basiliscus basiliscus, Biochemistry* 31:4547.
Sanchez, E. F., Diniz, C. R. and Richardson, M., 1991 The complete amino acid sequence of the haemorrhagic factor LHFII, a metalloproteinase isolated from the venom of the bushmaster snake *(Lachesis muta muta), FEBS Lett.* 282:178.
Scarborough, R. M., Rose, J. W., Hsu, M. A., Phillips, D. R., Fried V. A., Campbell, A. M., Hannizzi, L. and Charo, I. F., 1991 Barbourin: a GPIIb-IIIa-specific integrin antagonist from the venom of *Sistrurus m. barbouri, J. Biol. Chem.* 266:9359.
Shannon, J. D., Baramova, E. N., Bjarnason, J. B. and Fox, J. W., 1989 Amino acid Sequence of a *Crotalus atrox* venom metalloproteinase which cleaves type IV collagen and gelatin, *J. Biol. Chem.* 264:11575.
Takeya, H., Arakawa, M., Miyata, T., Iwanaga, S. and Omori-Satoh, T., 1989 Primary structure of H2-proteinase, a non-hemorrhagic metalloproteinase, isolated from the venom of habu snake, *Trimeresurus flavoviridis, J. Biochem. (Tokyo)* 106:151.
Takeya, H., Oda, K., Miyata, T., Omori-Satoh, T. and Iwanaga, S., 1990a The complete amino acid sequence of the high molecular mass hemorrhagic protein HRIB isolated from the venom of *Trimeresurus flavoviridis, J. Biol. Chem.* 265:16068.
Takeya, H., Onikura, A., Nikai, T., Sugihara, H. and Iwanaga, S., 1990b Primary structure of a hemorrhagic metalloproteinase, HT-2, isolated from the venom of *Crotalus ruber ruber, J. Biochem. (Tokyo)* 108:711.
Takeya, H., Kubo, J., Omori-Satoh, T. and Iwanaga, S., 1991 Snake venom platelet aggregation inhibitor, trigramin-like polypeptide, originates in the high molecular mass hemorrhagic protein, *Thomb. Haemostas.* 65:842.
Takeya, H., Nishida, S., Miyata, T., Kawata, S., Saisaka, Y., Morita, T. and Iwanaga, S., 1992a Coagulation factor X activating enzyme from Russell's viper venom (RVV-X), *J. Biol. Chem.* 267:14109.
Takeya, H., Nishida, S., Nishino, N., Makinose, Y., Omori-Satoh, T., Nikai, T., Sugihara, H. and Iwanaga, S., 1992b Primary structures of platelet aggregation inhibitors (disintegrins) autoproteolytically released from snake venom hemorrhagic metalloproteinases and new fluorogenic peptide substrates for these enzymes, *J. Biochem. (Tokyo)*, in press.
Takeya, H., Nishino, K., Miyata, T., Omori-Satoh, T. and Iwanaga, S., 1992c Snake venom hemorrhagic and non-hemorrhagic metalloendopeptidases, *Methods Enzymol.*, in press.

MASS SPECTROMETRY

RECENT ADVANCES IN PROTEIN SEQUENCING BY MASS SPECTROMETRY INTRODUCTION AND OVERVIEW

Klaus Biemann

Department of Chemistry
Massachusetts Institute of Technology
Cambridge, MA 02139

INTRODUCTION

During the past few years, developments in mass spectrometry have taken place that have had considerable impact on protein sequencing methodology. Fast atom (or ion) bombardment (FAB) ionization (Barber *et al.*, 1981) has made it possible to determine the molecular weights of peptides and small proteins (Mr < 15 kDa) with an accuracy of ± 0.01% (i.e., ± 0.1 to ± 1.5 Da), depending on molecular weight. More recently, the introduction of matrix-assisted laser desorption (MALD) (Karas and Hillenkamp, 1988) and electrospray (ES) (Meng *et al.*, 1988) ionization have extended the range to hundreds of kilodaltons with a similar accuracy (0.01 - 0.1%).

While molecular weight information alone can answer many structural questions directly, sequence information can be obtained by fragmenting the mono- or polyprotonated peptides produced by FAB (Biemann, 1987) or ES (Hunt *et al*, 1989; Barinaga *et al.*, 1989). This is accomplished by their collision-induced dissociation (CID) in a tandem mass spectrometer (MS/MS). All these mass spectrometric methods have the further advantage that they are applicable to mixtures, to peptides and proteins containing modified amino acids, or to those that are N-terminally blocked. A background in this area is provided by two recent reviews (Biemann, 1990; Biemann, 1992).

INFORMATION DERIVED FROM MOLECULAR WEIGHT MEASUREMENTS

Peptides

The accurate (± 0.3 Da or less) determination of the molecular weights of the peptides

Methods in Protein Sequence Analysis, Edited by K. Imahori
and F. Sakiyama, Plenum Press, New York, 1993

produced upon proteolytic or chemical cleavage of a protein can answer many questions. Foremost among them is the verification of a protein sequence that is derived either directly (by the Edman method) or indirectly (from a cDNA sequence). Based on that sequence, one can calculate the molecular weights of all peptides that would be produced using an enzyme of high specificity and compare these values with those obtained experimentally. Any discrepancy reveals an error in the original sequence. The earliest example of the correction of cDNA sequences involved a number of aminoacyl-tRNA synthetases from *E. coli* (Gibson and Biemann, 1983). More recently, the amino acid sequences of bovine serum albumin (BSA) were corrected by the detection of an additional tyrosine in position 156 (Hirayama *et al.*, 1990) which changed its molecular weight from 66,261 Da to 66,424 Da, a discrepancy that had been noticed from its ES spectrum. Another example is the difference of one dalton in the molecular weight of horse heart and sperm whale myoglobins, correct Mr = 16,951.5 and 17,199.9, respectively, due to aspartic acid instead of asparagine in position 122 (Zaia *et al.*, 1992). These corrections, although relatively minor, are important because these two proteins are widely used as calibration standards in MALD and ES mass spectrometry. In both cases, the new amino acid and its position were confirmed by the CID spectra of the appropriate proteolytic peptides.

The same approach can be used to characterize mutations and is thus widely applied to the identification of human hemoglobin mutants. The products of the proteolysis of the normal α–, β–,γ–, and δ-chains with various specific enzymes are now well established by mass spectrometry and a shift in mass of one of the peptides pinpoints the region where a mutation has occurred. In many cases, the mass difference alone identifies the site and nature of the mutation but, if there is more than one possibility, tandem mass spectrometry settles the question (Wada *et al.*, 1989).

Proteins

The accuracy and reliability of mass spectrometric molecular weight determinations of proteins are much higher than the conventional methods based on electrophoretic or chromatographic mobility, or ultracentrifugation. Thus, mass spectrometric molecular weights can be used to answer a host of questions or point out problems that are otherwise undetectable. The above case of BSA is one example. The difference in the expected and the mass spectrometrically determined M_r was +0.3%, clearly outside the error limit of ES-MS, thus indicating an error in the sequence or a modification of an amino acid.

A MALD or ES mass spectrum is also very useful in confirming the fidelity by which a cDNA sequence has been translated by a host cell into the proper gene-product. This is usually assumed to be the case (in part because of the lack of a suitably fast method to check) but either experimental errors or unexpected posttranslational modifications may have led to a different product. In the course of an investigation of the interaction of aspartate with the part of the aspartate chemoreceptor of *Salmonella typhimurium* that sits outside the membrane (Mowbray *et al.*, 1985), an N-terminal segment (amino acids 26-180, with an added methionine in position 25) was cloned and expressed (H.-P. Biemann and D.E. Koshland, Jr., unpublished work). The MALD spectrum of the product (Figure l) showed that none of the desired product Met(25)-(180) was present. The data indicated that it had been posttranslationally N-acetylated, as well as truncated to shorter sequences, mainly to (27)-(180) along with lesser amounts of (30)-(180), (32)-(180), and (33)-(180) (Mr calculated 17,436.6, 17,206.3, 16,922.9, 16,688.7, and 16,601.6, respectively). Since the purpose of the biosynthesis of the fragment was to obtain crystals for X-ray analysis on the three dimensional structure of this segment from the aspartate receptor, the mixture produced was obviously unsuitable for crystallization and any attempt thereof would have been a great waste of time. Considering the fact that it takes only a few picograms of protein and a few minutes to obtain the MALD spectrum, it is certainly worthwhile to incorporate such a test in any cloning and expression procedure. A conventional Edman degradation would probably have led to the conclusion that the product was pure (27)-(180), because the acetylated product would not have been reactive and the PTH-amino acids of the minor components could have been easily overlooked. It was finally possible to generate a product encompassing the segment (36)-(180), again preceded by a methionine. The MALD spectrum of this material exhibited a single peak corresponding to $(M + H)^+$ 16,348, which fits well the expected value of 16,355.4.

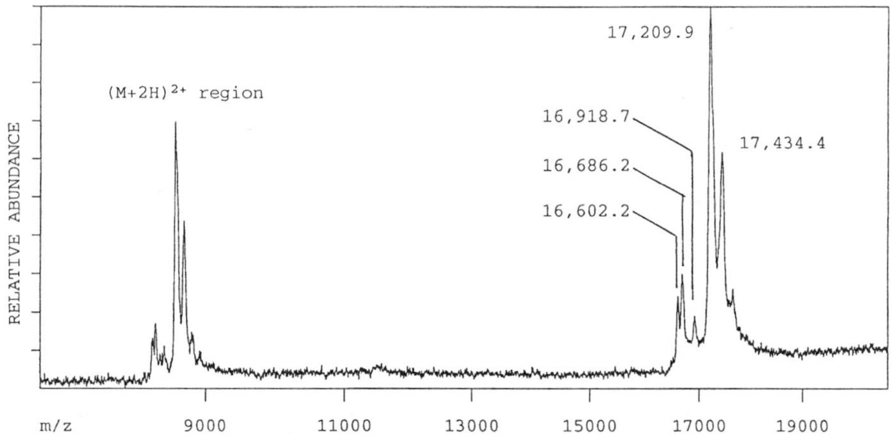

Figure 1. MALD spectrum of the protein produced by expression of a clone incorporating the segment of amino acids (26)-(180) of the aspartate receptor with Met at position 25. The product is clearly a mixture produced by unexpected posttranslational acetylation and N-terminal truncation (see text). The ions represent protonated peptides $(M + H)^+$ and their mass-to-charge value, *m/z*, coresponds to $M_r + 1$.

AMINO ACID SEQUENCING BY TANDEM MASS SPECTROMETRY

The ionization methods discussed above (FAB, MALD and ES) produce chiefly protonated peptides or proteins, $(M + H)^+$, but little or no fragmentation thereof. To derive structural information, fragment ions have to be produced - generally by collision of the $(M + H)^+$ ion of interest with an inert gas - and then separated in the second mass spectrometer of a tandem instrument. Presently, FAB is the most widely used ionization technique to generate the precursor ion, which is then mass selected in the first mass spectrometer (MS-1). As long as the resolving power of MS-1 is such that all components of a mixture can be separated, the sequence of each component peptide can be deduced from the CID spectrum recorded by the second mass spectrometer (MS-2). Furthermore, it is desirable to simplify the CID spectra by selecting as the precursor ion the species that contains only ^{12}C, because it results in a monoisotopic product ion spectrum.

Characteristics of CID Mass Spectra

Upon collision, fragmentation takes place along the peptide backbone (ions of type a_n, b_n, c_n, x_n, y_n, and z_n) and also at the side chain to form d_n and w_n ions, if the collisions take place at keV energies (Biemann, 1988) (see Scheme 1). Figure 2 represents a typical CID spectrum of a dodecapeptide. Complete sequences can be obtained for peptides as long as 20-25 amino acids and, in some cases, partial sequences can be derived from larger ones.

For protein sequencing, therefore, it is first necessary to cleave the material into peptides that are small enough to be suitable for this approach. This is generally done by the use of specific proteolytic enzymes or chemically. The peptide that gave the spectrum shown in Figure 2 was the product of the digestion of cytochrome *c* (horse) with trypsin. While the isobaric amino acids leucine and isoleucine (both of which create the same mass increment of 113 Da for any given backbone ion series) can be differentiated on the basis of d_n or w_n

Scheme 1 Nomenclature of high energy (keV) CID fragment ions.

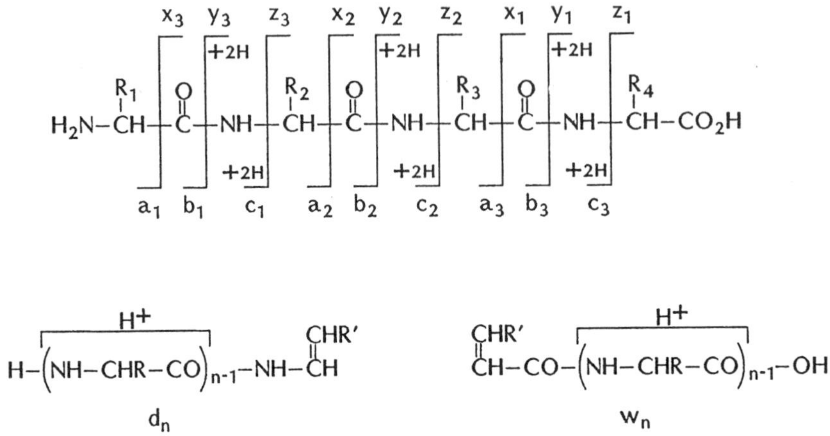

ions (w_5 and w_9 in Figure 2), the isobaric amino acids lysine and glutamine (mass increment 128 Da) require a different approach. One method for their differentiation is based on the specificity of the enzymes trypsin or Endo Lys-C, but a more definite assignment is possible by treatment of the peptide with acetic anhydride, which acetylates lysine but not glutamine. This is usually done at the peptide stage (i.e., after the proteolysis) but there are some advantages in acetylating the intact protein first (Annan and Biemann, 1992). Proteolysis leads then to peptides in which the lysines are already acetylated, giving rise to a mass increment of 170 Da, which is unique. A certain drawback is the fact that it is

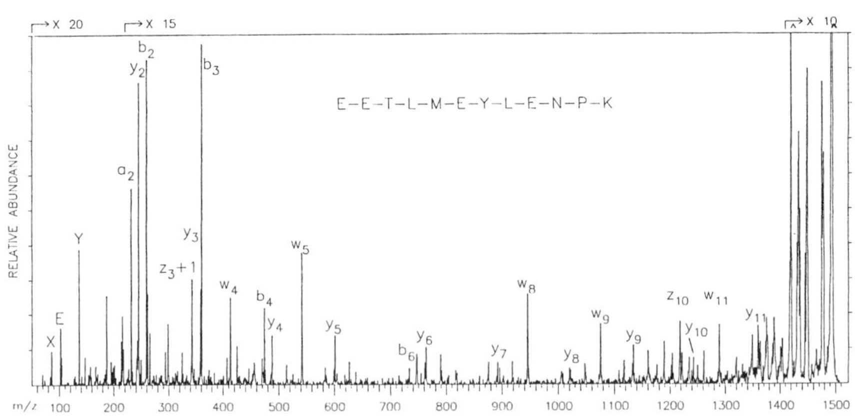

Figure 2. The CID spectrum of the $(M + H)^+$ ion = *m/z* 1,495.7 of a tryptic peptide (61-72) from horse heart cyctochrome *c*. Collision energy 7 keV. For the assignment of fragment ions, see Scheme 1. The capital letters X, E and Y denote the immonium ions $(NH_2 = CHR)^+$ of the amino acids Leu/Ile, Glu and Tyr. The *m/z* values of the peaks labeled w_5 and w_8 indicate that the 5th and 9th amino acids (counting from the C-terminus) are Leu rather than Ile (from Annan and Biemann, 1992; reproduced with permission).

difficult to completely acetylate all lysines in the native protein (unless it is quite small) (Riordan and Vallee, 1972) and thus producing some partially acetylated peptides. If there is only one lysine in the peptide, this has the inadvertent advantage that one will find acetylated and unacetylated peptide (which is easy to recognize by the 42 Da mass shift) and thus facilitates the interpretation. The situation becomes more complex if there are two lysines in the peptide. One may now encounter two isomeric monoacetylated peptides, which cannot be separated by MS-1 and lead to two superimposed CID spectra from MS-2. However, such a mixture spectrum can easily be interpreted in terms of the two isomeric structures, as demonstrated by Figure 3.

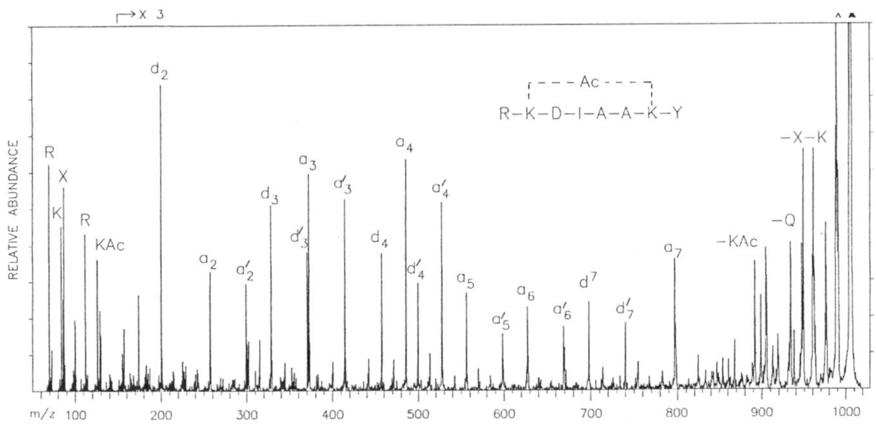

Figure 3. The CID spectrum of the two isomeric $(M + H)^+$ ions = m/z 1,006.6 of the chymotryptic peptides (139)-(146), monoacetylated at either 140 or 145, of sperm whale myoglobin. The peaks labeled a'n and d'n refer to the fragments due to the isomer acetylated at Lys-140 (from Annan and Biemann, 1992; reproduced with permission).

Cleavage at Cysteine

There are very few amino acid specific chemical peptide bond cleavages, the most common one is the conversion of methionine to homoserine by reaction with cyanogen bromide (Gross and Witkop, 1961). Because of the low frequency of occurrence of this amino acid in proteins, large fragments are obtained that are often not suitable for sequencing by tandem mass spectrometry without further enzymatic degradation. However, for proteins that are difficult to digest by proteolytic enzymes, prior cleavage with cyanogen bromide is advantageous.

Another specific chemical cleavage involves the cyanylation of cysteine with 2-nitro-5-thiocyanobenzoic acid (NTBC), followed by cyclization to a 2-iminothiazolidine-5-carboxylic acid derivative and accompanied by cleavage of the peptide bond N-terminal of the cysteine (Stark, 1977):

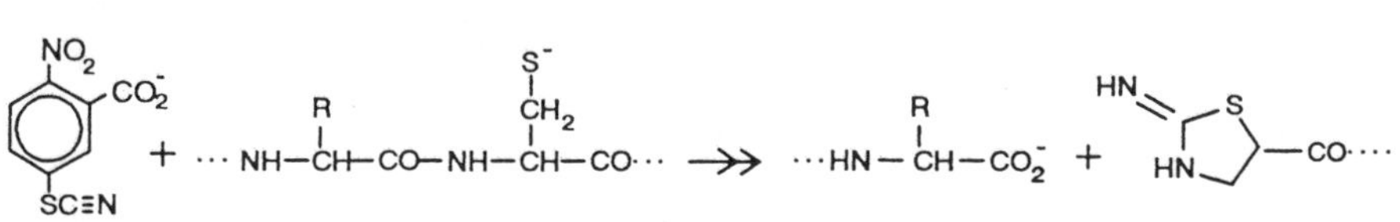

Because of the widely variable frequency of the occurrence of cysteine in proteins, the resulting fragments vary greatly in size. For example, the α-chain of human hemoglobin contains only a simple cysteine at position 104 and cyanylation produces, therefore, only two large fragments (1)-(103) and (104)-(141) (Zaia *et al*, 1991). In such cases, the utility of this reaction is like that of the cyanogen bromide cleavage at methionine.

For cysteine-rich proteins, cleavage at that amino acid resembles the enzymatic cleavage with a very specific enzyme. Unfortunately, the resulting peptides have a blocked N-terminus and are, therefore, not amenable to the Edman degradation, which is the reason why the "Stark reaction" has not been applied to protein sequencing. However, this feature is no hindrance to peptide sequencing by tandem mass spectrometry.

Sequencing of a Cysteine-Rich Protein by Tandem Mass Spectrometry

We have applied the reaction described above to the determination of the primary structure of a small protein, a protease inhibitor isolated from *Sarcophaga bullata* (Sugumaran *et al.*, 1985). The molecular weight was found to be 6,512.4, $(M + H)^+ = m/z$ 6,513.5 by FAB and 6,513.3 by MALD, which upon reduction and alkylation with iodoacetamide increased to 6,860.3, demonstrating the presence of six cysteines. Cleavage with NTBC resulted in 10 products, indicating the presence of some overlapping peptides with unreacted cysteine. The CID spectra of some of these revealed part of the sequence of the protein (Figure 4) and the

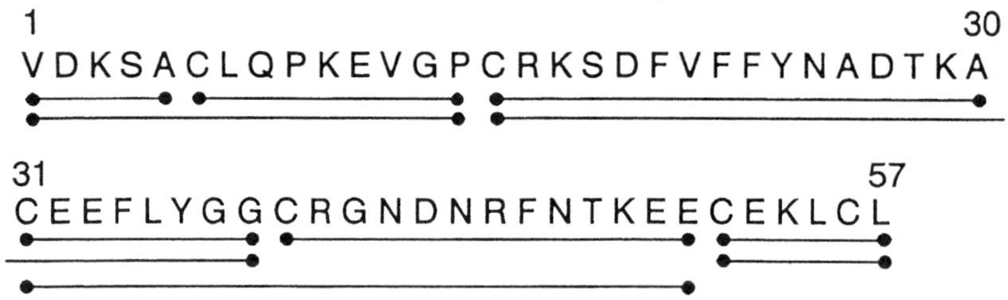

Figure 4. Amino acid sequence of the protease inhibitor from *S. bullata* determined by mass spectrometry. Peptides obtained by cleavage at cysteine are underlined.

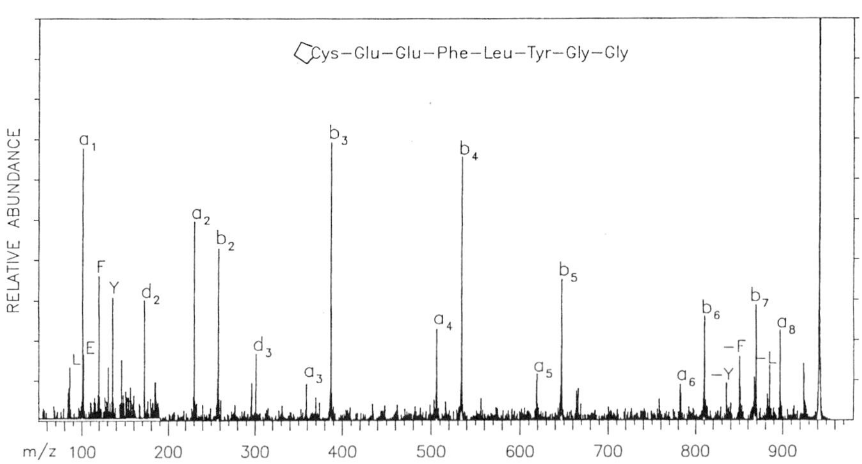

Figure 5. The CID mass spectrum of the $(M + H)^+$ ion = m/z 942.5, ultimately determined to represent the sequence (31)-(38). Ions C*, L, E, F, and Y are the immonium ions of cyclized cysteine (2-iminothiazolidine-4-carboxylic acid), leucine, glutamic acid, phenylalanine, and tyrosine, respectively. Ions -Y, -F, and -L are due to loss of the side chain of the respective amino acid from the protonated peptide molecule. The N-terminal cysteine, which has been converted to 2-iminothiazolidine-4-carboxylic acid, is indicated by the three corner symbol (from Papayannopoulos and Biemann, 1992; reproduced with permission).

remainder was deduced from digests with trypsin and chymotrypsin (Papayannopoulos and Biemann, 1992). The fragmentation of the peptide derivatives produced by cleavage at cysteine exhibits a complete a_n, and b_n, series, including b_1, and thus resembles those with an N-terminal acyl group (Figure 5).

CID Spectra of $(M + nH)^{n+}$ Ions Produced by ES Ionization

Singly and multiply protonated peptide ions produced by ES ionization can also be successfully dissociated by collision with a neutral gas. To date, most of these experiments have been carried out in triple-quadrupole mass spectrometers (TSQ) using either the infusion method (bleeding a solution of the peptide into the ES ion source) (Barinaga *et al.*, 1989; Hunt *et al.*, 1989) or by the direct coupling of the effluent from a liquid chromatograph to the ES ion source of the TSQ. A particularly elegant example of the latter approach is the identification of some of the short peptides that bind specifically to class I (Hunt *et al.*, 1992a) and class II (Hunt *et al.*, 1992b) major histocompatibility complexes. This immunologically significant finding is discussed elsewhere in this volume.

MATRIX-ASSISTED LASER DESORPTION ON A MAGNETIC MASS SPECTROMETER

While the ions formed by MALD ionization, which is a short-pulse process, are usually mass analyzed with a time-of-flight mass spectrometer, we have adapted the methodology to our JEOL HX110/HX110 tandem mass spectrometer (Annan *et al.*, 1992a) that is equipped with an integrating array detector (Hill *et al.*, 1991). Although the mass range of that instrument is limited to < *m/z* 14,000, the high sensitivity of MALD and its lower discrimination among certain peptides in complex mixtures have advantages over FAB ionization. On the other hand, the resolution of a double-focusing mass spectrometer is a distinct advantage over a linear TOF instrument.

The beam of an excimer laser generating 12 ns pulses at 353 nm was deflected and focused onto the sample probe of the conventional FAB-ion source that is located between MS-1 and MS-2 of the tandem instrument and fixed at a rate of 5-20 Hz. The signal was integrated for one second periods. A resolution of 5800 (FWHM) was obtained for the peptide substance P, $(M + H)^+ = m/z$ 1,347.7 (monoisotopic). At higher mass, the resolution is limited mainly by the width of the isotopic cluster of the protein molecules but is still superior to that of TOF spectrometers. Good signals have been obtained with cytochrome *c* (Mr 12,359), horse myoglobin (Mr 16,952) and, at reduced accelerating voltage, with trypsinogen (Mr 23,980) at the picomole (2-8) level. Because of the high tolerance of MALD for salts and buffers, the use of this ionization method for the direct analysis of proteolytic digests of proteins with a magnetic mass spectrometer may have an advantage over FAB (Annan *et al.*, 1992a). It has also been used in the ion source of MS-1 of our tandem mass spectrometer to produce CID spectra of peptides (Annan *et al.*, 1992b).

Acknowledgements. The author is indebted to his many former and present associates, especially R. S. Annan, J. E. Biller, C. E. Costello, K. M. Downard, J. A. Hill, P. Juhasz, H. J. M. Köchling, I. A. Papayannopoulos, and J. Zaia for their contributions to this work. Also to H.-P. Biemann and M. Sugumaran for providing the samples of the aspartate receptor fragments and the protease inhibitor, respectively. The work in the author's laboratory was supported by grants from the National Institutes of Health (RR00317, GM05472 and ES01640).

REFERENCES

Annan, R. S. and Biemann, K., 1992 The utilily of N-peracetylation of proteins for their structure determination by mass spectrometry, *J. Am. Soc. Mass Spectrom.*, in press.

Annan, R. S., Köchling, H. J., Hill, J. A. and Biemann, K., 1992a Matrix-assisted laser desorption using a fast atom bombardment ion source and a magnetic mass spectrometer, *Rapid Commun. Mass Spectrom.* 6:298.

Annan, R. S., Köchling, H. J., Biller, J. E., Hill, J. A. and Biemann, K., 1992b Matrix-assisted laser desorption ionization in a magnetic deflection mass spectrometer, *in*: Proceedings of the 40th ASMS Conference on Mass Spectrometry and Allied Topics, Washington, DC.

Barber, M., Bordoli, R. S., Sedgwick, R. D. and Tyler, A. N., 1981 Fast atom bombardment of solids (F.A.B.): a new source for mass spectrometry, *J. Chem. Soc. Chem. Commun.* 325.

Barinaga, C. J., Edmonds, C. G., Udseth, H. R. and Smith, R. D., 1989 Sequence determination of multiply charged peptide molecular ions by electrospray-ionization tandem mass spectrometry, *Rapid Commun. Mass Spectrom.* 3:160.

Biemann, K., 1987 The growing role of mass spectrometry, *in*: Methods in Protein Sequence Analysis 1986, Walsh, K.A. ed., Humana Press, Clifton, NJ.

Biemann, K., 1988 Contributions of mass spectrometry to peptide and protein structure, *Biomed Environ. Mass Spectrom.* 16:99.

Biemann, K., 1990 Sequencing of peptides by tandem mass spectrometry upon high energy collision-induced dissociation, *Methods Enzymol.* 193:455.

Biemann, K., 1992 Mass spectrometry of peptides and proteins, *Annu. Rev. Biochem.* 61:977.

Gibson, B. W. and Biemann, K. 1984 Strategy for the mass spectrometric verification and correction of the primary structures of proteins deduced from their DNA sequences, *Proc. Natl. Acad. Sci. USA* 81:1956.

Gross, E. and Witkop, B., 1961 Selective cleavage of the methionyl peptide bonds in ribonuclease with cyanogen bromide, *J. Am. Chem. Soc.* 83:1510.

Hill, J. A., Biller, J. E. and Biemann, K., 1991 A variable dispersion array detector for a tandem mass spectrometer, *Int. J. Mass Spectrom. Ion Processes* 111:1.

Hirayama, K., Akashi, S., Furuya, M. and Fukuhara, K., 1990 Rapid confirmation and revision of the primary structure of bovine serum albumin by ESIMS and FRIT-FAB LC/MS, *Biochem. Biophys. Res. Commun.* 173:639.

Hunt, D. F., Zhu, N. Z. and Shabanowiz, J., 1989 Oligopeptide sequence analysis by collision-activated dissociation of multiply charged ions, *Rapid Commun. Mass Spectrom.* 3:122.

Hunt, D. F., Henerson, R. A., Shabanowitz, J., Sakaguchi, K., Michel, H., Sevilir, N., Cox, A. L., Appella, E. and Engelhard, V. H., 1992a Characterization of peptides bound to the class I MHC molecule HLA-A2.1 by mass spectrometry, *Science* 255:1261.

Hunt, D. F., Michel, H., Dickinson, T. A., Shabanowitz, J., Cox, A. L., Sakaguchi, K., Appella, E., Grey, H. M. and Sette, A., 1992b Peptides presented to the immune system by the murine class II major histocompatibility complex molecule I-A^d, *Science* 256:1817.

Karas, M. and Hillenkamp, F., 1988 Laser desorption ionization of proteins with molecular masses exceeding 10,000 daltons, *Anal. Chem.* 60:2299.

Meng, C. K., Mann, M. and Fenn, J. B., 1988 Of protons or proteins, *Z. Phys. D.* 10:361.

Mowbray, S. L., Foster, D. L. and Koshland, Jr., D. E., 1985 Proteolytic fragments identified with domains of the aspartate chemoreceptor, *J. Biol. Chem.* 260:11711.

Papayannopoulos, I. A. and Biemann, K., 1992 Amino acid sequence of a protease inhibitor isolated from *Sarcophaga bullata* determined by mass spectrometry, *Protein Sci.* 1:278.

Riordan, J. F. and Vallee, B. L., 1972 Acetylation, *Methods Enzymol.* 25:494.

Stark, G. R., 1977 Cleavage at cysteine after cyanylation, *Methods Enzymol.* 47:129.

Sugumaran, M., Saul, S. J. and Ramesh, N., 1985 Endogenous protease inhibitors prevent undesired activation of prophenolase in insect hemolymph, *Biochem. Biophys. Res. Commun.* 132:1124.

Wada, Y., Matsuo, T. and Sakurai, T., 1989 Structure elucidation of hemoglobin variants and other proteins by digit-printing method, *Mass Spectrom. Rev.* 8:379.

Zaia, J., Papayannopoulos, I. A. and Biemann, K., 1991 Utility of a cysteine-specific cleavage reaction in the structure determination of proteins by laser desorption time-of-flight and tandem mass spectromelry, *in*: Proceedings of The 39th ASMS Conference on Mass Spectrometry and Allied Topics, Nashville, TN.

Zaia, J., Annan, R. S. and Biemann, K., 1992 The correct molecular weight of myoglobin, a common calibrant for mass spectrometry, *Rapid Commun. Mass Spectrom.* 6:32.

SEQUENCE ANALYSIS OF PEPTIDES PRESENTED TO THE IMMUNE SYSTEM BY CLASS I AND CLASS II MHC MOLECULES

Donald F. Hunt[1], Jeffrey Shabanowitz[1], Hanspeter Michel[1], Andrea L. Cox[1], Tracey Dickinson[1], Theresa Davis[1], Wanda Bodnar[1], Robert A. Henderson[2], Noelle Sevilir[2], Victor H. Engelhard[2], Kazuyasu Sakaguchi[3], Ettore Appella[3], Howard M. Grey[4] and Alessandro Sette[4]

[1]*Department of Chemistry, University of Virginia Charlottesville, VA 22901*
[2]*Department of Microbiology and Beirne Center for Immunological Research, University of Virginia School of Medicine, Charlottesville, VA 22901*
[3]*Laboratory of Cell Biology, National Cancer Institute, NIH Bethesda, MD 20892 and*
[4]*Cytel Corporation, 3525 John Hopkins Court, San Diego CA, 92121 USA*

INTRODUCTION

Cytotoxic T lymphocytes (CTL) are an arm of the immune system concerned with recognition of host cells that express new antigens as a result of viral infection. CTL do not recognize new antigens directly, but only as short peptides bound to cell surface class I molecules of the major histocompatability complex (MHC) (Germain, 1986; Monaco, 1992; Morrison *et al.*, 1983). It is thought that a small fraction of newly synthesized self and viral proteins are degraded into small peptides in the cytoplasm, and these are then transported into a subcellular compartment, the endoplasmic reticulum, where they become bound to a groove on the top of class I molecules, and are subsequently transported to the cell surface for presentation to cytotoxic lymphocytes. Class I molecules are glycoproteins consisting of light (12 kDa) and heavy (47 kDa) chains. Each cell presents up to 10^5 copies of six different class I molecules. Since the cell is also synthesizing several thousand different proteins, each type of class I molecule is expected to present several thousand peptide fragments to the immune system at any particular moment. CTL bind to the class I molecules and check the nature of the peptides being presented. If they are of foreign origin, the CTL become activated, multiply, and lyse the infected cell. If only self peptides are presented, the CTL remain inactive and continue their surveilance of other cells in the neighborhood.

In the second stage of the immune response to a viral infection, antibodies secreted by B-cells bind to and inactivate viral proteins dumped into the extracellular medium as a result of cell lysis. Labelled proteins are injested or endocytosed by macrophage cells that then degrade the foreign proteins to peptides in endosomal compartments, and present them back on the cell surface in conjunction with class II molecules (Germain, 1986; Monaco, 1992; Morrison *et al.*, 1983). The latter glycoproteins consist of two 30 kDa chains. T-helper cells recognize the peptide fragments of foreign proteins presented on the surface of class II molecules, become activated, multiply, and secrete cytokines that stimulate multiplication of B-cells, and the synthesis of more antibody. Memory B-cells primed to attack the virus if it is encountered again in the future are also created as a result of the above process.

Although the above description is somewhat oversimplified, it is clear that presentation of peptides derived from disease state associated proteins plays a critical role in activating an immune response. Sequence analysis of these peptides is a necessary first step in developing vaccines or other agents that modulate the immune response to particular diseases. The task is a challenging one because the peptide fragments characteristic of the disease state will be present in a complex mixture dominated by peptide fragments derived from normal cellular proteins. If several thousand peptides are presented, the concentration of any individual peptide is likely to be in the range 10 - 1,000 fmol/ 10^8 cells. Extensive fractionation of these mixtures by high performance liquid chromatography (HPLC) has made it possible to sequence a small number of self peptides by Edman degradation (Chicz *et al.*, 1992; Jardetzky *et al.*, 1991; Rudensky *et al.*, 1991). Described here is the use of microcapillary HPLC in combination with electrospray ionization/tandem mass spectrometry to sequence peptides presented by both class I and class II molecules (Henderson *et al.*, 1992; Hunt *et al.*, 1992a; 1992b).

Class I, HLA-A2.1 molecules were purified by immunoprecipation from $3x10^9$ cells of the human B lymphoblastoid cell line, ClR-A2.1 (Hunt *et al.*, 1992a). Associated peptides were released by acid extraction and separated from HLA-A2.1 and antibody by filtration. Aliquots of the resulting material corresponding to 10^8 cells were then analyzed by microcapillary HPLC in conjunction with electrospray ionization/tandem mass spectrometry on Finnigan TSQ-70 and TSQ-700 instruments (Hunt *et al.*, 1991). Microcapillary HPLC columns were constructed from 70 cm of 50 micron (i.d.), 200 micron (o.d.) fused silica. The last 10 cm of the column were packed with C-18 absorbent (10 micron silica particles). Peptides were eluted with a 15 min linear gradient of 0-80% acetic acid (0.1 M)/acetonitrile. Flow rate through the column was 250-500 nl/min. Sheath flow for the electrospray ionization was 2-3 μl/min of 75:25 methanol : 0.1 M acetic acid.

Shown in Figure la is the total ion chromatogram generated from the analysis of peptides released from the MHC molecule, HLA-A2.1. Results from the identical experiment performed on an equal number of cells that do not express the HLA-A2.1 molecule are shown in Figure lb. No peptides were obtained in this latter experiment. Mass spectra were acquired every 2 s as the peptides eluted from the HPLC column. Most of the peptide signals were observed during the three minute time period corresponding to scans 100-190. More than 200 distinct ions having a signal-to-noise ratio (S/N) of at least 2 were observed in the individual mass spectra. Known quantities of synthetic peptides added to cell extracts facilitated quantitation of peptides and defined the lower limit of detection (S/N = 2) as 30 fmol in the above experiment. Of the 200 peptides detected, 10% were present at 150-600 fmol/10^8 cells; 90% were found at the 30-150 fmol level. Recent experiments with larger amounts of material isolated from the class I molecule, B-

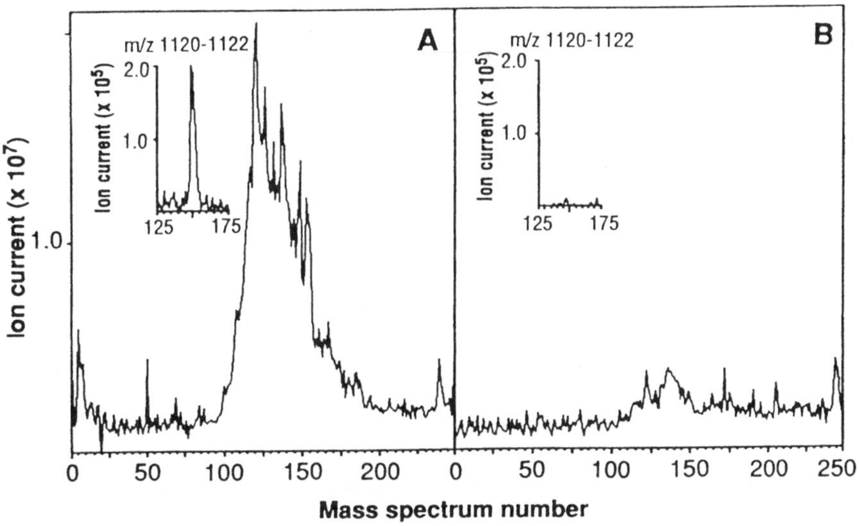

Figure 1. Total ion chromatograms of HLA-A2.1 associated peptides obtained from an equal number of ClR cells that are HLA-A2.1 positive (A) and negative (B), respectively.

7, indicate that as many as a thousand additional peptides are probably present at or just below the above detection limit, 10-30 fmol.

Sequence information on individual peptides in the above mixture was obtained by repeating the microcapillary HPLC experiment and subjecting $(M+H)^+$ or $(M+2H)^{+2}$ ions to the process of collision activated dissociation (CAD) in the triple quadrupole mass spectrometer (Hunt *et al.*, 1986). Shown in Figure 2a is the CAD mass spectrum recorded on the $(M+H)^+$ ions at m/z 1121 observed in scans 145-155 (Figure 1, insert). To obtain this data, the first mass filter of the triple quadrupole instrument, quadrupole 1, is set to pass all ions within a 6 mass unit window centered around the $(M+H)^+$ ion at m/z 1121. On transmission to the collision chamber, a bent-quadrupole mass filter (quadrupole 2) operated in the rf only mode, the $(M+H)^+$ ions suffer 10-100 low energy (15-40 eV) collisions with an Ar present at a total pressure of 3 mtorr. During this process, translational energy is converted to vibrational energy and the peptide $(M+H)^+$ ions suffer fragmentation, more or less randomly, at the various amide bonds along the backbone of the linear peptide chain. Elimination of small molecules such as water and ammonia also occurs from the side chains of some amino acids. A collection of neutrals and ions is produced by the above processes. Neutrals are pumped away by the vacuum system and the charged fragments are transmitted to a third mass filter, quadrupole 3, that separates them according to mass. To maximize ion transmission (sensitivity), quadrupole 3 is also usually operated at less than unit resolution. Sample quantity employed to record the spectrum in Figure 2a was 100-300 fmol. The data shown in Figure 2b was recorded with quadrupole 3 operating at near unit resolution and a sample size of 2-4 pmol. With a little experience, both spectra can be interpreted with equal facility. Accordingly, most spectra in our laboratory are recorded as in Figure 2a in order to minimize sample consumption.

To generate the spectrum in Figure 2a, the mass range from 50-1121 Da was scanned 6 times, data was acquired in the centroid mode, and the resulting spectra were then summed. Total instrument time to acquire the data was 12 s. Signals for each of the fragment ions in the spectrum appear in clusters both because the data system is attempting to resolve signals that are more than 1 mass unit wide and because the clusters contain contributions from ^{13}C and other isotopes in each fragment.

					98	261	390	507	
102	215	401	500	615	712	875	1004	1121	b_n
Thr	Lxx	Trp	Val	Asp	Pro	Tyr	Glu	Val	
1121	1020	907	721	622	507	410	247	118	y_n

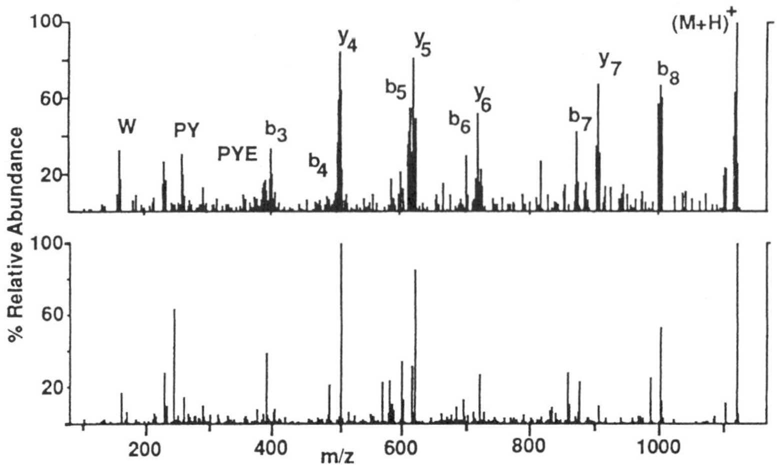

Figure 2. CAD mass spectra of the $(M+H)^+$ ion at m/z 1121 observed in scans 145-155 in Figure la. The top spectrum was recorded on a sample quantity of 100-300 fmol with quadrupole 3 operating with a four mass unit window. The bottom spectrum required 2 pmol of sample and is recorded at approximately unit resolution.

The sequence of amino acids in the sample is deduced by substracting m/z values for peaks that occupy the same relative positions within two cluster patterns and are separated by at least 57 and no more than 186 mass units, the residue masses of the lightest and heaviest amino acids. Presence of additional signal multiplets corresponding to the loss of ammonia (17 Da), water (18 Da), and carbon monoxide (28 Da) from the principal fragment ions often facilitates recognition of the next highest member of a particular fragment ion series.

Predicted masses, m/z values, for ions of type y for the sequence deduced from the spectrum are shown below the structure in Figure 2a. These ion types all contain the C-terminal residue plus one, two, three, four, etc. additional residues. Those observed in the spectrum are underlined. Subtraction of m/z values for any two fragments that differ by a single amino acid, NHCH(R)CO, generates a value that specifies the mass and thus the identity of the extra residue in the larger fragment. In the present example, ions of type y facilitate identification of residues 3, 4 and 5 in the peptide chain.

Predicted fragment ions of type b in Figure 2a are shown above the structure. Those observed in the spectrum are underlined. These ions all contain the amino terminal residue plus one or more additional residues. Subtraction of m/z values for any two fragments of type b that differ by a single amino acid, NHCH(R)CO, generates a value that identifies the extra residue present in the larger fragment. In the present example, ions of type b confirm the identity of residues 4 to 9 in the deduced sequence. Fragments resulting from multistep cleavage of the peptide chain at Pro are labeled with the single letter code for the amino acid residues contained within the fragment (Hunt *et al.*, 1986). CAD spectra recorded on an additional aliquot of the sample mixture that had been subjected to a single cycle of Edman degradation confirmed the order of the first two amino acids as Thr and

Leu/Ile, respectively. Since the amino acids, Leu and Ile, have identical masses, they cannot be differentiated on the triple quadrupole instrument and are specified in the structure as Lxx. Assignment of residue 2 as Leu in this peptide was established by adding synthetic peptides containing Leu or Ile at this position to the cell extract and then demonstrating that only the Leu containing peptide coeluted from a microcapillary HPLC column with the naturally processed species.

Examples of other sequences deduced from HLA-A2.1 associated peptides are shown in Table 1. All but peptides 3 and 4 are nonamers that contain either Leu or Ile at position 2. Residue 9 is one of four small hydrophobic aliphatic residues, Leu, Ile, Val or Ala. The remaining residues are highly variable among the various peptides.

The crystal structure of the HLA-A2 molecule shows five pockets, designated A through F, in the peptide binding groove (Saper *et al.*, 1991). Residues lining the A pocket form hydrogen bonds to the amine terminus. The side chain of residue 1 points toward the solvent exposed surface. Pocket B is lined with hydrophobic residues and accomodates the hydrophobic side chain observed on residue 2. Pocket D is also hydrophobic but large enough to accomodate many different side chains on residue 3. Side chains of residues 4 and 8 point outward from the groove and are fully accessible to the cytotoxic T-cell. Side chains of residues 5 and 6 point back and forth across the binding groove and are partially accessible to the T-cell as well. The MHC molecule binds the side chain of residue 7. Pocket F of HLA-A2.1 contains residues that form hydrogen bonds to the carboxy termimls plus hydrophobic residues that readily acommodate the small hydrophobic side chains observed on residue 9.

In the crystal structure of the HLA-B27 molecule (Madden *et al.*, 1992), acidic residues are located at the bottom of pockets B and F. Accordingly, these two pockets bind

Table 1. Partial listing of data for peptides extracted from HLA A2.1 molecules.

Peptide	(M+H)+ (m/z)	Signal x10⁵	Yield (fmol)	Sequence[1]	Database Match
1	786	9.1	137	SXPSGGSGV	-
2	898	11.9	179	XXDVPTAAV	LLDVPTAAV[4]
3	1011	6.0	90	XXXDVPTAAV	LLLDVPTAAV[4]
4	1211[2]	21.9	320	XXXDVPTAAVQA	LLLDVPTAAVQA[4]
5	930	4.9	74	GXVPFXVSV	-
6	954	11.9[3]	179	SXXPAXVEX	SLLPAIVEL[5]
7	999	3.5	53	SXXVRAXEV	-
8	1037[2]	10.0	150	KXNEPVXXX	-
9	1038[2]	39.4[3]	591	YXXPAXVHX	YLLPAIVHI[6]
10	1121	6.4[3]	320	TLWVDPYEV	TLWVDPYEV[7]

[1]X refers to Leu or Ile, two amino acids of identical mass that can not be distinguished on the triple quadrupole mass spectrometer.
[2]Observed as a doubly charged, $(M+2H)^{2+}$ ion at half this m/z value.
[3]Signal observed for 1 pmol of synthetic peptide standards corresponding to peptides 6, 9, and 10 added into the sample afforded signals of 60 x 10s, 70 x 10^5 and 20 x 10^5, respectively; background noise gave a signal of 1 x 10^5
[4]Gamma interferon induced, human protein, IP-30.
[5]β-Regulatory subunit of human protein phosphatase 2A.
[6]Human nuclear protein p68.
[7]Growth factor induced protein, TIS21.

positively charged side chains strongly and the peptides presented by the HLA-B27 molecule generally have Arg at position 2 and a basic residue, Lys or Arg at position (Rudensky *et al.*, 1991). Thus, positions 2 and 9 appear to be anchor residues that strongly influence which peptides are presented by a particular class I molecule. If four of the five residues in positions 4 through 8 can be any of twenty common amino acids, the number of different nine residue peptides presented by a single class I molecule could be as high as 160,000.

Note that two of the entries in Table 1, peptides 3 and 4, contain 10 and 12 residues, respectively. Mutant cells, defective in antigen processing, present a simplified mixture of peptides dominated by seven major species, including peptides 3 and 4, that each represent between 1% and 21% of the total (Henderson *et al.*, 1992). All seven of these peptides are derived from signal sequences that mediate translocation of the parent proteins into the endoplasmic reticulum. This finding suggests that proteolysis of signal peptide domains in the endoplasmic reticulum constitutes a second mechanism for processing and presentation of peptides for association with class I molecules (Henderson *et al.*, 1992).

Table 2. Data for nine of peptides associated with the class II molecule, I-A^d.

Peptide No.	(M+H)+ m/z	Yield (pmol)[4]	Peptide Sequence[5]	Residue Length	Protein Source[6]
11	1887[1]	9	**WANLMEKIQASVATNPI**	17	Apo-E
12	1774[1]	1	**WANLMEKIQASVATNP**	16	Apo-E
13	1900[2]	6	**DAYHSRAIQVVRARKQ**	16	Cys-C
14	1677[1]	5	**ASFEAQGALANIAVDKA**	17	H2-Edα
15	1606[1]	4	**ASFEAQGALANIAVDK**	16	H2-Edα
16	2089[3]	5	**EEQTQQIRLQAEIIQAR**	17	Apo-E
17	1960[3]	2	**EQTQQIRLQAEIIQAR**	16	Apo-E
18	1989[3]	5	**KPVSQMRMATPLLMRPM**	17	Ii
19	1926[1]	3	**VPQLNQMVRTAAEVAGQX**	18	TF Recp.

[1]Observed a doubly charged, $(M+2H)^{2+}$ ion at half this m/z value.
[2]Observed as a quadruply charged ion, $(M+4H)^{4+}$, at one fourth this m/z value.
[3]Observed as a triply charged, $(M+3H)^{3+}$ ion at one third this m/z value.
[4]Picomole yields in 1/30 of each HPLC fraction were determined on the basis of 3 x 10^6 counts per picomole observed for several synthetic peptides corresponding to entries in the table.
[5]Deduced amino acid sequences are shown in bold, critical binding residues to the class II molecules are underlined.
[6]Protein sources were located by computer searches of gene and protein sequence data banks. ApoE, mouse apolipoprotein-E; CysC, rat cystatin-C; TF recp., rat transferrin receptor protein; I-E^d α–chain of mouse class II MHC protein, I-E^d; Ii, mouse class II invariant chain.

Peptides presented to the immune system in combination with the class II molecule, I-A^d, have also been analyzed by the combination of reverse phase HPLC and microcapilary HPLC\electrospray ionization\tandem mass spectrometry (Hunt *et al.*, 1992b). Class II I-A^d molecules were purified from cells of the murine B cell lymphoma A20-1.11 by affinity chromatography. Peptides were released from the class II molecules on treatment with acid, separated from protein by filtration through an Amicon filter, and fractionated by reverse phase HPLC. Aliquots from fractions 1 to 37 were subjected to further fractionation by microcapillary HPLC and eluted directly into the electrospray ionization source of a triple quadrupole mass spectrometer.

The complexity of the mixture of peptides presented by the I-A^d, class II molecule is considerably greater than that observed associated with the class I molecule, HLA-A2.1. Signals for 664 peptides were detected in the HPLC chromatogram. As many as 2,000 other peptides could be present at levels at or near the detection limit of the instrument, 30 fmol, (S/N = 2). Sequences for nine of the peptides bound to I-A^d have been determined and are shown in Table 2. All are 16-18 residues long, exhibit ragged N- and C-termini, and contain a six residue binding motif that is variably placed in the peptide sequence. Protein precursors for these peptides are all located either on the cell surface or in the extracellular medium and, therefore, expected to participate in the class II processing pathway.

Truncated class I peptides lose their binding ability whereas one or more residues can be removed from class II molecules without destroying activity. Binding constants for class II molecules are in the 5-500 nM range and the length of the naturally processed peptides is not minimal. The above findings are consistent with a processing mechanism that involves binding of large peptide fragments that are then proteolytically trimmed to peptides, the final length of which is determined by the nature of the proteolytic system rather than by optimal binding parameters of the class II molecule. The data suggest that the binding groove on the I-A^d molecule is open at both ends.

Future research will focus on the development of methods that facilitate sequence analysis of a few disease associated antigens presented to the immune system along with the complex mixture of self peptides. Vaccines or other therapeutic agents against viral infection, cellular transformation, and autoimmune diseases should result from this effort.

REFERENCES

Chicz, R. M., Urban, R. G., Lane, W. S., Gorga, J. C., Stern, L. J., Vignali, D. A. A. and Strominger, J. L., 1992 Predominant naturally processed peptides bound to HLA-DRl are derived from MHC-related molecules and are heterogeneous in size, *Nature* 358:764.

Germain, R. N., 1986 The ins and outs of antigen processing and presentation, *Nature* 322:687.

Henderson, R. A. Michel, H., Sakaguchi, K., Shabanowitz, J., Appella, E., Hunt, D. F. and Engelhard, V. H., 1992 HLA-A2.1 associated peptides from a mutant cell line: A second pathway of antigen presentation, *Science* 255:1264 .

Hunt, D. F., Alexander, J. E., McCormack, A. L., Martino, P. A., Michel, H., Shabanowitz, J., Sherman, N., Moseley, M. A., Jorgenson, J. W. and Tomer, K. B., 1991 Mass spectrometric methods for protein and peptide sequence analysis, *in*: Techniques in Protein Chemistry II, Villafranca, J. J. ed., Academic Press, N.Y.

Hunt, D. F., Henderson, R. A., Shabanowitz, J., Sakaguchi, K., Michel, K., Sevilir, N., Cox, A., Appella, E. and Engelhard, V. H., 1992a Characterization of peptides bound to the class I MHC molecule, HLA-A2.1 by mass spectrometry, *Science* 255:1261.

Hunt, D. F., Michel, H., Shabanowitz, J., Cox, A. L., Sakaguchi, K., Appella, E., Grey H. and Sette, A., 1992b Peptides presented to the immune system by the murine class II MHC molecule, IAd, *Science* 256:1817.

Hunt, D. F., Yates, III, J. R., Shabanowitz, J., Winston, S. and Hauer, C. R., 1986 Protein sequencing by tandem mass spectrometry, *Proc. Natl. Acad. Sci. USA* 83:6233.

Jardetzky, T. S., Lane, W. S., Robinson, R. A., Madden D. R. and Wiley, D. C., 1991 Identification of self peptides bound to purified HLA-B27, *Nature* 353:326.

Madden, D. R., Gorga, J. C., Strominger J. L.and Wiley, D. L., 1991 The structure of HLA-B27 reveals nonamer self-peptides bound in an extended conformation, *Nature* 353:321.

Monaco, J. J., 1992 A molecular model of MHC Class I restricted antigen processing, *Immunol. Today* 13:173.

Morrison, L. A., Lukacher, A. E., Braciale, V. L., Fan, D. P. and Braciale, T. J., 1983 Differences in antigen presentation to MHC class I and class II restricted influenza virus specific cytolytic T lymphocyte clones, *J. Exp. Med.* 163:903.

Rudensky, A. Y., Preston-Hurlburt, P., Hong, S.-C., Barlow, A. and Janeway, C. A. Jr., 1991 Sequence analysis of peptides bound to MHC class II molecules, *Nature* 353:622.

Saper, M. A., Bjorkman P. A. and Wiley, D. C., 1991 Refined structure of the human histocompatibility antigen HLA-A2 at 2.6 angstrom resolution, *J. Mol. Biol.* 219:277.

HIGH PRECISION MASS SPECTROMETRY WITH THE ION TRAP MASS SPECTROMETER

R. Graham Cooks, Kathleen A. Cox and Jon D. Williams

Department of Chemistry, Purdue University
West Lafayette, IN 47907, U.S.A.

INTRODUCTION

The quadrupole ion trap was first described by Wolfgang Paul (Paul *et al.*, 1959). It was developed as a means of performing gas chromatography/mass spectrometry by Stafford and coworkers (Stafford *et al.*, 1984) who introduced a mass selective method of ion ejection which provided a convenient method of scanning a mass spectrum. This instrument, introduced commercially in 1983, utilized electron impact ionization to convert gaseous samples into ions. Refinements such as chemical ionization (Brodbelt *et al.*,1984), automatic gain control (Stafford *et al.*, 1987) and axial modulation (Grabau *et al.*, 1983) increased the dynamic range and sensitivity of the ion trap for the analysis of ions which have a mass-to-charge ratio < 650 Da/charge.

Further improvements were necessary in order to obtain structural information on biological compounds using the quadrupole ion trap. Injection of ions generated externally to the trap by Cs^+ ion bombardment (Kaiser *et al.*, 1991) and laser desorption (Glish *et al.*, 1989) and electrospray (VanBerkel *et al.*, 1990) ionization methods, which are well-suited to ionize peptides, was demonstrated. After the resulting peptide ions are accelerated into the trap, they are contained through collisions with helium buffer gas. The trapped ions can be mass-analyzed using resonant ejection (Kaiser *et al.*, 1991) which extends the mass-to-charge range by resonantly exciting ions with an external dipole field causing them to become unstable, and hence to be ejected from the trap and detected. Using this method, CsI cluster ions having m/z > 80,000 Da/charge (Kaiser *et al.*, 1991) and molecular ions of proteins such as egg albumin (m/z 43,300 Da/charge) (Bier *et al.*,1992) have been observed. The ability to resonantly excite ions in a mass-selective fashion also forms the basis of collision induced dissociation, an experiment in which parent ions gain a sufficient amount of energy to dissociate. The resulting product ions are mass analyzed and detected in a tandem mass spectrometry (MS/MS) experiment. Peptide molecular ions yield characteristic fragment ions from which sequence information can be obtained on low pmole to femtomole amounts of sample (Kaiser *et al.*, 1990; Cox *et al.*, 1992). MS/MS experiments can be extended to multiple dissociation steps (MS^n) (Schwartz *et al.*, 1990). After the initial dissociation of the molecular ion, all fragments but one are ejected and this fragment becomes the precursor for the next dissociation step. The amount of structural information obtained is significantly enhanced over that obtained in a single dissociation step (MS/MS experiment).

Recently, developments have occurred which further increase the ability of the ion trap to solve peptide sequencing problems. Techniques to perform parent and neutral loss scans (Johnson *et al.*,1991) as well as increased mass resolution for both MS and MS/MS experiments (Schwartz *et al.*, 1991; Williams and Cooks *et al.*, 1991) have been demonstrated. The increased resolution has made it easy to monitor small mass changes such as those associated with disulfide bond reduction and the distinction of H_2O versus NH_3 loss in a fragmentation spectrum (Cox *et al.*, 1992). However, current demands on ion trap technology in the analysis of biomolecules include the ability to make highly accurate molecular weight measurements on the molecular ion and fragments and so improve the quality of the structural information available on higher mass compounds. This paper focuses on two ongoing projects which are aimed at further enhancing these capabilities: (i) the use of mixed target gases as a means of depositing greater energies into precursor ions during collision-induced dissociation (CID) and (ii) internal calibration methods for improved mass measurement accuracies on higher molecular weight compounds in MS and MS/MS experiments.

EXPERIMENTAL

A description of the prototype Finnigan ion trap mass spectrometer (ITMS) fitted with a Cs^+ ion gun is given elsewhere (Kaiser *et al.*, 1991). Leucine enkephalin (average MW 555.6) and gramicidin S (average MW 1141.5) were used as model compounds for mass measurement accuracy and energy deposition studies. Commercial peptide samples were dissolved in 1:1 MeOH:H_2O and dispensed in 1 μl aliquots onto a gold-plated stainless steel probe tip in concentrations of 54 pmol/μl and 10.7 pmol/μl respectively. The peptide samples were then mixed with 1 μl of 1:1 glycerol:thioglycerol matrix on the probe tip. An aqueous solution of CsI was used to provide calibrant ions for high mass accuracy. Internal calibration was achieved by simultaneously injecting ions of the sample and the calibrant by using a split probe (Williams *et al.*, 1992) on which both the sample and calibrant were placed but physically separated. Helium buffer gas pressures were in the range 1 to 2 mTorr. For energy deposition studies, 1.3 x 10^{-5} torr Ar and Xe were admixed individually with 2 mTorr of He buffer gas.

Once desorbed from the probe tip by Cs^+ bombardment, ions were accelerated to kinetic energies ranging from 6 to 10 eV and injected into the trap. Mass analysis of the trapped ions was achieved by mass-selective instability using axial modulation for increased sensitivity and resolution in the case of leucine enkephalin and resonant ejection with a mass range extension factor of 1.98 in the analysis of gramicidin S. In this latter case, a 15 $V_{p\text{-}p}$ supplementary AC voltage was applied to the end-caps at a frequency of 186.0 kHz. As a result of the 1.98-Fold mass range extension, the rate of the mass-selective instability scan increased from 5555 Da/sec to 11000 Da/sec. To improve mass resolution, the scan rate was slowed to 1100 Da/sec by attenuating by a factor of 10 the output of the 12 bit accuracy mass analysis control DAC, which determines the amplitude of the 1.1 MHz RF voltage applied to the ring electrode. An appropriate offset voltage was applied to place each mass analyzed calibrant and peptide ion at an apparent m/z value of 374.5 ± 0.1 Da/charge to perform peak matching (Williams *et al.*, 1992) which is necessary for high resolution calibration. A moving two-point calibration was used to analyze the data.

In MS/MS experiments, isolation and dissociation stages were added to the scan sequence prior to the mass analysis stage. Isolation of leucine enkephalin was achieved by a forward RF scan to eliminate all ions of lower mass-to-charge than the protonated molecular ion. A "reverse-then-forward" RF scan (Cox *et al.*, 1992) was used to isolate ions due to the protonated molecule of gramicidin S to eliminate higher and lower mass ions. The AC voltage was turned off at several points in the RF ramp (see Figure 1 below) to allow calibrant ions (m/z 652, 912 and 1172) to remain trapped. Also, an additional activation stage was added to partially dissociate m/z 1172 in order to produce calibrant ions below m/z 650 since these were eliminated during the RF ramp. Ion activation was achieved by applying a small supplementary AC voltage (typically 1 to 3 V_{p-p} at frequencies from 78 to 120 kHz) to the end-caps for 20 to 30 ms.

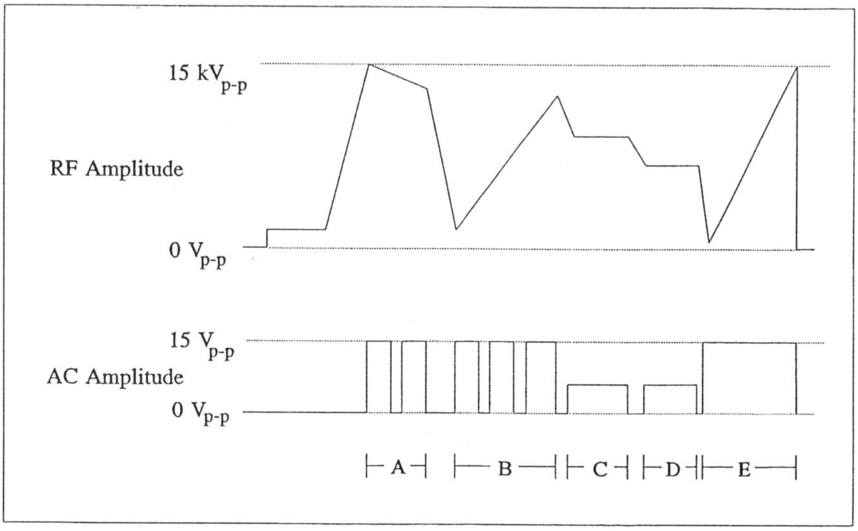

Figure 1. Scan diagram for co-injection experiment: A) Reverse isolation to eliminate all masses higher than gramicidin S MH^+ (m/z 1141) with an AC voltage notch to retain calibrant ion m/z 1172, B) Forward isolation to eliminate masses lower than gramicidin MH^+ with notches for calibrant ions m/z 652, 912 and 1172, C) MS/MS period for dissociation of calibrant ion m/z 1172, D) MS/MS period for dissociation of gramicidin S (MH^+), E) RF scan for detection of ions.

MIXED GAS STUDIES

The ion trap is normally operated in the presence of approximately one millitorr of helium to aid in the translational cooling of injected ions through collisions. Helium is also used as a target gas during collision-induced dissociation (CID) where selected ions are resonantly excited and collide inelastically with the buffer gas, causing dissociation into structurally characteristic fragments (Kelley *et al.*, 1985). However, compounds of increasing molecular weight require the deposition of higher internal energies into the precursor ion in order to fragment. Access to higher energy fragmentation pathways can be achieved in the ion trap through MS^n experiments in which selected fragment ions produced during CID are subsequently dissociated into smaller fragments (Louris *et al.*, 1990). Recent data on the highly ordered CsI cluster ions indicates that greater energies can be deposited directly into the original precursor ion when a small amount of a heavier target gas is added to the helium already present (Morand *et al.*, 1992). The heavier noble gas species provides a more massive target for collisions, so allowing greater energy transfer. Evidence is also available on the inorganic cluster ions to show that both precursor and fragment ions are better trapped in the presence of heavy targets allowing higher mass precursors to be analyzed and much lower mass fragments to be produced and detected (Morand *et al.*, 1992).

Biomolecular systems are considerably more complex than the well-ordered CsI system. However, examination of the small peptide, leucine enkephalin (MW 556) indicates that the presence of heavy target gases does allow the deposition of higher energies in the precursor ion upon CID. Leucine enkephalin has previously been described as a thermometer molecule, producing a characteristic high energy fragment, b_3 (m/z 278) and a low energy fragment, y_2 (m/z 279) (Alexander and Boyd, 1989). The ratio of these fragment ion intensities serve as a relative gauge of energy deposition into the precursor molecular ion. Figure 2 shows that the addition of argon to the helium does indeed result in increased access to the higher energy pathway. However, when xenon is added to the target gas, the high energy pathway is not accessed as efficiently, suggesting that optimum conditions for dissociation are dependent on the mass of both the target and the sample. The phenomenon of accessing more energetic pathways with a heavier target can also be observed in the CID spectrum of the cyclic peptide, gramicidin S (MW 1141). Two abundant fragments observed in the CID spectrum are m/z 831 and its subsequent water loss fragment, m/z 813. Since water loss from the fragment requires more energy deposition than formation of m/z 831 itself, a relative gauge of energy deposition can be achieved by evaluating the ratio of the intensities of these two fragment ions. Figure 3 shows a) the entire CID spectrum of gramicidin S and b) the effect of adding a heavier target gas on the fragment ions m/z 813 and m/z 831 under otherwise identical CID conditions. At a low CID voltage, 1.1 $V_{p\text{-}p}$, the addition of argon provides sufficient energy to invert the ratio of the two fragment ions, providing a method for preferentially accessing higher energy fragmentation pathways under mild CID conditions.

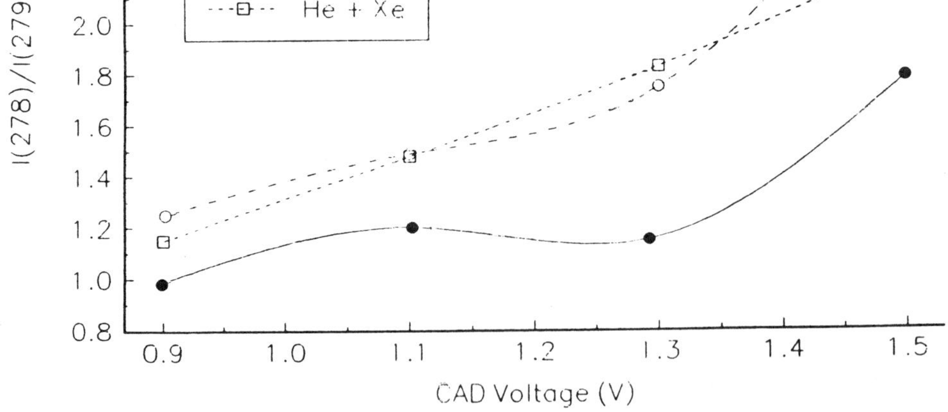

Figure 2. Examination of the b_3/y_2 ratio (m/z 278/279) of leucine enkephalin as a function of excitation voltage in the presence of different target gases; b_3 is a higher energy fragment and y_2 is a lower energy fragment.

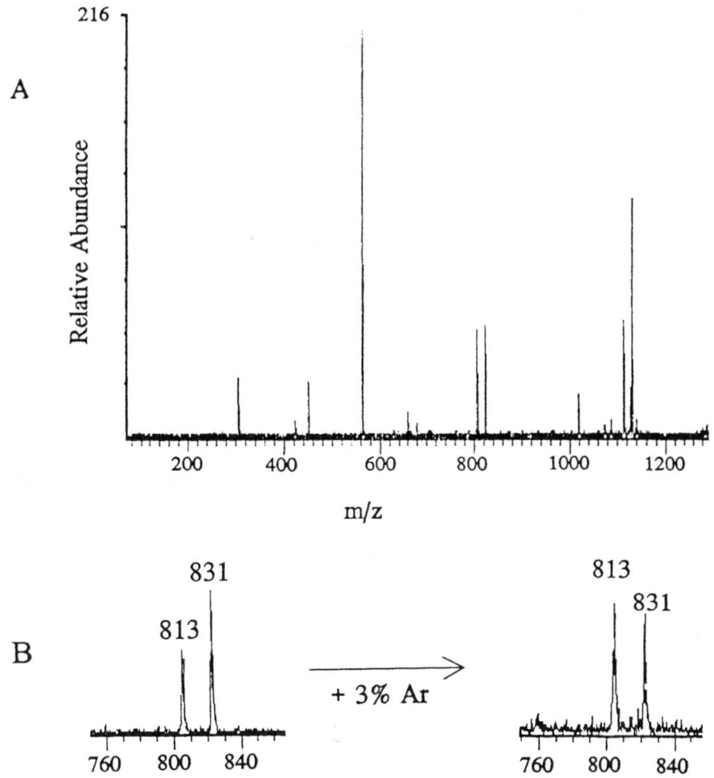

Figure 3. A) Typical CID spectrum of gramicidin S. B) Effect of addition of 6 x 10^{-5} torr of argon along with helium on the abundance of m/z 831 and 813. Resonant ac voltage was 1.1 V_{p-p} in both cases.

HIGH RESOLUTION USING INTERNAL CALIBRATION

Mass accuracy measurements in the ion trap for ions of high m/z ratios are typically < 0.1% when external calibration and low resolution experiments are performed (Cox *et al.*, 1992). Recently, this limit has been improved to < 0.010% by using i) a split probe tip to allow both analyte and calibrant ions to be desorbed by the primary ion beam, injected and trapped simultaneously to provide internal calibration and ii) a peak matching technique which enables high resolution mass spectra to be calibrated (Williams *et al.*, 1992). In this report, the first results of using these methods to assign accurate m/z values to fragment ions in MS/MS spectra are reported.

CsI cluster ions were co-injected from a split probe tip with gramicidin S to serve as calibrant ions. After isolation (Figure 1, steps A and B), two separate MS/MS stages were employed. In the first stage (Figure 1, step C), approximately 75% of the $Cs_3I_2^+$ ions were fragmented to produce Cs_2I^+ and so provide calibration ions down to m/z 392. Gramicidin S molecular ions were resonantly excited to produce fragment ions during the second MS/MS step (Figure 1, step D). Figure 4 displays the results recorded under low resolution conditions from this co-injection MS/MS experiment.

The high resolution experiment was done by slowing the mass-selective instability scan rate (Figure 1, stage E) from 11000 to 1100 Da/sec to provide better than unit resolution throughout the observed m/z range. Mass assignments were then made using peak matching (Williams *et al.*,1992). This involved using a moving two point calibration covering three linear functions relating known m/z values for adjacent CsI cluster ions to input voltages that control the amplitude of the RF at which ion ejection occurs. This produced calibrated m/z ranges: m/z 392.7 - 652.5, m/z 652.5 - 912.3 and m/z 912.3 - 1172.1. The input voltages at which gramicidin S and its fragment ions were ejected were placed in the appropriate calibrated m/z regions to yield the data shown in Table 1. These data display a dramatic improvement in mass measurement accuracy when compared with previous results (Cox *et al.*, 1992).

Further refinements such as i) choosing calibrant ions which have smaller mass differences than the CsI clusters and ii) improving the stability of the electronics should yield mass measurement errors of less than 10 ppm for all detected ions.

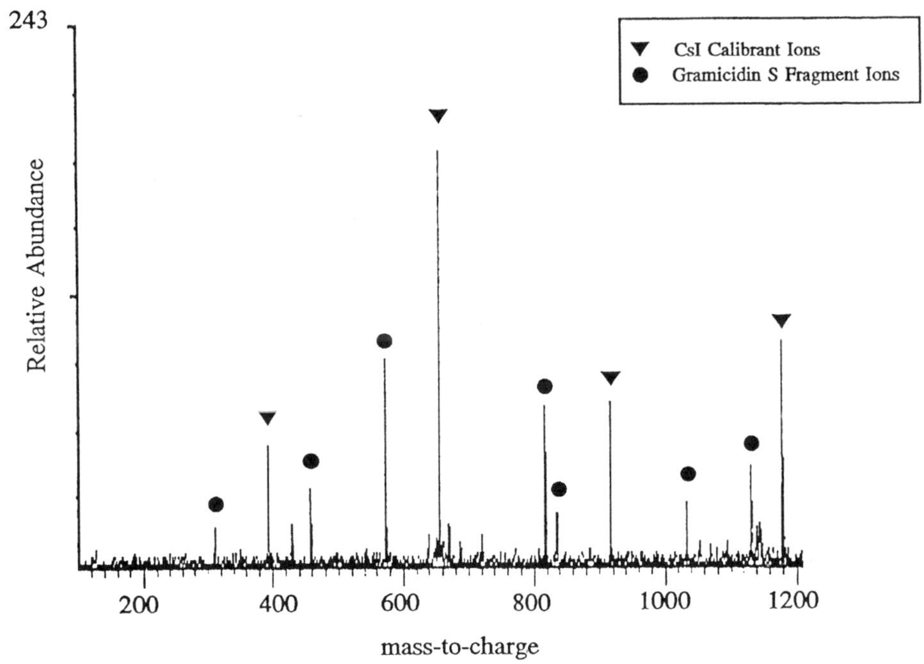

Figure 4. Low resolution co-injection MS/MS experiment of gramicidin S and CsI cluster ions: calibrant ions (▼) peptide fragment ions (●).

Table 1. Calibration for Monoisotopic (^{12}C) Gramicidin S Fragment Ions.

Theoretical m/z	Calibrated m/z[a]	Δ m/z	Mass Accuracy (ppm)
571.3608	571.31	0.0468	82
813.5026	813.55	-0.0494	60
831.5132	831.51	0.0015	2
1027.634	1027.6	0.0260	25
1141.715	1141.7	-0.0040	4

[a]Values were calculated using a moving two-point calibration.

REFERENCES

Alexander, A. J. and Boyd, R. K., 1989 Experimental investigation of factors controlling the collision-induced dissociation spectra of peptide ions in a tandem hybrid mass spectrometer I. Leucine Enkephalin, *Int J. Mass Spectrom. Ion Proc.* 90:211.

Bier, M. E., Schwartz, J. C., Jardine, I. and Stafford, G. C., 1992 Matrix-assisted laser desorption quadrupole ion trap mass spectrometry, Proc. of the 40th ASMS Conference of Mass Spectrometry and Allied Topics, Washington, DC.

Brodbelt, J. S., Louris, J. N. and Cooks, R. G., 1984 Chemical ionization in an ion trap mass spectrometer, *Anal. Chem.* 59:1278.

Cox, K. A., Williams, J. D., Cooks, R. G. and Kaiser, R. E. Jr., 1992 Quadrupole ion trap mass spectrometry: current applications and future directions for peptide analysis, *Biol. Mass Spectrom.* 21:226.

Glish, G. L., Goeringer, D. E., Asano, K. G. and McLuckey, S. A., 1989 Laser desorption mass spectrometry and MS/MS with a three dimensional quadrupole ion trap, *Int. J. Mass Spectrom. Ion Proc.* 94:15.

Johnson, J. V., Pedder, R. E. and Yost, R. A., 1991 Implementation of parent and neutral loss scanning and selected reaction monitoring on a quadrupole ion trap mass spectrometer, Proc. of the 39th ASMS Conference of Mass Spectrometry and Allied Topics, Nashville, TN, 530.

Kaiser, R. E. Jr., Cooks, R. G., Syka, J. E. P. and Stafford, G. C. Jr., 1990 Collisionally activated dissociation of peptides using a quadrupole ion trap mass spectrometer, *Rapid Commun. Mass Spectrom.* 4:30.

Kaiser, R. E. Jr., Cooks, R. G., Stafford, G. C. Jr., Syka, J. E. P. and Hemberger, P. H., 1991. Operation of a quadrupole ion trap mass spectrometer to achieve high mass/charge ratios, *Int. J. Mass Spetrom. Ion Proc.* 106:79.

Kelley, P. E., Stafford, G. C., Syka, J. E. P., Reynolds, W. E., Louris, J. N., Amy, J. W. and Todd, J. F. J., 1985 New advances in the operation of the ion trap mass spectrometer, proceedings of the 33rd ASMS Conference on Mass Spectrometry and Allied Topics, San Diego, CA, 707.

Louris, J. N., Brodbelt, J. S., Cooks, R. G., Glish, G. L., VanBerkel, G. J. and McLuckey, S. A., 1990 Ion isolation and sequential stages of mass spectrometry in a quadrupole ion trap mass spectrometer, *Int. J. Mass Spectrom. Ion Proc.* 96:117.

Morand, K. L., Cox, K. A. and Cooks, R. G., 1992 Efficient trapping and collision-induced dissociation of high mass cluster ions using mixed target gases in the quadrupole ion trap, *Rapid Commun. Mass Speclrom.* 8:520.

Paul, W., Reinhard, H. and von Zahn, V., 1959 Das elektrische massenfilter als Massenspektrometer und Isotopentrenner, *Z. Phys.* 156:1.

Schwartz, J. C., Wade, A. P., Enke, C. G. and Cooks, R. G., 1990 A systematic delineation of scan modes in multidimensional mass spectrometry, *Anal. Chem.* 62:1809.

Schwartz, J. C., Syka, J. E. P. and Jardine, I., 1991 High resolution on a quadrupole ion trap mass spectrometer, *J. Am. Soc. Mass Spectrom.* 2:198.

Stafford, G. C., Kelley, P. E., Syka, J. E. P., Reynolds, W. E. and Todd, J. F. J., 1984 Recent improvements in and analytical applications of advanced ion trap technology, *Int. J. Mass Spectrom. Ion Proc.* 60:85.

Stafford, G. C., Taylor, D. M., Bradshaw, S. C., Syka, J. E. P. and Urich, M., 1987 Enhanced sensitivity and dynamic range on an ion trap mass spectrometer with automatic gain control (AGC), Proc. of the 35th ASMS Conference of Mass Spectrometry and Allied Topics, Denver, CO, 775

VanBerkel, G. J., Glish, G. L. and McLuckey, S. A., 1990 Electrospray ionization combined with ion trap mass spectrometry, *Anal. Chem.* 62:1284 .

Weber-Grabau, M., Kelley, P. E., Bradshaw, S. C. and Hoekman, D. J., 1988 Advances in MS/MS analysis with the ion trap mass spectrometer, Proc. of the 36th ASMS Conference of Mass Spectrometry and Allied Topics, San Francisco, CA, 1106 .

Williams, J. D., Cox, K. A., Cooks, R. G., Kaiser, R. E. and Schwartz, J. C., 1991 High mass-resolution using a quadrupole ion trap mass spectrometer, *Rapid Commun. Mass Spectrom.* 5:327.

Williams, J. D. and Cooks, R. G., 1992 Improved accuracy of mass measurements with a quadrupole ion trap mass spectrometer, *Rapid Commun. Mass Spectrom.* 8:524.

ELECTROSPRAY-MASS SPECTROMETRY, AN EMERGING METHODOLOGY FOR ELUCIDATING STRUCTURE-FUNCTION RELATIONSHIPS OF PROTEINS

Kenneth A. Walsh, Lowell H. Ericsson
Katheryn Resing and Richard S. Johnson

Department of Biochemistry SJ-70, University of Washington
Seattle, WA, 98195, U.S.A.

INTRODUCTION

For 20 years, the MPSA conferences have focussed on improving the technologies available for determining amino acid sequences of proteins. During that time the sensitivity and speed of the techniques have each increased by six orders of magnitude; new technologies have filled databases with descriptions of the detailed nature of proteins, illustrated by the subtle beauty of their three-dimensional structures. From linear sequences we now seek the substructural domains and informational motifs that reveal clues to ancestry, function and regulation. Interplay of the methods of protein chemistry and molecular biology leads to redesigned proteins, and to descriptions of complex cellular events in discreet molecular terms.

This remarkable increase in our level of understanding has its roots in a punctuated series of methodological breakthroughs, in part stimulated by conferences such as the MPSA series. *New methodologies continue to be catalysts for the growth of understanding of the molecular basis of biological function.* In the last 3-4 years it has become evident that new techniques in mass spectrometry have introduced major new strategies with which structure/function questions can be approached (Carr *et al.*, 1991; Chait and Kent, 1992; Covey *et al.*, 1988; Fenn *et al.*, 1990). Both Electrospray ionization Mass Spectrometry (ES-MS) and Matrix-Assisted Laser Desorption Time-of-Flight analysis offer access to molecular weights of proteins with accuracies approaching 99.99%. This is easily sufficient to detect the differences frequently observed between the mass predicted by a cDNA-derived sequence and that observed in the expressed protein. A particularly powerful technique involves direct ES-MS analysis of tryptic peptides as they elute from an HPLC. ES-MS of peptides in instruments with tandem MS capability allows direct sequence analysis and placement of posttranslational modifications. We have applied ES-MS to a variety of problems in protein chemistry in our laboratory, and are pleased with its

Methods in Protein Sequence Analysis, Edited by K. Imahori
and F. Sakiyama, Plenum Press, New York, 1993

flexibility and adaptability. We use a Sciex model API-III instrument, which has a nebulization-assisted "Ionspray" source, and three quadrupoles, one of which serves as a collision cell. The range of applications of this instrumentation to proteins and peptides during its first year in our laboratory is suggested by the list in Table 1.

Table 1. Typical applications of ES-MS to proteins.

1) Accurate mass determination of proteins permits:
   - Comparison with cDNA sequence.
   - Detection of posttranslational modifications.
   - Detection of isoprotein mixtures.

2) Mass analysis of a synthetic peptide tests for:
   - Fidelity of synthesis.
   - Removal of blocking groups.

3) Coupled HPLC/ES-MS of a proteolytic digest of a protein of *known* sequence yields a "fingerprint" wherein particular peptide sequences may be located by their predicted masses, and putative posttranslationally modified peptides are suggested by their anomalous masses.

4) Coupled HPLC/ES-MS of a proteolytic digest of a protein of *unknown* sequence provides a ready source of pure peptide ions that can be sequenced by tandem MS for comparison with databases or for the design of oligonucleotide probes.

5) Collision-induced dissociation (CID) of a peptide may yield:
   - Its amino acid sequence.
   - The site of its posttranslational modification.
   - The nature of the posttranslational modification.

EXAMPLES OF THE APLICATION OF ES-MS TO PROTEINS

Examples of the power of ES-MS are found in recent reports of our detection of double bonds within fatty acids covalently attached to the N-termini of two proteins of the visual cycle, recoverin and transducin (Dizhoor *et al.*, 1992; Neubert *et al.*, 1992). Evidence for the presence of fatty acids was found in the extra mass of the isolated proteins when compared with those predicted by their cDNA sequences. However, comparison of the M_r of native recoverin with that of a recombinant form lacking fatty acylation suggested that the native protein weighed 2 Da less than it would if the fatty acid were saturated. Following tryptic digestion of native recoverin, four different N-terminal tetrapeptides were isolated and three were proven, by CID in the triple quadrupole MS, to differ by the presence or absence of double bonds in a 14-carbon myristoyl group attached to the N-terminal glycyl residue. The fourth peptide was shown to contain a 12-carbon lauric acid (saturated) rather than myristic acid. Similar analysis of the alpha subunit of transducin revealed the same distribution of saturated and unsaturated fatty acids. In both cases, the principal evidence for the presence of the double bonds was the finding that the mass of the protein or of a derived peptide was smaller by two daltons than expected.

Since the ES-MS methodology is accurate enough to give evidence of a double bond in a protein of M_r 23,000, it is clear that comparison of the M_r of any native protein with that predicted from its cDNA should provide initial clues to the mass of endogeneous posttranslational modifications in the isolated protein. Since virtually all proteins are posttranslationally modified (cf. Krishna and Wold, this volume), this approach has general utility. The accuracy of the methodology is sufficient to suggest the nature of the modification. However, as with the study of fatty acylations, detailed tandem MS analysis of isolated peptides is then necessary for a complete characterization. Unfortunately, this simple strategy has been confounded more than once by inaccurate information in a sequence database. This can arise directly from sequencing errors, from analysis of the cDNA sequence of a minor variant (or a reverse transcriptase artefact), or from a simple lapse in proofreading.

The sensitivity of the instrumentation for sequence analysis of peptides by CID in the tandem ES-MS mode is approximately the same as for conventional microsequencing by the Edman degradation. Collection of MS data is orders of magnitude more rapid, but their interpretation requires experience and emerging software aids. These contrasting methodologies of sequence analysis are complementary to each other in the sense that segments which resist analysis by one technique tend to be the most readily analyzed by the other. Edman degradations typically become less reliable as the C-terminus is approached, whereas by tandem MS techniques the N-termini are more likely to be problematic. Hydroxyamino acids do not present the problems in MS that they do in extended Edman degradations. Triple quadrupole MS does not distinguish between the

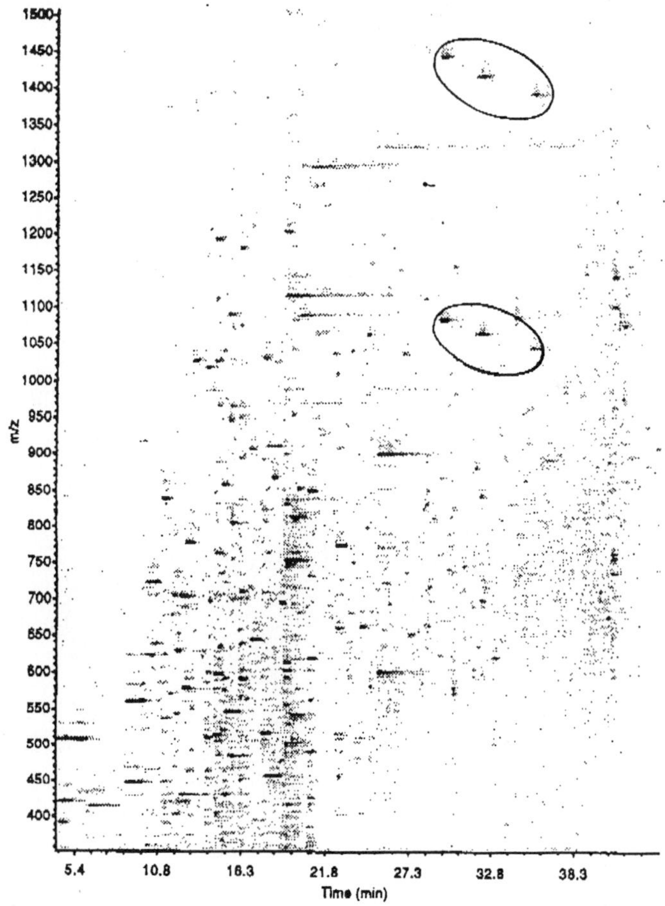

Figure 1. On-line coupled HPLC/MS analysis of 10% of the elution stream of a tryptic digest of 250 pmol of the epidermal protein profilaggrin (Resing *et al.*, in preparation) on a 2 mm ID Hypersil C18 column (Hewlett Packard) eluted with an acetonitrile gradient. The distribution of each peptide ion in the eluate stream is indicated by a horizontal line with a single mass/charge (m/z) ratio.

isobaric residues Gln and Lys or the isotopic residues Leu and Ile (and isobaric HO-Pro), whereas these are readily recognized during Edman degradations. Mass spectrometric analysis uniquely appears to provide an internal check of the validity of a sequence in that the final result must agree with the observed mass. However an astonishing number of sequences can fit any initial mass and the corrrect interpretation of CID data may depend upon recognition of minor fragment ions. Recently we have described a software system designed to draw on the contrasting strengths of the two techniques to solve the sequences of peptides in unresolved mixtures (Johnson and Walsh, 1992).

In our hands, perhaps the most powerful methods for locating peptides for confirmation of sequence, and for selecting peptides of anomalous mass for further characterization, is ES-MS analysis coupled directly to an eluate stream, during HPLC separation of a tryptic digest of a protein of known sequence. Figure 1 illustrates such an LC/MS "fingerprint", wherein each horizontal band defines both the mass of a peptide and its elution behaviour. Each mass is then compared with lists of peptides predicted from the known sequence. A mass that does not fit predictions is worthy of further study, usually by reexamining the peptide in a fraction collected by splitting the original HPLC eluate stream. By using a triple quadrupole instrument for this reexamination, it does not matter if other peptides contaminate the fraction because the first quadrupole serves as the final purification step, the CID quadrupole yields information only from the selected ion, and the third quadrupole separates the collision induced fragments exclusively from the peptide of interest.

An LC/MS fingerprint is simpler and more informative than it may first appear because several ions of different m/z values may correspond to a single peptide in several charged forms. This is evident in Figure 1, where two ions happen to elute in an anomalously broad manner at a coincident time (25-26 min). The m/z of one (591) is precisely 2/3 that of the other, and one represents the doubly charged ion, whereas the other is triply charged. This particular protein contains many phosphoserine residues, where the sites tend to be clustered and incompletely phosphorylated. This is apparent within the circled group of peptide ions of m/z 1040-1080 in Figure 1, where a set of three ions form a staircase-like pattern differing from each other in m/z by 20 daltons. Since these three ions each have a charge of +4, the difference in mass corresponds to 80, precisely that of a phosphoryl group. The three ions are thus recognized as di-, mono-, and un-phosphorylated forms of the same peptide. A similar set of three ions of m/z 1386-1440, displayed in the same fractions represents simply the triply charged ions of the same peptides. These six ions correspond to a single peptide locus in the protein, and their pattern of elution times and m/z values draws attention to their phosphopeptide character.

Alternative methods of locating other specifically modified peptides are being developed and some have already been described. For example, one can compare by LC/MS fingerprints of phosphorylated and dephosphorylated or glycosylated and deglycosylated proteins (Carr *et al.*, 1991). One can seek parent ions that release a fragment of characteristic size in the collision cell.

In current studies of intact proteins, they are usually electrosprayed from acidic aqueous solutions, where positively charged ions predominate. Recently, there have been several reports that specific non-covalent binding of ligands may be observed when selected proteins are electrosprayed at neutral pH (Ganem *et al.*, 1991; Katta and Chait, 1991); we have reported that calmodulin retains calcium ions, but not magnesium ions under these condition (Walsh *et al.*, 1992). The general utility of these observations is not yet clear. For example, ligand loss could be expected due to changes in pH or conformation during the electrospray ionization and drying process. Hence, while the observation of a retained ligand may be of biological interest, evidence of this type indicating a lack of binding is probably inconclusive.

The power of this technology for the elucidation of protein structure and structure/function relationships is only beginning to be generally apparent. Small proteins can be sequenced in days, not months. Blocked proteins are sequenced as easily as their unblocked counterparts. These methods surely represent the best current technology for identifying posttranslational modifications in subnanomole amounts of protein, and for quickly identifying discrepancies between the mass of an expressed protein and that predicted by its cDNA.

REFERENCES

Carr, S. A., Hemling, M. E., Bean, M. F. and Roberts, G. D., 1991 Integration of mass spectrometry in analytical biotechnology, *Anal. Chem.* 63:2802.

Chait B. T. and Kent, S. B. H., 1992 Weighing naked proteins: practical, high-accuracy mass measurement of peptides and proteins, *Science* 257:1885.

Covey, T. R., Bonner, R. P., Shushan, B. I. and Henion, J., 1988 The determination of protein, oligonucleotide and peptide molecular weights by ion spray mass spectrometry, *Rapid Commun. Mass Spectrom.* 2:249

Dizhoor, A. M., Ericsson, L. H., Johnson, R. S., Kumar, S., Olshevskaya, E., Zozulya, S., Neubert, T. A., Stryer, L., Hurley, J. B. and Walsh. K. A., 1992 The NH_2 terminus of retinal recoverin is acylated by a small family of fatty acids, *J. Biol. Chem.* 267:16033.

Fenn, J. B., Mann, M., Meng, C. K. and Wong. S. F., 1990 Electrospray ionization - principles and practice, *Mass Spectrom. Rev.* 9:37.

Ganem, B. , Li, Y. T. and Henion, J. D., 1991 Observation of noncovalent enzyme-substrate and enzyme-product complexes by ion-spray mass spectrometry, *J. Am. Chem. Soc.* 113:7818.

Johnson R. S. and Walsh, K. A., 1992 Sequence analysis of peptide mixtures by automated integration of Edman and mass spectrometric data, *Protein Science* 1:1083.

Katta, V. and Chait, B. T., 1991 Observation of the heme-globin complex in native myoglobin by electrospray-ionization mass spectrometry, *J. Am. Chem. Soc. 113:8535.*

Neubert, T. A., Johnson, R. S., Hurley, J. B. and Walsh. K. A., 1992 The rod transducin α subunit amino terminus is heterogeneously fatty acylated, *J. Biol. Chem.* 267:18274.

Resing, K. A., Johnson, R. S. and Walsh, K. A., Characterization of protease processing sites during conversion of rat profilaggrin to filaggrin, in preparation.

Walsh, K. A. Ericsson, L. H. and Johnson. R. S., 1992 Application of electrospray ionization-mass spectometry to proteins, *Proc. Japan Soc. Biomed, Mass Spectrom., Kanazawa,* Sept. pp.113.

A GENERAL STRATEGY FOR THE USE OF MASS SPECTROMETRIC MOLECULAR WEIGHT INFORMATION IN PROTEIN PURIFICATION AND SEQUENCE DETERMINATION

Peter Roepstorff and Peter Højrup

Department of Molecular Biology, Odense University
DK 5230 Odense M, Denmark

INTRODUCTION

At the MPSA 1988 in Berlin we presented a general strategy for the combined use of mass spectrometric molecular weight information and automatic Edman degradation for sequence determination of proteins (Roepstorff *et al.*, 1989). Among the mass spectrometric methods available then, i.e. plasma desorption mass spectrometry (PDMS) and fast atom bombardment mass spectrometry (FABMS), we had chosen PDMS because this method in our experience yielded better sensitivity and higher practical mass range than FABMS. The mass accuracy although not as good as for FABMS was in most cases adequate. PDMS had further advantages over FABMS in simplicity in sample preparation, instrument operation and data interpretation. This has been of utmost importance for our routine use of mass spectrometry in protein sequencing. In the past years we have sequenced several proteins using this strategy (Andersen *et al.*, in press; Andreassen *et al.*, submitted; Højrup *et al.*, 1986; 1991; Klarskov *et al.*, 1989; Mikkelsen *et al.*, 1989; Parello *et al.*, in preparation; Talbo *et al.*, 1990).

The appearance of two new ionization methods, matrix assisted laser desorption ionization (MALDI)(Karas *et al.*, 1988) and electrospray ionization (ES)(Fenn *et al.*, 1989), overcame some of the shortcomings of PDMS and FABMS. The achievable mass ranges exceed 300 and 200 kDa respectively. Mass accuracy of 100 ppm or better over the complete mass range can be obtained with ESMS, and sensitivity in the low femtomole range can be obtained with MALDIMS. These new performances have considerably widened the perspectives for the use of mass spectrometry in combination with traditional methods in protein purification and sequence determination. Having access to all the ionization methods in our laboratory, we have accordingly modified our strategy to take optimum advantage of the specific features of each method.

Methods in Protein Sequence Analysis, Edited by K. Imahori and F. Sakiyama, Plenum Press, New York, 1993

APPLICATION OF MS IN PROTEIN PURIFICATION

Purification of a protein from a complex mixture requires localization of the protein of interest in the fractions collected in the different purification steps. This is frequently performed by analysis of each fraction by tedious and often expensive assays based on biological activity, use of specific antibodies, or 2D-PAGE. A major project in our laboratory is studies of the structural proteins from locust cuticle. The hard exocuticle contains more than 100 different proteins and the soft endocuticle approximately 20 as indicated by 2D-PAGE. Preparative isolation of the proteins prior to sequence determination is performed by a combination of gel chromatography, ion exchange chromatography, and RP-HPLC. As no biological or immunological assays are available, it has until recently been necessary to monitor the fractions collected in the different purification procedures by 2D-PAGE. After installation of MALDI and ESMS in our laboratory we have tested the use of these methods to monitor the purification procedures. An example is the purification of the abdominal protein no. 4 (Abd 4) from the desert locust, *Schistocerca gregaria*. Based on the sequence of the corresponding protein from the migratory locust, *Locusta migratoria* (Talbo *et al.*, 1990), its approximate molecular weight was known to be close to 13,000. Monitoring the HPLC purification by MALDIMS localizes the Abd 4 protein to the fraction for which the spectrum is shown in Figure 1. Further purification is indicated by the presence of another recently sequenced cuticle protein, Abd 5 (MW 8,898)(Jespersen *et al.*, in preparation).

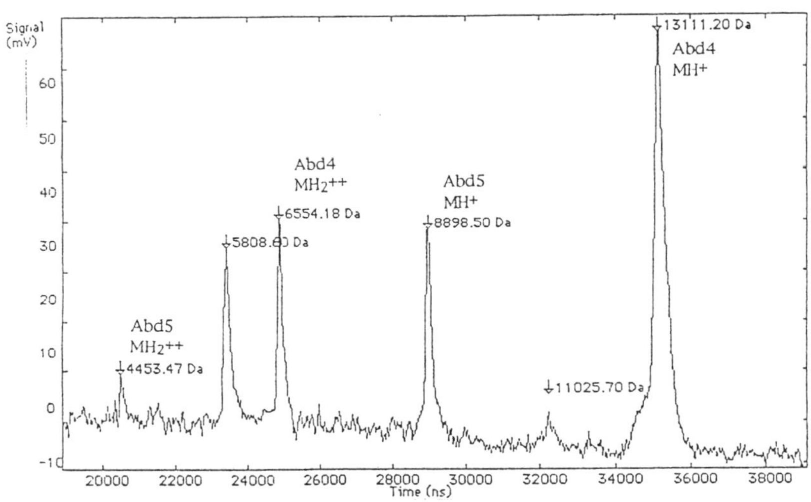

Figure 1. MALDI spectrum of an HPLC fraction obtained during purification of protein Abd 4 (MW 13,111) from the cuticle of desert locust. The fraction is still contaminated with Abd 5 (MW 8,998) and some minor components. The peak at MW 5,808.6 is human insulin used for calibration.

Recently, we have also been involved in purification and sequencing of a specific ribosomal protein from the ciliate, *Tetrahymena termophila* (Palm *et al.*, in preparation). Also here mass spectrometry, in this case ESMS, was used to monitor the purification as the only alternative would have been 2D-PAGE. The molecular weight informations were found very useful, not only to locate the appropriate fractions, but also for optimization of the purification strategy.

The advantages of using mass spectrometry to monitor protein purification include high speed (a few minutes per fraction), high specificity due to the precision of the molecular weight information, and low direct analysis cost. However, a precondition for its use is that the molecular weight of the protein of interest is known. This is often the case because the MW information is one of the first parameters we use to characterize a protein in the initial analytical studies. MALDIMS is favoured in the early stages of protein purification because it can tolerate the presence of salts, buffers, and denaturing agents such as urea. It is also more effective than ESMS for analysis of crude protein mixtures. The strengths of ESMS are detection of minor components, microheterogeniety, and high mass accuracy. It is consequently optimally used in the final purification steps and for precise molecular weight determination of the purified protein.

STRATEGY FOR SEQUENCING UNKNOWN PROTEINS COMBINING MASS SPECTROMETRY AND AUTOMATIC EDMAN DEGRADATION

Once the protein is purified and its molecular weight determined, the strategy given in Figure 2 is followed. If the protein is not N-terminally blocked it is first submitted to automatic Edman degradation for as many cycles as possible.

Based on the amino acid composition a suitable enzymatic or chemical cleavage procedure is selected to produce fragments of reasonable size. A small scale analytical cleavage procedure is performed and time-course monitored by mass spectrometry. The purpose of this step is to check if the chosen cleavage procedure yields appropriately sized peptides and also to identify partial cleavage products which allow alignment of the peptides based on the molecular weight information. To minimize sample consumption in this step MALDIMS seems preferable because it is excellent for peptide mixture analysis and can be performed on subpicomole sample amounts. Thus, a complete time course monitoring of an enzymatic protein digestion can be performed starting with one picomole of protein.

Once the optimum cleavage conditions are established, a preparative digest is performed and the peptides separated by HPLC. Each peptide fraction is analysed by mass spectrometry, amino acid analysis if sufficient sample is available, and submitted to automatic Edman degradation. In our laboratory PDMS is preferred at this stage because of its simplicity, excellent compatibility with the solvents used for HPLC, and the possibility to perform further studies on the sample after mass spectrometric analyses e.g., reduction, alkylation, enzymatic cleavage, removal of the sample for further chromatography, and/or transfer to the PVDF membrane used in the sequencer (Jespersen *et al.*, in press)

The molecular weight determination at this stage serves several purposes. By adding the molecular weights of the peptides and comparing with the molecular weight of the intact protein, peptides which account for the complete sequence can be selected for sequencing. Possible losses of single amino acids or small peptides in the purification procedure will also be disclosed. The molecular weights of the peptides give an estimate of the number of cycles to perform in the subsequent automatic sequencing and confirms the sequencing result once this is concluded. Often ambiguities encountered in the identification of the PTH amino acid derivatives can be resolved directly based on the molecular weight. The presence of modified amino acid residues is revealed and their nature frequently determined from the combined sequencing and molecular weight information.

Failure to sequence the peptides to the end is immediately disclosed from the MW information. If only one or a few residues are missing, a carboxypeptidase digestion of the peptide monitored by mass spectrometry frequently yield the missing information. This may when PDMS is used be performed directly on the remaining nitrocellulose bound sample (Klarskov *et al.*, 1989). If carboxypeptidase digests are carried out on large peptides or intact proteins, ESMS is preferable because the high mass accuracy is needed for unambiguous identifications. This was recently used for determination of the two C-terminal residues in a 13,423 Da protein from the locust cuticle. According to MW and amino acid analysis data a histidyl and a leucyl residue seemed to be missing. The mass accuracy of ESMS in this molecular weight range is sufficient to distinguish the removal of a leucyl residue (residue weight 113) from that of Asn (114) or Asp (115) (Roepstorff *et al.*, in preparation). If several residues are missing, subdigestion of the relevant peptides followed by separation and MW determination allows immediate identification of the peptides containing the missing residues. These can then be sequenced. Alternatively the gaps can be located and filled out in the final verification procedure using a cleavage procedure with a different specificity. It is also our experience that sequencing of simple peptide mixtures often is possible when the molecular weights of the components are known.

At this stage most of the sequence will be established. The sequence of N-terminally blocked peptides, precise localization of post translational modifications, and disulfide bridging may remain to be settled. For N-terminally blocked peptides the combined MW and amino acid analysis data normally identifies the nature of the blocking group

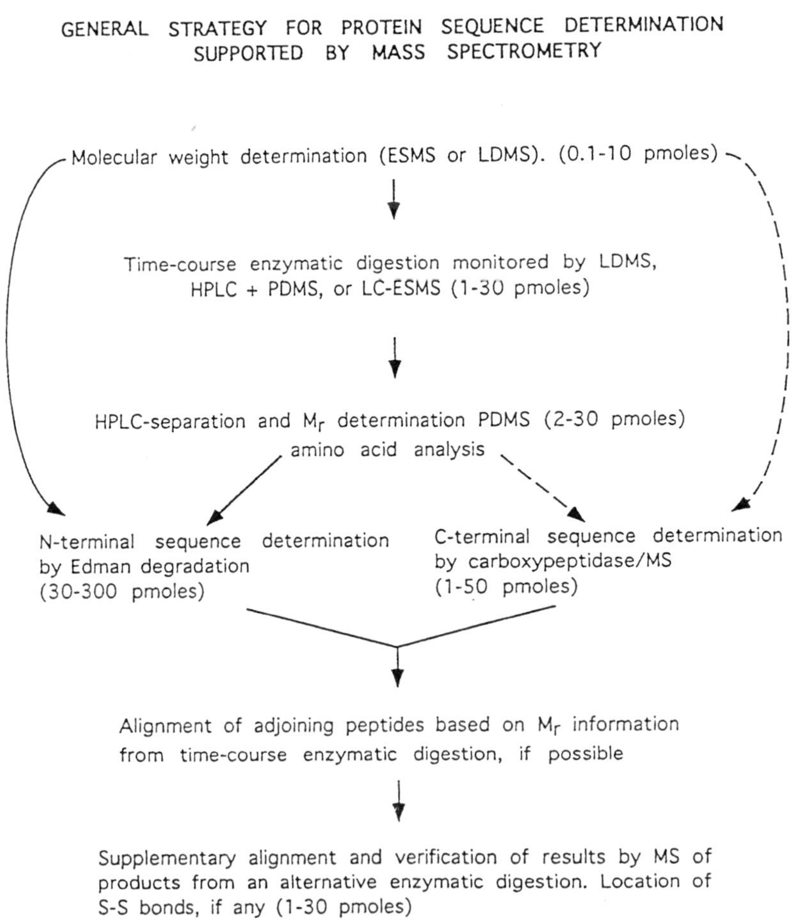

Figure 2. General strategy for protein sequence determination by combined Edman degradation and mass spectrometry. The total sample use for mass spectrometric analysis is 50-200 picomoles.

(Jespersen *et al.*, in preparation; Roepstorff *et al.*, in preparation). In the case of a blocked cuticular protein a pyroglutamic acid residue was indicated. This was confirmed by MW determination by ESMS after treatment of the intact protein with pyroglutamate amino peptidase. The sequence from the second residue was then established by Edman degradation (Jespersen *et al.*, in preparation). For an N-terminally blocked parvalbumin the acetylated N-terminal peptide was immediately identified from MW and amino acid analysis data. Subdigestion of this peptide with trypsin allowed sequencing of most of the peptide leaving only two possibilities for the first three residues, AcMAK or AcAMK. Cleavage of the intact protein with cyanogen bromide followed by analysis by ESMS confirmed the latter (Roepstorff *et al.*, in preparation). If such simple procedures are not sufficient to sequence the N-terminally blocked peptides, direct mass spectrometric sequencing is possible using either MS/MS and collision induced dissociation (CID)(Biemann, 1990) or PDMS after derivatization by permethylation (Talbo and Roepstorff, 1990). Post translational modifications can frequently be exactly located by subdigestion followed by determination of the molecular weight of the resulting peptides. If not sufficient direct mass spectrometric sequencing can be applied. Identification of the disulfide bonds using mass spectrometric peptide mapping of the unreduced protein followed by reduction of the disulfide bonds is now a standard procedure (e.g. Sørensen *et al.*, 1990). This can conveniently be combined with the last verification step in Figure 2.

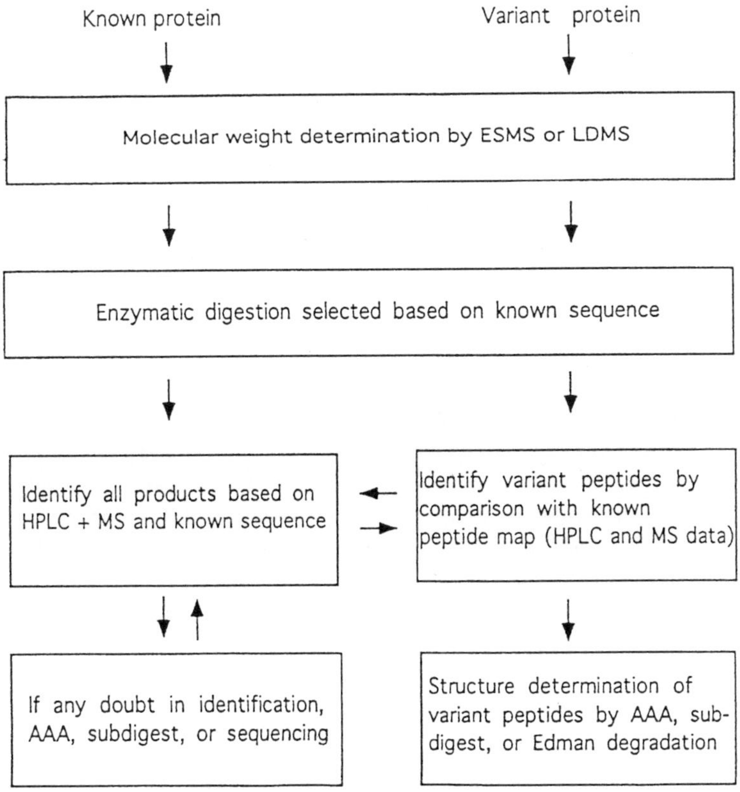

Figure 3. Strategy for structure determination of proteins which are variants of proteins with known sequence. The left column can be simulated if the protein is not available but known only from database information.

IDENTIFICATION AND SEQUENCE OF KNOWN OR PARTIALLY KNOWN PROTEINS

Often a protein considered unknown upon sequence determination shows up already to be known or to be strongly related to known proteins. The information may be available in either protein or in DNA sequence databases. We have developed a program to translate the database information into molecular weights of the proteins as well as a number of peptide MW tables corresponding to chemical or enzymatic cleavage of the proteins. Acquisition of accurate protein molecular weights and/or selected peptide molecular weights is very fast and highly sensitive with modern mass spectrometric methodology. In our experience searches in the above mentioned transformed databases are highly efficient to identify a protein (Mann *et al.*, submitted). Especially, the use of data sets derived for a variety of specific enzymatic cleavages of the proteins has proven to be highly tolerant to database errors and/or short-comings such as missing information about post translational modifications. Thus, variants as well as modifications of the proteins in the databases can be identified and often also the region or exact point of variation/modification.

In general the structure of a protein recognized to be a variant of a protein with known sequence can be determined using the scheme given in Figure 3. Identification of all common peptides is based on MW determinations and only the deviating peptides need further examination. Often subdigests combined with MW determinations and/or amino acid analysis are sufficient for assignment of the structure. In cases where sequencing is necessary it can be performed either by Edman degradation or, because the sequence is partially known and interpretation of the spectra therefore simple, by CID-MS/MS. We have used this or a simplified scheme to determine the structure of hemoglobin variants (Jensen *et al.*, 1991), isoforms of fatty acid binding proteins (Unterberg *et al.*, 1990), and interspecies variation for cuticular proteins (Jespersen *et al.*, in preparation).

CONCLUSION

With the introduction of MALDI and ESMS molecular weight information of both proteins and peptides can be obtained with high precision and high sensitivity. These two techniques are now preferable for analysis of intact proteins while PDMS is still advantageous in routine peptide analysis. The inclusion of mass spectrometric molecular information in protein purification and primary structure studies increases the speed, sensitivity and specificity. In sequencing projects it reduces time consuming and costly automatic sequencing. For studies of variant proteins it may entirely eliminate the need for Edman degradation. Most important, the inclusion of mass spectrometry answers questions that might not be solved otherwise and provides independent confirmatory evidence.

Acknowledgements. G. Talbo is acknowledged for providing the result in Figure 1. Economical support for the described studies has been obtained from The Danish Technical Science and Natural Science Research Councils, Villum Kann Rasmussens Foundation and the Danish Biotechnology Programme through our participation in the Danish Protein Engineering Center.

REFERENCES

Andersen, J., Andersen, S. O., Højrup, P. and Roepstorff, P. Primary structure of a protein from the cuticle of the migratory locust, *locusta migratoria. Insect Biochem. Molec. Biol.*, in press.

Andreassen, L., Højrup, P., Andersen, S. O. and Roepstorff, P. Combined plasma desorption mass spectrometry and Edman degradation applied to simultaneous sequence determination of isoforms of structural proteins from cuticle of *Locusta migratoria, Biochem. J.* submitted.

Biemann, K., 1990 Sequencing peptides by Tandem mass spectrometry and high-energy collision induced dissociation, *Methods in Enzymol.* 193:455.

Fenn, J. B., Mann, M., Meng, C. K., Wong, S. F. and Whitehouse, C. M., 1989 Electrospray ionization for mass spectrometry of large biomolecules, *Science* 246:64.

Højrup, P., Andersen, S. O. and Roepstorff, P., 1986 Primary structure of a structural protein from the cuticle of the migratory locust, *Locusta migratoria, Biochem. J.* 236:713.

Højrup, P., Gerola, P., Mikkelsen, J., Hansen, H. F., El-Shahed, A., Knudsen, J., Roepstorff, P. and Olson, J. M., 1991 The primary structure of the bacteriochlorophyll c-binding protein from chlorosomes of *Chlorobium limicola f. thiosulfatophilum, Biochem. Biophys. Acta* 1077:220.

Jensen, O. N., Højrup, P. and Roepstorff, P., 1991 Plasma desorption mass spectrometry as a tool in characterization of abnormal proteins. Application to variant human hemoglobins, *Anal. Biochem.* 199:175.

Jespersen, S., Højrup, P., Andersen, S. O. and Roepstorff, P. The primary structure of an endocuticle protein from two locust species, *Locusta migratoria* and *Schistocerca gregaria*, determined by a combination of mass spectrometry and automatic Edman degradation, in preparation.

Jespersen, S., Talbo, G. and Roepstorff, P. Optimization of sample recovery from the nitrocellulose support used in plasma desorption mass spectrometry and its use for multiple analysis of insulin, *Biol. Mass Spectrom.*, in press.

Karas, K. and Hillenkamp, F., 1988 Laser desorption ionization of proteins with molecular masses exceeding 10,000 Da, *Anal. Chem.* 60:2299.

Klarskov, K., Breddam, K. and Roepstorff, P., 1989 C-terminal sequence determination of peptides degraded with carboxypeptidases with different specificities and analyzed by 252 Cf plasma desorption mass spectrometry, *Anal. Biochem.* 180:28.

Klarskov, K., Højrup, P., Andersen, S. O. and Roepstorff, P., 1989 Plasma desorption mass spectrometry as an aid in protein sequence determination. Application of the method on a cuticular protein from the migratory locust, *Biochem. J.* 262:923.

Mann, M., Højrup, P. and Roepstorff, P. Use of mass spectrometric molecular weight information to identify proteins in sequence databases, *Proc. Nat. Acad. Sci. USA*, submitted.

Mikkelsen, J., Højrup, P., Nielsen, P. F., Roepstorff, P. and Knudsen, J., 1987 Amino acid sequence of acyl-CoA-binding protein from cow liver, *Biochem. J.* 245:857.

Palm, L., Kristiansen, K. and Højrup, P. The phosphorylated ribosomal protein in *Tetrahymena thermophila* is homologous to mammalian ribosomal protein S4, in preparation.

Parello, J., Garrod, S., Klarskov, K., Etienne, G., Blancuzzi, Y. and Roepstorff, P. Primary structure of parvalbumin from the leopard shark, *Triakis sanifasciata*, in preparation.

Roepstorff, P., Klarskov, K. and Højrup, P., 1989 Strategy for the use of plasma desorption mass spectrometry in protein sequence analysis, *in*: Methods in Protein Sequence Analysis 1988, Wittmann-Liebold, B., ed., Springer Verlag, Berlin.

Roepstorff, P., Rafn, K., Etienne, G. and Parello, J. Primary structure of parvalbumin 1 from commun carp (*Cyprinus Carpio*), in preparation.

Roepstorff, P., Skou, L., Højrup, P. and Andersen, S. O. Primary structure of protein Lm7 from the exocuticle of *Locusta migratoria*, in preparation.

Sørensen, H. H., Thomsen, J., Bayne, S., Højrup, P. and Roepstorff, P., 1990 Strategies for determination of disulfide bridges in proteins using plasma desorption mass spectrometry, *Biomed. Environm. Mass Spectrom.* 19:713.

Talbo, G. and Roepstorff, P., 1990 Sequence determination of N-acetylated-N, O-permethylated peptides by plasma desorption mass spectrometry, *Biomed. Environm. Mass Spectrom.* 19:589.

Talbo, G., Højrup, P., Rahbek-Nielsen, H., Andersen, S. O. and Roepstorff, P., 1990 Determination of the covalent structure of an N- and C-terminally blocked glycoprotein from endocuticle of *locusta migratoria.* Combined use of plasma desorption mass spectrometry and

Edman degradation to study post-translationally modified proteins, *Eur. J. Biochem.* 195:495.

Unterberg, C., Börchers, T., Højrup, P., Roepstorff, P., Knudsen, J. and Spener, F., 1990 Cardiac fatty acid binding protein. Isolation and characterization of the mitochondrial fatty acid binding protein and its structural relationship with the cytocolic isoforms, *J. Biol. Chem.* 265:16255.

DETERMINATION OF POSTTRANSLATIONAL MODIFICATIONS BY MASS SPECTROMETRY

Toshifumi Takao and Yasutsugu Shimonishi

Institute for Protein Research, Osaka University
Yamadaoka 3-2, Suita, Osaka 565, Japan

INTRODUCTION

Mass spectrometry has become the most promising tool for analysis of the structures of posttranslationally modified peptides and proteins, which is essential for understanding their biological functions or metabolic regulations. The increased value of this method should be ascribed to the development of several soft-ionization techniques. Gas chromatography-mass spectrometry (GC-MS) has also been successfully used for determination of modified structures of proteins (Carr and Biemann, 1984; Farnsworth *et al.*, 1989). However, GC-MS usually requires large amounts of peptides and their chemical derivatization for their vaporization. Chemical treatments may change or destroy unstable functional groups. Recently, for elucidating posttranslational modifications or unknown components in proteins, we and others have measured the accurate masses of peptides and proteins by fast atom bombardment (FAB) (Andrews and Dixon, 1989; Biemann, 1990; Carr and Biemann, 1984; Fukada *et al.*, 1990; Takao *et al.*, 1990), electrospray (ESI) (Jardine, 1990; Kokame *et al.*, 1992; Ohguro *et al.*, 1991) and laser desorption ionization (LDI) (Hillenkamp and Karas, 1990) mass spectrometry, procedures that do not involve any chemical derivatization process. A modification can be determined directly by measurement of the absolute molecular weight only or in combination with other methods such as collision-induced dissociation in tandem mass spectrometry (MS/MS) and chemical synthesis.

In this paper, we describe the determination of novel posttranslational modifications of proteins, such as those of the α- and γ-subunits of the photoreceptor GTP-binding protein (transducin) by FAB and ESI mass spectrometry.

EXPERIMENTAL SECTION

Sample Preparation

The α- and γ-subunits of transducin (Tα and Tγ, respectively) were purified from freshly

Methods in Protein Sequence Analysis, Edited by K. Imahori
and F. Sakiyama, Plenum Press, New York, 1993

excised bovine retinas as described previously (Fukada *et al.*, 1989). Purified Tα and Tγ were digested with *Achromobacter lyticus* protease I (API) (Wako Pure Chemical Industries) and *Staphylococcus aureus* protease V8 (Miles), respectively. The digests were loaded on a reversed-phase C_{18}- or phenyl-column (4.6 x 150 mm; Nacalai Tesque). Fragments were eluted with a linear gradient of acetonitrile in 0.1 % trifluoroacetic acid and their absorbances at 214 and 280 nm were monitored simultaneously. The N-terminal nonapeptide of Tα (Gly-Ala-Gly-Ala-Ser-Ala-Glu-Glu-Lys) and the C-terminal pentapeptide of Tγ (Leu-Lys-Gly-Gly-Cys) were synthesized in an automated peptide synthesizer (9050 PepSynthesizer, Milligen or Model 430A, Applied Biosystems). The former peptide was acylated by treatment with myristoyl or lauroyl succinimide prior to deprotection (Kokame *et al.*, 1992). The latter peptide was *S*-farnesylated at the C-terminal Cys residue by treatment with farnesyl bromide (Kitada *et al.*, 1979). The resulting peptides were purified by reversed-phase HPLC and identified by amino acid analysis and FAB mass spectrometry.

Mass Spectrometry

FAB mass spectra were recorded in a JMS-HX100 double-focusing mass spectrometer equipped with an FAB ion source (JEOL, Ltd.). Peptides (0.1~0.5 nmol) dissolved in water-acetonitrile (1:1) were placed on the tip of a stainless steel probe and mixed with a liquid matrix (a mixture of dithiothreitol (DTT) and dithioerythritol (DTE), 5:1 (w/w)). The samples were bombarded with xenon atoms accelerated by a 6 keV potential. Accurate masses were measured on a rotatable probe with a dual-target at the top (JEOL MS-DTM) as described (Takao *et al.*, 1990). The sample and standard compounds (polyethyleneglycol 400 or 1,540) were analyzed alternately on the probe and their spectra were stored separately in a JMA-DA5000 data acquisition system. Then mass calibration of the sample from the standard peaks was carried out. The elemental composition for the accurate mass was calculated with an error of less than 10 millimass units .

MS/MS spectra were obtained in a JEOL JMS-HX/HX110A four-sector tandem mass spectrometer equipped with a collision cell and a JMA-DA7000 data acquisition system. The precursor ions were selected with a resolving power such that only the monoisotopic ions were transmitted into the collision cell operated at 8 kV potential. The amount of collision gas (argon) introduced into the cell was adjusted to reduce the intensity of the precursor ions by half. Then the daughter ion spectra were acquired on an array detector (10% dispersion) by stepping up an electric and magnetic field keeping the ratio of B_2/E_2 constant in MS2, where the magnetic field (B_2) was calibrated with a salt mixture (Sato *et al.*, 1987).

The ESI mass spectrum was measured in a JEOL JMS-SX102 mass spectrometer equipped with an ESI ion source (Analytica of Branford, CT). Purified Ta was dissolved in 30μl of H_2O/acetic acid/methanol (5:1:4) and infused into the ion source at a flow rate of lμl/min.

Determination of Posttranslational Modifications

The scheme for determination of the structure of a posttranslational modification in a protein by mass spectrometry is shown in Figure 1. Direct measurement of the molecular weight of a protein by ESI-MS may lead to estimation of a posttranslational modification. Furthermore, the structure can be exhaustively analyzed by accurate FAB mass measurement or tandem mass spectrometry of a peptide with a modification derived from the protein. The residual accurate mass of the modification, which is readily calculated by subtracting the mass of the peptide portion identified by amino acid or sequence analysis from the observed accurate mass, provides the possible elemental composition of the modification. Moreover, tandem mass spectrometric analysis of the peptide provides the structure of the modification in more detail.

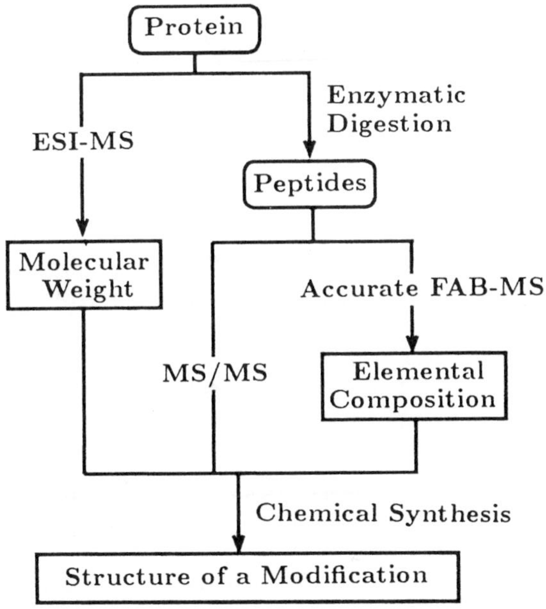

Figure 1. Scheme for determination of a posttranslational modification by mass spectrometry.

Finally, the structure of the modification is confirmed by comparison with that of a chemically synthesized amino acid or peptide including the modification.

RESULTS AND DISCUSSION

Transducin, a GTP-binding protein involved in the visual transduction process in vertebrate photoreceptor cells, consists of subunits α with a GTP-binding site, β and γ, whose primary structures have been studied extensively (Hurley *et al.*, 1984a; Ovchinnikov *et al.*, 1985; Sugimoto *et al.*, 1985; Tanabe *et al.*, 1985). Transducin is associated with the rod outer segment disk membrane and serves as an intermediate between receptor (rhodopsin) and effector (cyclic GMP phosphodiesterase) proteins in the visual transduction process. On interaction with light-activated rhodopsin (metarhodopsin II), GDP-bound Tα changes to a GTP-bound form that can activate cyclic GMP phosphodiesterase until it hydrolyzes its own GTP to a GDP-bound form.

Recently, Fukada *et al.* (1989) found that bovine Tβγ can be separated into two components, biologically inactive (Tβγ-l) and active (Tβγ-2) forms, with different γ-subunits, Tγ-l and Tγ-2, respectively, which can facilitate the interaction of Tα with metarhodopsin II or its localization on the disk membrane. The ESI-MS of intact Tγ (8315.7 or 8330.2 Da) (Ohguro *et al.*, 1991) and the accurate FAB-MS of the C-terminal *S. aureus* protease V8 fragments (681.438 and 695.453) (Fukada *et al.*, 1990) revealed the unique elemental compositions of the modified structures, strongly suggesting farnesylation of the C-terminal cysteine residue and partial methyl-esterification of its α-carboxyl group (Figure 2).

For confirmation of these structures of the substituents on the cysteine residue, we synthesized a farnesylated C-terminal pentapeptide (Leu-Lys-Gly-Gly-Cys) and its methyl-esterified form, and compared the synthetic peptides with the C-terminal V8 fragments of

Tγ by reversed-phase HPLC. The synthetic peptides had the same retention times as the natural ones (data not shown). These results demonstrate that the mature γ-subunit of transducin is modified by a farnesyl group at the C-terminal cysteine residue and that its C-terminal α-carboxyl group is partly methyl-esterified (Fukada *et al.*, 1990).

```
1                                   10                                      20
Pro-Val-Ile-Asn-Ile-Glu-Asp-Leu-Thr-Glu-Lys-Asp-Lys-Leu-Lys-Met-Glu-Val-Asp-Gln-

                                    30                                      40
Leu-Lys-Lys-Glu-Val-Thr-Leu-Glu-Arg-Met-Leu-Val-Ser-Lys-Cys-Cys-Glu-Glu-Phe-Arg-

                                    50                                      60
Asp-Tyr-Val-Glu-Glu-Arg-Ser-Gly-Glu-Asp-Pro-Leu-Val-Lys-Gly-Ile-Pro-Glu-Asp-Lys-

                                    70
Asn-Pro-Phe-Lys-Glu-Leu-Lys-Gly-Gly-Cys-OH (OMe)
```

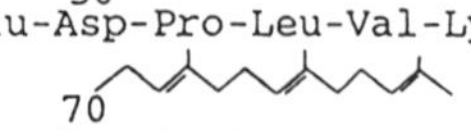

Figure 2. Posttranslational modification of the γ-subunit of photoreceptor GTP-binding protein (transducin). The C-terminal *S. aureus* protease V8 fragment is indicated by a solid bar.

The N-terminal glycine residue of Tα was reported to be blocked by an acyl group other than a myristoyl group (Hurley *et al.*, 1984b), which is a well-known posttranslational modification occurring at the N-terminal glycine and has been found in several kinds of proteins (Aitken *et al.*, 1982; Carr *et al.*, 1982; Chow *et al.*, 1987; Henderson *et al.*, 1983; Marchildon *et al.*, 1984; Ozols *et al.*, 1984; Schultz *et al.*, 1985) including the α-subunits of some G-proteins (Buss *et al.*, 1987; Schultz *et al.*, 1987).

ESI-MS of bovine Tα showed multiply-charged ion signals (Figure 3), from which the molecular weight of Tα was calculated to be 40,049 Da (SD: 7.3). The molecular mass was 215 Da higher than that deduced from the cDNA (Tanabe *et al.*, 1985) (39,834). Since the N-terminal amino acid of Tα was blocked, this finding suggested that the N-terminus had

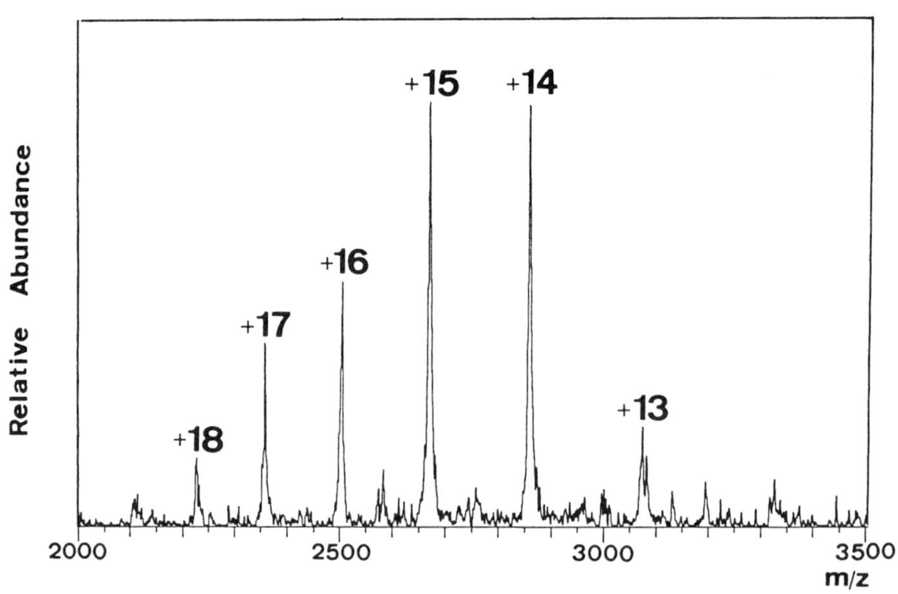

Figure 3. ESI mass spectrum of intact Tα (Kokame *et al.*, 1992).

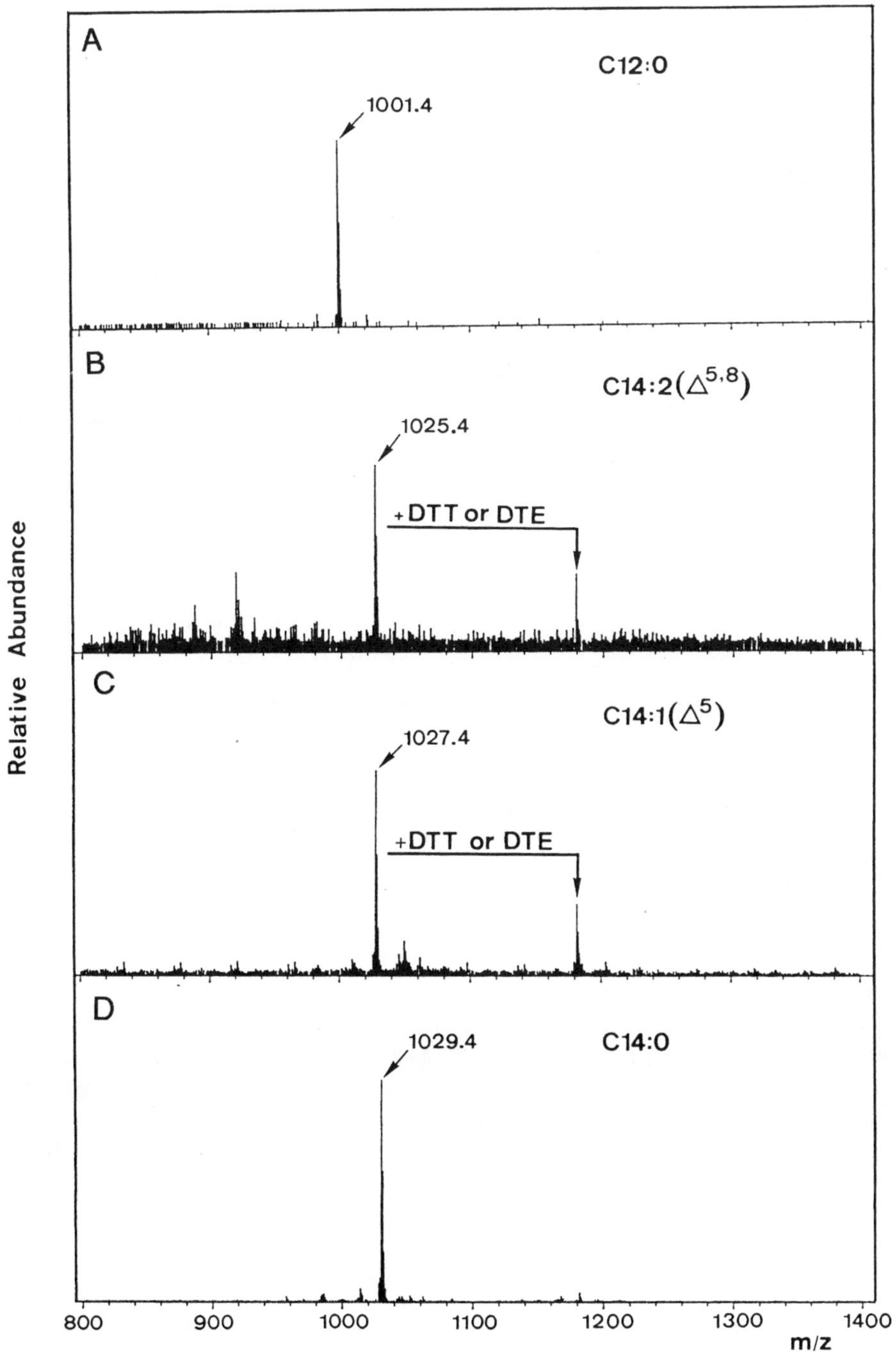

Figure 4. FAB mass spectra of the N-terminal peptides with heterogeneous lipid modifications isolated from the *A. lyticus* protease I digest of Tα.

undergone some unknown modification, the mass of which nearly agreed with that (211.2) of a myristoyl group. For elucidation of the N-terminal structure, Tα was digested with API and the proteolytic fragments isolated by reversed-phase HPLC (data not shown) were subjected to amino acid analyses and FAB-MS. The result unexpectedly revealed four kinds of peptides, which had the same amino acid compositions corresponding to that of the N-terminal API peptide (Gly-Ala-Gly-Ala-Ser-Ala-Glu-Glu-Lys) of Tα and whose FAB mass spectra showed intense signals at m/z=1001.4, 1025.4, 1027.4 and 1029.4 (Figures 4A-4D). In addition, the ion signals of the adducts of dithiothreitol or dithioerythritol used as a matrix were observed to be ascribed to the presence of olefinic double bonds (Figures 4B and 4C). These values, together with the mass (818.4) calculated from the sequence of the N-terminal nonapeptide, predicted four kinds of fatty acids, C12:0, C14:2, C14:1 and C14:0. Moreover, all the peptides were refractory to Edman degradation like intact Tα, indicating that these fatty acids are linked via an amide bond to the N-terminal glycine of the nonapeptide.

To verify the structure, we chemically synthesized the N-lauroylated (C12:0) and N-myristoylated (C14:0) nonapeptides of Tα and compared them with the N-terminal API peptides. These synthetic peptides were identified with the API fragments with masses of 1001.4 and 1029.4, respectively.

For identification of positions of the double bonds in C14:2 and C14:1, the N-terminal API peptides were completely digested with carboxypeptidase Y to yield the acyl glycine. Negative-ion FAB tandem mass spectrometric analyses of the acyl glycine in the digests (C14:2-Gly and C14:1-Gly) indicated highly specific fragmentations (Tomer *et al.*, 1983) which made it possible to assign the locations of the double bonds, (Δ^5, Δ^8) and (Δ^5) in C14:2 and C14:1, respectively (Takao *et al.*, in preparation). These results demonstrate that the N-terminal Gly of the mature Tα is modified by heterogeneous fatty acids consisting of laurate (C12:0), myristate (C14:0) and unsaturated C14:1 and C14:2.

The modified structures determined in the present study, the C-terminal farnesylation of Tγ and the N-terminal lipid modification of Tα, are presently assumed to mediate formation of the Tα/Tβγ complex or its membrane localization (Fukada *et al.*, 1990; Fukuda *et al.*, 1992; Kokame *et al.*, 1992; Ohguro *et al.*, 1991) and should be helpful in understanding the structure-function or structure-protein recognition characteristics of the visual transduction process.

CONCLUSION

Novel posttranslational modifications of the photoreceptor GTP-binding protein, which are essential for biological activity in a visual transduction process, could be determined by mass spectrometry. The present study demonstrates that soft-ionization mass spectrometry can easily provide clear information on the mass of a posttranslational modification, and that this method is more efficient for elucidation of the structure when used in combination with accurate FAB mass measurement or tandem mass spectrometric analysis.

Acknowledgements. We are grateful to Dr. Toru Yoshizawa (Department of Applied Physics and Chemistry, University of Electro-Communications), Dr. Yoshitaka Fukada and Mr. Koichi Kokame (Department of Biophysics, Faculty of Science, Kyoto University) and Drs. Toyoaki Akino and Hiroshi Ohguro (Department of Biochemistry, Sapporo Medical College) for their cooperation. This work was partly supported by a Grant-in-Aid from the Ministry of Education, Science and Culture of Japan.

REFERENCES

Aitken, A., Cohen, P., Santikarn, S., Williams, D. H., Calder, A. G., Smith, A. and Klee, C. B., 1982 Identification of the NH_2-terminal blocking group of calcineurin B as myristic acid, *FEBS Lett.* 150:314.

Andrews, P. C. and Dixon, J. E., 1989 Application of fast atom bombardment mass spectrometry to posttranslational modifications of neuropeptides, *Methods Enzymol.* 168:72.

Biemann, K., 1990 Utility of extract mass measurements, *Methods Enzymol.* 193:295.

Buss, J. E., Mumby, S. M., Casey, P. J., Gilman, A. G. and Sefton, B. M., 1987 Myristoylated α subunits of guanine nucleotide-binding regulatory proteins, *Proc. Natl. Acad. Sci. USA* 84:7493.

Carr, S. A. and Biemann, K., 1984 Identification of posttranslationally modified amino acids in proteins by mass spectrometry, *Methods Enzymol.* 106:29.

Carr, S. A., Biemann, K., Shoji, S., Parmelee, D. C. and Titani, K., 1982 *n*-Tetradecanoyl is the NH_2-terminal blocking group of the catalytic subunit of cyclic AMP-dependent protein kinase from bovine cardiac muscle, *Proc. Natl. Acad. Sci. USA* 79:6128.

Chow, M., Newman, J. F. E., Filman, D., Hogle, J. M., Rowlands, D. J. and Brown, F., 1987 Myristylation of picornavirus capsid protein VP4 and its structural significance, *Nature* 327:482.

Farnsworth, C. C., Wolda, S., Gelb, M. H. and Glomset, J. A., 1989 Human lamin B contains a farnesylated cysteine residue, *J. Biol. Chem.* 264:20422.

Fukada, Y., Ohguro, H., Saito, T., Yoshizawa, T. and Akino, T., 1989 βγ-Subunit of bovine transducin composed of two components with distinctive γ-subunits, *J. Biol. Chem.* 264:5937.

Fukada, Y., Takao, T., Ohguro, H., Yoshizawa, T., Akino, T. and Shimonishi, Y., 1990 Farnesylated γ-subunit of photoreceptor G protein indispensable for GTP-binding, *Nature* 346:658.

Fukuda, M., Takao, T. Hidaka, Y. Matsuda, T., Fukada, Y., Akino, T., Yoshizawa, T. and Shimonishi, Y., 1992 Synthesis and biological function of the C-terminal farnesylated peptide of the γ-subunit of photoreceptor GTP-binding protein, *Peptide Chemistry* 1991, 259, Suzuki, A., ed., Protein Research Foundation, Osaka.

Henderson, L. E., Krutzsch, H. C. and Oroszlan, S., 1983 Myristyl amino-terminal acylation of murine retrovirus proteins: An unusual posttranslational protein modification, *Proc. Natl. Acad. Sci. USA* 80:339.

Hillenkamp, F. and Karas, M., 1990 Mass spectrometry of peptides and proteins by matrix-assisted ultraviolet laser desorption/ionization, *Methods Enzymol.* 193:280.

Hurley, J. B., Fong, H. K. W., Teplow, D. B., Dreyer, W. J. and Simon, M. I., 1984a Isolation and characterization of a cDNA clone for the γ subunit of bovine retinal transducin, *Proc. Natl. Acad. Sci. USA* 81:6948.

Hurley, J. B., Simon, M. I., Teplow, D. B., Robishaw, J. D. and Gilman, A. G., 1984b Homologies between signal transducing G proteins and *ras* gene products, *Science* 226:860.

Jardine, I., 1990 Molecular weight analysis of proteins, *Methods Enzymol.* 193:441.

Kitada, C., Fujino, M., Kamiya, Y., Sakurai, A., Tamura, S., Takahashi, N., Tsuchiya, E., Abe, K. and Fukui, S., 1979 Synthesis of rhodotorucine A, the inducing factor of mating tube formation of *Rhodosporidium toruloides*, *Experientia* 35:1275.

Kokame, K., Fukada, Y., Yoshizawa, T., Takao, T. and Shimonishi, Y., 1992 Lipid modification at the N terminus of photoreceptor G-protein α-subunit, *Nature* 359:749.

Marchildon, G. A., Casnellie, J. E., Walsh, K. A. and Krebs, E. G., 1984 Covalently bound myristate in a lymphoma tyrosine proten kinase, *Proc. Natl. Acad. Sci. USA* 81:7679.

Ohguro, H., Fukada, Y., Takao, T., Shimonishi, Y., Yoshizawa, T. and Akino, T., 1991 Carboxyl methylation and farnesylation of transducin γ-subunit synergistically enhance its coupling with metarhodopsin II, *EMBO J.* 10:3669.

Ovchinnikov, Y. A., Liplin, V. M., Shuvaeva, T. M., Bogachuk, A. P. and Shemyakin, V. V., 1985 Complete amino acid sequence of γ-subunit of the GTP-binding protein from cattle retina, *FEBS Lett.* 179:107.

Ozols, J., Carr, S. A. and Strittmatter, P. J., 1984 Identification of the NH_2-terminal blocking group of NADH-cytochrome b_5 reductase as myristic acid and the complete amino acid sequence of the membrane-binding domain, *J. Biol. Chem.* 259:13349.

Sato, K., Asada, T., Ishihara, M., Kunihiro, F., Kammei, Y., Kubota, E., Costello, C. E., Martin, S. A., Scoble, H. A. and Biemann, K., 1987 High-performance tandem mass spectrometry: calibration and performance of linked scans of a four-sector instrument, *Anal. Chem.* 59:1652.

Schultz, A. M., Henderson, L. E., Oroszlan, S., Garber, E. A. and Hanafusa, H., 1985 Amino terminal myristylation of the protein kinase p60*src*, a retroviral transforming protein, *Science*

227:427.

Schultz, A. M., Tsai, S.-C., Kung, H.-F., Oroszlan, S., Moss, J. and Vaughan, M., 1987 Hydroxylamine-stable covalent linkage of myristic acid in Goα, a guanine nucleotide-binding protein of bovine brain, *Biochem. Biophys. Res. Commun.* 146:1234.

Sugimoto, K., Nukada, T., Tanabe, T., Takahashi, H., Noda, M., Minamino, N., Kangawa, K., Matsuo, H., Hirose, T., Inayama, S. and Numa, S., 1985 Primary structure of the β-subunit of bovine transducin deduced from the cDNA sequence, *FEBS Lett.* 191:235.

Takao, T., Kokame, K., Fukada, Y., Yoshizawa, T. and Shimonishi, Y. in preparation.

Takao, T., Yoshino, K., Suzuki, N. and Shimonishi, Y., 1990 Analysis of post-translational modifications of proteins by accurate mass measurement in fast atom bombardment mass spectrometry, *Biomed. Environ. Mass Spectrom.* 19:705 .

Tanabe, T., Nukada, T., Nishikawa, Y., Sugimoto, K., Suzuki, H., Takahashi, H., Noda, M., Haga, T., Ichiyama, A., Kangawa, K., Minamino, N., Matsuo, H. and Numa, S., 1985 Primary structure of the α-subunit of transducin and its relationship to *ras* proteins, *Nature* 315:242.

Tomer, K. B., Crow, F. W. and Gross, M. L., 1983 Location of double bond position in unsaturated fatty acids by negative ion MS/MS, *J. Am. Chem. Soc.* 105:5487.

POST-TRANSLATIONAL MODIFICATIONS

POST-TRANSLATIONAL MODIFICATIONS OF PROTEINS

Radha G. Krishna and Finn Wold

Department of Biochemistry and Molecular Biology
University of Texas Medical School
Houston, Texas 77225, U.S.A.

INTRODUCTION

The purpose of this chapter is to attempt to compile a list of the amino acid derivatives known to exist in proteins in a manner that may be of use to protein chemists concerned with protein sequencing and with the prediction and elucidation of the complete covalent structure of proteins. With the many reviews of the area of post-translational modifications of proteins written in the course of the last 15 years (Alix and Hays, 1983; Krishna and Wold, 1992; 1993; Uy and Wold, 1977; Whitaker, 1977; Wold, 1981; 1983), there is little justification for just another review; however, with the rapidly increasing use of mass spectrometry in the elucidation of protein structure (Biemann, 1992), and with mass spectrometry as the most obvious tool toward recognizing and identifying unusual amino acids in proteins, it has been suggested that a list of most of the known derivatives along with their molecular masses might be of some use, especially to the individuals at this meeting.

The main content of this chapter consequently. is a list of known post-translational derivatives with the associated mass changes involved in converting the primary amino acids to the secondary amino acid derivatives. As expected, the attempt to fully document the derivatives in the present format became impractical; the references alone would have exceeded the space allotment by a factor of 2. To obtain the needed documentation of individual derivatives, interested readers are referred to past reviews (Alix and Hayes, 1983; Krishna and Wold, 1993; Uy and Wold, 1977; Whitaker, 1977; Wold, 1981), especially the most recent one (Krishna and Wold, 1993), which contains the titles of the individual reports.

MASS CHANGES ASSOCIATED WITH POST-TRANSLATIONAL MODIFICATIONS

A current listing of secondary amino acids is presented in Table 1 with its footnotes. Part A of the table lists the various derivatives with their respective mass contributions,

Methods in Protein Sequence Analysis, Edited by K. Imahori
and F. Sakiyama, Plenum Press, New York, 1993

Table-1. A. Mass changes associated with post-translational modifications in proteins*.

THE N-TERMINUS: H_2N–

N^{α}1. N-Formyl-	28
N^{α}2. N-Acetyl-	42
N^{α}3. N-Myristoyl-	210
N^{α}4. N-Aminoacyl-[a]	
N^{α}5. N-α-Ketoacyl- (See S9)[a]	
N^{α}6. N-Methyl-	14
N^{α}7. N-Pyrrolidone carboxyl-	-17
N^{α}8. N-Glucuronyl-	176
N^{α}9. N-Glucosyl-[a,b] (Aminoketose)	162

CYSTEINE: HS–CH_2– [103]

C1. Cystine[c]	-2
C2. Selenocysteine[#] (See S10)	
C3. S-γ-Glutamyl-[c]	-18
C4. S-Palmitoyl-[a]	238
C5. S-(Hexosyl)$_{2\text{-}3}$[a]	324 - 486
C6. S-(sn-1-Glyceryl)-	74
C7. S-(sn-1-Dipalmitoyl-glyceryl)-[a]	524
C8. S-(2-Histidyl)-[c]	-2
C9. S-Heme[a]	617
C10. S-Phycocyanobilin[a]	587
C11. S-(6-Flavin [FAD])-	784
C12. S-Coenzyme A	454
C13. S-Farnesyl-	206
C14. S-Geranylgeranyl-	276
C15. S-(3-Tyr)[c]	-2
C16. Dehydroalanine	-2
C17. Lysinoalanine	-2
C18. Lanthionine	-34
C19. S-(ADP-ribosyl)-	541

ARGININE: H_2N–C(=NH)–NH–CH_2–CH_2–CH_2– (ω) [156]

R1. N-(ADP-ribosyl)-	541
R2. N-Methyl-	14
R3. N^{ω}-Dimethyl-	28
R4. N^{ω}, $N^{\omega'}$-Dimethyl-	28
R5. Ornithine	-42
R6. Citrulline	1

HISTIDINE: 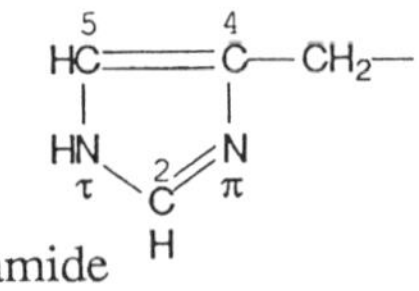 [137]

H1. Diphthamide	107
H2. N^{τ}-(ADP-ribosyl) diphthamide-	648
H3. N-Phosphoryl-	80
H4. N^{π}-Methyl-	14
H5. 4-Iodo-	126
H6. N^{τ}- and N^{π}-(8α-Flavin)	784

LYSINE: H_2N–CH_2(ε)–CH_2(δ)–CH_2–CH_2– [128]

K1. N-Acetyl-	42
K2. N-Lipoyl-	188
K3. N-Biotinyl-	226
K4. N-Ubiquitinyl-[d]	
K5. N-Phosphoryl-	80
K6. N-Pyridoxyl-	229
K7. N-Methyl-	14
K8. N-Dimethyl-	28
K9. N-Trimethyl-	42
K10. N-(4-NH_2-2-OH-butyl)- (Hypusine)	87
K11. N-Glucosyl-[a,b] (Aminoketose)	162
K12. Allysine	-1
K13. δ-Hydroxy-	16
K14. δ-Glycosyloxy-[a,b]	177
K15. δ-Hydroxy-allysine	15
K16. Desmosine[c]	-58
K17. Syndesine[c]	13

SERINE: HO–CH_2– [87]

S1. O-Phosphoryl-	80
S2. O-Acetyl-	42
S3. O-pantetheinephosphoryl-	324
S4. O-γ-Glutamyl-[c]	-18
S5. O-Methyl-	14
S6. O-Glycosyl-[a,b]	162
S7. O-(GlcNAc-1-phosphoryl)-	266
S8. Alanino (τ- or π-histidine)[c]	-2
S9. Pyruvoyl-	16
S10. Selenocysteine[#]	64

Table-1A (continued)

THREONINE: 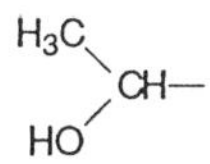	[101]
T1. O-Phosphoryl-	80
T2. O-Glycosyl-[a,b]	162
T3. O-Methyl-	14
METHIONINE: $H_3C-S-CH_2-CH_2-$	[131]
M1. Sulfoxide	16
PHENYLALANINE: 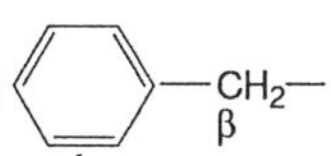	[147]
F1. β-Glycosyloxy [a,b]	177
TYROSINE: 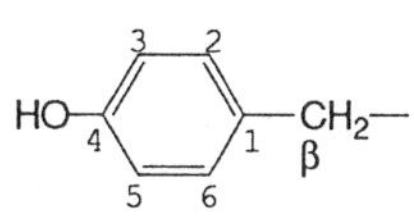	[163]
Y1. 3-Iodo-	126
Y2. 3-Bromo-[e]	78, 80
Y3 3-Chloro-[e]	34, 36
Y4. 3,5-Diiodo-	252
Y5. 3,5-Dibromo-[e]	156, 158, 160
Y6. 3,5-Dichloro-[e]	68, 70, 72
Y7. 3,5,3'-Triiodothyronine	470
Y8. 3,5,3',5'-Tetraiodothyronine	596
Y9. 3,4-Dihydroxy-Phe (DOPA)	16
Y10. 3,4,6-Trihydroxy-Phe (TOPA)	32
Y11. O-Phosphoryl-	80
Y12. O-Sulfonyl-	80
Y13. O-Adenylyl-	329
Y14. O-Uridylyl-	306
Y15. O-(8α-Flavin [FAD])-	783
Y16. β-Glycosyloxy-[a,b]	177
Y17. 3,3'-Bityr[c]	-2
Y18. 3,3';5,5'-TerTyr[c]	-4
Y19. IsodiTyr[c]	-2
TRYPTOPHAN: 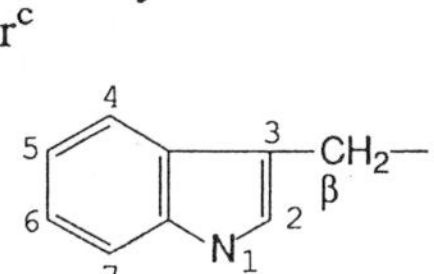	[186]
W1. β-Hydroxy-	16
W2. 6,7-Dione	30
W3. 2,4-BisTrp-6,7-dione[c]	28
PROLINE: 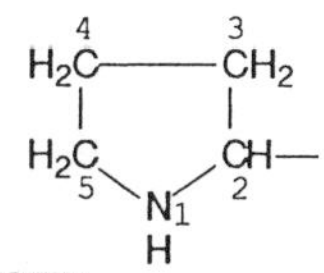	[97]
P1. 3-Hydroxy-	16
P2. 4-Hydroxy-	16
P3. 3,4-Dihydroxy-	32
P4. 4-Glycosyloxy- (pentosyl, C_5)[a,b]	147
P5. 4-Glycosyloxy- (hexosyl, C_6)[a,b]	177
GLUTAMINE: $H_2N-C(=O)-CH_2(\gamma)-CH_2-$	[128]
Q1. Glutamate	1
Q2. N^{ε}-(γ-Glutamyl)-lysine [c]	-17
Q3. N^{α}-(γ-Glutamyl)-Glu	130
Q4. N^{α}-(γ-Glutamyl)-Glu_2	259
Q5. N^{α}-(γ-Glutamyl)-Glu_3	388
ASPARAGINE: $H_2N-C(=O)-CH_2(\beta)-$	[114]
N1. Aspartate	1
N2. N-Methyl-	14
N3. N-Glycosyl-[a,b]	892 - 2770
N4. N^{ε}-(β-Aspartyl)-lysine[c]	-17
GLUTAMATE: $HO-C(=O)-CH_2(\gamma)-CH_2-$	[129]
E1. O-(ADP-ribosyl)-	541
E2. γ-Carboxy-	44
E3. O-Methyl-	14
ASPARTATE: $HO-C(=O)-CH_2(\beta)-$	[115]
D1. β-Carboxy-	44
D2. β-Hydroxy-	16
D3. O-Phosphoryl-	80
THE C-TERMINUS: $-C(=O)-OH$	
C^{α}1. O-Methyl-	14
C^{α}2. O-(ADP-ribosyl)-	541
C^{α}3. -Amide	-1
C^{α}4. -(N^{α}-Tyr)	163
C^{α}5. -(N-Ethanolamine-glycan-phosphoinositides)[a]	>1907

Table-1. B. Mass changes from A. listed in increasing order.

Mass Change	Compounds
-58	K16
-42	R5
-34	C19
-18	C3,S4
-17	N$^{\alpha}$7,Q2,N4
-4	Y18
-2	C1,C8,C16,C17,C18,S8,Y17,Y19
-1	K10,C$^{\alpha}$3
1	Q1,N1,R6
13	K17
14	N$^{\alpha}$6,S5,H4,R2,K7,T3,N2,E3,C$^{\alpha}$1
15	K15
16	K13,W1,P1,P2,D2,S9,M1,Y9
28	N$^{\alpha}$1,K8,R3,R4,W3
30	W2
32	P3,Y10
34 (+36)	Y3
42	N$^{\alpha}$2,S2,K1,K9
44	E2,D1
64 (+62 & more)	S10
68 (+70, 72)	Y6
74	C6
78 (+80)	Y2
80	S1,H3,K5,T1,Y11,Y12,D3
107	H1
126	Y1,H5
130	Q3
147	P4
156 (+158, 160)	Y5

Mass Change	Compounds
162	N$^{\alpha}$9,S6,T2,K11,
163	C$^{\alpha}$4
176	N$^{\alpha}$8
177	F1,Y16,P5,K14,
188	K2
206	C14
210	N$^{\alpha}$3
226	K3
229	K6
238	C4
252	Y4
259	Q4
266	S7
276	C15
306	Y14
324	S3,C5
329	Y13
388	Q5
454	C12,C13
470	Y7
486	C5
524	C7
541	C20,H2,R1,E1,C$^{\alpha}$2
587	C10
596	Y8
617	C9
648	H2
784	C11,Y15,H6
892-2770	N3
1907	C$^{\alpha}$5

*The reported mass changes represent the mass of the radical(s) added or removed in forming the derivatized amino acid. The residue mass for each of the primary amino acids is given [in brackets]. To illustrate: the protein-bound serine residue (87) is modified by the dihydrogenphosphoryl radical (80) to yield protein-bound dihydrogenphospho-Ser (167); upon hydrolysis of the peptide bonds, the free phospho-Ser (185) is obtained. In cases where two or more residues are involved in the reaction, all the partners must be considered: 2 Cys residues (2 x 103) give up 2 H (-2) to yield a cystine residue (204) which after hydrolysis becomes free cystine (2 x 18 +204 = 240); 1 Gln residue (128) reacts with 1 Lys residue (128) to release NH_3 (-17) and form γ-glutamyl-Lys (128 +128 - 17 = 239) and after hydrolysis (275). For all compounds except some halogen derivatives (see [e]), the reported values are derived from the average atomic mass, and the final sum has been rounded off to the nearest integer. A number of derivatives (e.g. lanthionine, citrulline, syndesine, etc.) have trivial names that do not give any information about their structure and origin; their mass changes should nevertheless serve as a first indication of their possible existance.

Table-1 (continued)

#Based on recent studies (Stadtman, 1991) it seems clear that selenocysteine needs no longer be considered as a secondary, derivatized amino acid. With both a specific codon and tRNA, it should now probably be accepted as the 21st primary, genetically encoded amino acid.

[a] There are several possible derivatives in this category.

[b] The glycosylated amino acid derivatives are particularly complex to record because of the heterogeneity that may exist at each glycosylation site. In the table this problem has been handled in different ways: the common O- and N-linked glycans are reported as a mass range selected to cover the majority of known derivatives; in the case of the glycosyloxy derivatives of Pro, Lys, Tyr and Phe the mass of a single hexosyloxy radical (177), and in the case of Pro, also of a single pentosyloxy radical (147) has been listed. It should be emphasized that more than one sugar may well be present in all these derivatives. In matching experimental mass values to the ranges given, it is a matter of assembling the appropriate composition of radical masses for hexoses (162), N-acetylhexosamines (203), deoxyhexoses (146), pentoses (132) and sialic acids (291). Thus a typical O-linked glycan from yeast, Man_2-[Ser/Thr], would have a radical mass of (2 x 162) = 324, and one from animals, Sia-Gal-GalNAc-[Ser/Thr], a mass of 291 + 162 + 203 = 656. Similarly, a typical N-linked oligomannose structure, Man_6-$GlcNAc_2$-[Asn], would be (6 x 162) + (2 x 203) = 1378, a hybrid structure, $GlcNAc_2$--Man_5-$GlcNAc_2$--[Asn], (5 x 162) + (4 x 203) = 1622, and a complex structure, Sia_3-Gal_3-$GlcNAc_3$-Man_3-(Fuc-$GlcNAc_2$-)-[Asn], (3 x 291) + (6 x 162) + (5 x 203) + 146 = 3006. A number of structures from proteoglycans containing hexuronic acids and sulfated sugars have not been included in establishing the mass-ranges in the table.

[c] Two or more amino acid residues are involved in these derivatives, formed by condensation or oxidative coupling, in which small molecules like H_2O, NH_3, H_2 are eliminated to yield a product mass less than the sum of the masses of the starting material. Since these derivatives constitute crosslinks, their characterization is likely to start with a set of covalently linked peptides, confounding the sequencing results in a very characteristic and usually recognizable manner.

[d] No mass has been listed for ubiquitinated Lys. The mass of the ubiquitin radical is 8566, and an intact ubiquitinated protein should have a mass increase of 8548, 17096, 25644 etc. depending on how many ubiquitin units are attached. In isolating the derivatized peptide or amino acid, however, the ubiquitin moiety will be just as sensitive to proteolysis and chemical degradation as the modified protein, and a variety of derivatives might be expected from different treatments.

[e] In the case of Br- and Cl-derivatives, masses have been reported accounting for the two most abundant isotopes of each, ^{79}Br, ^{81}Br, ^{35}Cl and ^{37}Cl.

and part B lists the mass changes in increasing order and identifies the derivatives in part A associated with each mass change. The various footnotes are included to illustrate the rationale for the presentation and justify the shortcuts and simplifications needed to compile the data in this particular form. It should be emphasized that compilations like these are never complete; some derivatives are inadvertently omitted, and new ones appear regularly.

In presenting just the list of derivatives, the most exciting features of post-translational modification reactions have been totally omitted from consideration. Thus, the very complex and intriguing questions of the biological functions of the different derivatives might have been discussed in great length, in spite of the fact that for many of the derivatives we do not really have any clear reasons for their existence. Similarly, the exquisite specificity by which a single residue or a single peptide bond may be selected for modification reactions that do not affect any of the thousands of other similar structures is well worthy of serious consideration both as an illustration of the nearly infinite number of unique structural features that can be expected in proteins and as a means of determining the specificity features that will permit us to predict that a given modification will take place. With the steadily increasing number of protein sequences deduced from cDNA sequences, it will clearly be of interest to be able to identify the modification sites even before the protein itself has been purified. At present it is not possible to make such predictions; consensus structures have been established that allow one to distinguish between structures that may be modified and those that will definitely not be, but they do not permit one to unequivocally state that a given structure will be modified. Most of these points have been discussed in past reviews and are summarized in the most recent one.

Acknowledgements: We are grateful to Dr. Kenneth A. Walsh for making the suggestion of the format of this paper. Aspects of the work discussed were supported by grants from the USPHS (GM 31305) and the Robert A. Welch Foundation (AU-0916 and AU-0009).

REFERENCES

Alix, J.-H. and Hayes, D., 1983 Why are macromolecules modified post-synthetically?, *Biol. Cell* 47:139.

Biemann, K., 1992 Mass spectrometry of peptides and proteins. *Ann. Rev. Biochem.* 61:977.

Krishna, R. G. and Wold, F., 1992 Post-translational modifications: unique amino acids in proteins, *in*: Frontiers and New Horizons in Amino Acid Research, Takai, K., ed., Elsevier, Amsterdam.

Krishna, R. G. and Wold, F., 1993 Post-translational modification of proteins. *Adv. Enzymology and Related Areas of Mol. Biol.* 67:in press.

Stadtman, T. C., 1991 Biosynthesis and functioning of selenocysteine-containing enzymes, *J. Biol. Chem.* 266:16257.

Uy, R. and Wold, F., 1977 Posttranslational covalent modification of proteins, *Science* 198:890.

Whitaker, J. R., 1977 *in*: Food Proteins, Improvement through Chemical and Enzymatic Modification, Feeney, R. E. and Whitaker, J. R., eds., Am. Chem. Soc., Washington DC.

Wold, F., 1981 *In vivo* chemical modification of proteins, *Ann. Rev. Biochem.* 50:783.

Wold, F., 1983 Posttranslational protein modifications: perspectives and prospectives, *in*: Posttranslational Covalent Modifications of Proteins for Function, Johnson, C. B., ed., Academic Press, New York.

N-TERMINAL ACETYLATION OF MUTATIONALLY ALTERED FORM OF ISO-1-CYTOCHROMES *c* IN NORMAL AND *nat1*⁻ STRAINS DEFICIENT IN THE MAJOR N-TERMINAL ACETYL TRANSFERASE OF THE YEAST *SACCHAROMYCES CEREVISIAE*

Fred Sherman[1,2], Richard P. Moerschell[2,3], Susumu Tsunasawa[3], and Rolf Sternglanz[4]

Departments of [1]Biochemistry and [2]Biophysics, School of Medicine and Dentistry, Rochester, New York 14642
[3]Institute for Protein Research, Osaka University, Suita, Osaka 565, Japan
[4]Department of Biochemistry, State University of New York, Stony Brook, New York 11794

INTRODUCTION

The two co-translational processes, cleavage of N-terminal methionine and N-terminal acylation, are by far the most common modification events, affecting nearly all proteins. Protein synthesis initiates with formylmethionine in prokaryotes, mitochondria and chloroplasts, whereas proteins initiate with methionine in the cytosol of eukaryotes. The formyl group is removed from prokaryotic proteins by a deformylase, resulting in methionine at N-termini. The N-terminal methionine is cleaved from the nascent chains of most prokaryotic and eukaryotic proteins by methionine aminopeptidase. Subsequently, N-terminal acetylation can occur on certain proteins, either containing or lacking the initiator methionine. This N-terminal acetylation occurs on over one-half of eukaryotic proteins but only rarely on prokaryotic proteins (for reviews see Driessen *et al.*, 1985; Kendall *et al.*, 1990)

N-terminal acetylation of the majority of eukarotic proteins occurs co-translationally, usually at the site where there are between 20 and 50 residues extruding from the ribosome. Less often, acetylation occurs after translation is completed on N-termini of special proteins such as actins from some but not all organisms (Sheff and Rubenstein, 1989; Cook *et al.*, 1991); or at internal sites after specific proteolytic processing, as in cases of peptide hormones (Symth and Zakariam, 1980; Chappell *et al.*, 1986; Glembotski, 1982a,b). Apparently, co-translational and post-translational acetylation involve different sets of

Methods in Protein Sequence Analysis, Edited by K. Imahori and F. Sakiyama, Plenum Press, New York, 1993

acetyltransferases having different specificities. Also, post-translational acetylation of each protein or group of peptides involves a different acetylation system.

THE ISO-1-CYTOCHROME *c* SYSTEM

We have systematically investigated N-terminal processing with the iso-1-cytochromes *c* system of the yeast *Saccharomyces cerevisiae*. The nuclear gene, *CYC1*, encoding the mitochondrial protein iso-1-cytochrome *c* comprises one of the most thoroughly studied gene-protein systems of eukaryotes. We wish to emphasize that, in contrast to the import

Table 1. Examples of the creation of altered iso-l-cytochromes *c* by transforming the *cyc1-31* strain with synthetic oligonucleotides and selecting for functional transformants.

Met cleavage	Acetylation	Sequence	Allele
		-1 1 2 3 4 5 6 (Met)**Thr**-Glu-Phe-Lys-Ala-Gly- ATA ATG ACT GAA TTC AAG GCC GGT	*CYC1*$^+$
		ATA ATG ACT GAA TA- AAG GCC GGT	*cyc1-31*
0	0	ATA ATA ATG TTG TTC TTG GCC GGT Met-**Leu**-Phe-Leu-Ala-Gly-	*CYC1-850*
+	0	ATA ATA ATG GGT TTC TTG GCC GGT (Met)**Gly**-Phe-Leu-Ala-Gly-	*CYC1-841*
0	+	ATA ATA ATG GAA TTC TTG GCC GGT Ac-Met-**Glu**-Phe-Leu-Ala-Gly-	*CYC1-853*
+	+	ATA ATG TCT GAA TTC TTG GCC GGT (Met)**Ser**-Glu-Phe-Leu-Ala-Gly- Ac	*CYC1-793*

Amino acid sequences of the N-terminal region of the iso-l-cytochromes *c* are presented along with the corresponding DNA sequences of *CYC1* alleles. The penultimate residues are denoted in boldface. The *cyc1-31* mutant completely lacks iso-l-cytochrome *c* because of the frameshift and TAA nonsense mutation altered iso-l-cytochromes *c* with four types of amino termini are illustrated, without (0) and with (+) cleavage of the N-terminal methione and without (0) and with (+) N-terminal acetylation. (Adapted from Moerschell *et al.*, 1990)

of most if not all other mitochondrial proteins, the import of cytochrome *c* does not involve an N-terminal signal and does not require a membrane potential as an energy source. Because the N-terminal region of iso-l-cytochrome *c* is dispensable for its biosynthesis and function, N-terminal processing can be freely investigated with essentially any alteration. Thus, altered forms of iso-l-cytochrome *c* proved to be ideally suited for investigating the specificity of methionine aminopeptidase and N-terminal acetylation.

Over the years, numerous altered forms of iso-l-cytochrome *c* were generated in various studies (Tsunasawa *et al.*, 1985). Of more importance, we have described a convenient procedure for producing specific alteration of genomic DNA by transforming yeast directly with synthetic oligonucleotides (Moerschell *et al.*, 1988, 1991; Yamamoto *et al.*, 1992). This procedure is easily carried out by transforming a defective *cyc1* mutant and selecting for revertants that are at least partially functional. The oligonucleotide used for transformation contains a sequence that corrects the defect and produces additional alterations at nearby sites. This technique ideally suited for producing a large number of specific alterations that change a completely nonfunctional allele to at least a partially functional form. The selection procedure used with *cycl* mutants allows recovery of altered iso-l-cytochromes *c* with even less than 1% of the normal activity. This procedure is illustrated with the *cycl-31* mutant (Table l) that has been used to construct various single and multiple amino acid alterations at the N-terminal region of iso-l-cytochrome *c* (Moerschell *et al.*, 1990).

N-TERMINAL ACETYLATED ISO-l-CYTOCHROMES *c*

Proteins susceptible to N-terminal acetylation have a variety of different N-terminal sequences, with no simple consensus motif, although serine, alanine and methionine are the most common N-termini residues (Driessen *et al.*, 1985). Attempts to decipher the rules by which eukaryotic proteins are co-translationally acetylated have involved examination of published amino acid sequences (Spencer, 1984; Driessen *et al.*, 1985; Persson *et al.*, 1985), systematic studies of altered proteins *in vivo*, especially iso-l-cytochrome *c* (Tsunasawa *et al.*, 1985; Moerschell *et al.*, 1990), and systematic studies of synthetic peptides *in vitro*,

Table 2. N-terminal acetylated and related nonacetylated forms of altered iso-l-cytochromes *c*.

Ac-Ser-Glu-Phe-Leu-	Ac-Met-Ile-Arg-Ile-	Ac-Met-Met-Asn-Ser-
Ser-Phe-Lys-Ala-	Ac-Met-Ile-Lys-Phe-	Ac-Met-Asn-Asn-Ser-
	Met-Ile-Thr-Glu-	Ac-Met-Asn-Asn-Met-
Ac-Gly-Glu-Phe-Leu-	Met-Ile-Glu-Phe-	Met-Met-Ile-Met-
Gly-Phe-Leu-Ala-		Met-Asn-Glu-Lys-
	Ac-Met-Glu-Phe-Leu-	Met-Asn-Gln-Phe-
Ac-Ala-Glu-Phe-Leu-	Ac-Met-Asp-Phe-Leu-	Met-Asn-Lys-Phe-
Ala-Phe-Leu-Ala-	Met-Gln-Phe-Leu-	

% Acetylation	Sequence	% Acetylation	Sequence
100%	Ac-Met-Glu-Phe-Leu-	55%	Ac-Met-Glu-Phe-Lys-
100%	Ac-Met-Asp-Phe-Leu-	67%	Ac-Met-Asp-Pro-Leu-
79%	Ac-Met-Asn-Phe-Leu-		
0%	Met-Etc-Phe-Leu-		
0%	Etc-Phe-Leu-		

Top: Various N-terminal acetylated and related nonacetylated iso-l-cytochromes *c*. Bottom: Results of the series Met-*All*-Phe-Leu-, having all 20 residues. (Etc denote the other 20 residues.) These and other results established that Ac-Met-Glu- and Ac-Met-Asp- is sufficient, but not necessary for amino-terminal acetylation in eukaryotes, and at the degree of acetylation is influenced by other nearby residues, such as Lys and Pro.(Adapted from Tsunasawa, Stewart and Sherman, (1985); Moerschell *et al.*, (1990); and R. P. Moerschell, S. Tsunasawa, and F. Sherman, unpublished results).

especially with the rat liver enzyme (Yamada and Bradshaw, 1991b). Specific alterations of iso-l-cytochrome *c* can be used to systematically define possible consensus sequences *in vivo*, thereby obtaining information that could not be accessible from just examining protein data bases.

In normal yeast strains, the N-terminal methionine of iso-l-cytochrome *c* is cleaved and the newly exposed threonine is not acetylated. However, certain mutant proteins having alterations in the N-terminal region were acetylated (Table 2, Top). In some sequences, only two or three specific residues appear to specify acetylation. In spite of the diversed eukaryotic sequences that are acetylated, one simple pattern emerged in the first systematic study by Moerschell *et al.* (1990), which included all 20 residues in the Met-*All*-Phe-Leu-

N-terminus (Table 2, Bottom). In this series, only Met-Glu- and Met-Asp- were completely acetylated, whereas Met-Asn- was partially acetylated. We believe that Met-Glu- and Met-Asp- are members of the class of Met-Acidic- proteins, whereas Met-Asn-Phe- may be a member of a different class that may include Ac-Met-Asn-Asn- and possibly other iso-l-cytochromes *c* (Table 2, Top). Thus, Met-Acidic- termini are sufficient, but not necessary to direct methionine acetylation because some Met-Ile- and Met-Met-iso-l-cytochromes *c* are also acetylated. Furthermore, the diminished acetylation of Met-Asp-Pro- suggests that antepenultimate proline residues inhibit acetylation in addition to methionine cleavage. Acetylation of Met-Acidic- termini also may be diminished by basic residues in the terminal region, as exemplified by Ac-Met-Glu-Phe-Lys-.

On going studies with altered forms of iso-l-cytochromes *c* are aimed at deciphering other consensus sequences for acetylation. However, these studies are hampered by the complexity of N-terminal acetylation, which we believe is due in part to the existence of multiple N-terminal acetyltransferases, each acting on different patterns of N-terminal sequences (see below).

nat1⁻ STRAINS DEFICIENT IN AN N-TERMINAL ACETYL TRANSFERASE

We have reported that the *NAT1* and *ARD1* genes encode an N-terminal acetyltransferase in the yeast *Saccharomyces cerevisiae*. Both *nat1*⁻ and *ard1*⁻ mutants are unable to carry out N-terminal acetylation of a subset of normally acetylated proteins (Mullen *et al.*, 1989). The

Table 3. Percent N-terminal acetylation of altered iso-l-cytochromes *c* from *NAT1*$^{+}$ and *nat1-3* strains.

		% Acetylation	
		NAT1^{+}	*nat1-3*
CYC1-793	ATA ATG TCT GAA TTC TTG GCC Ac-(Met)Ser-Glu-Phe-Leu-Ala-	100%	0%
CYC1-RAT11	AAA ATG GGT GAT GTT GAA AAA Ac-(Met)Gly-Asp-Val-Glu-Lys-	70%	0%
CYC1-872	ATA ATA ATG AAC AAC TTG GCC Ac-Met-Asn-Asn-Leu-Ala-	100%	~95%
CYC1-848	ATA ATA ATG GAC TTC TTG GCC Ac-Met-Asp-Phe-Leu-Ala-	100%	100%
CYC1-853	ATA ATA ATG GAA TTC TTG GCC Ac-Met-Glu-Phe-Leu-Ala-	100%	100%

Methionine residues cleaved before acetylation are denoted by (Met). *CYC1-RAT11* encodes rat cytochrome *c* (Clements *et al.*, 1989). (From R. P. Moerschell, S. Tsunasawa, J. R. Mullen, R. Sternglanz, and F. Sherman, unpublished results.)

nat1⁻ mutant was originally uncovered by screening a collection lethal temperature-sensitive mutants for protein acetyltransferase activity *in vitro*. Furthermore, regions of the *NAT1* gene and the chloramphenicol acetyltransferase genes of bacteria have limited but significant homology. The previously identified *ard1*⁻ mutant was first suspected to be related to *nat1*⁻ because of certain similar phenotypes. In addition to lacking acetyltransferase activity, strains with mutations in either *NAT1* or *ARD1* exhibit slower growth, derepression of the silent mating type locus *HMLα*, and failure to enter Go. Some of these findings were confirmed by Lee *et al.* (1989a,b), who denoted *nat1* as *aaa1*. Also, a 20-fold increase of acetyltransferase activity occurred when both *NAT1* and *ARD1* were concomitantly overexpressed, but not when either gene was individually overexpressed (Mullen *et al.*, 1989). Furthermore, studies with epitoped tagged derivatives demonstrated that both *NAT1* and *ARD1* proteins are required for enzyme activity *in vitro* (Park and Szostak, 1992). Thus, the *NAT1* and *ARD1* proteins function together as an N-terminal acetyltransferase.

Analysis of two-dimensional gels revealed that at least 20 proteins, and possibly many more (Mullen *et al.*, 1989), or approximately 20% of the different species of soluble proteins (Lee *et al.*, 1989c), were affected in *nat1⁻* strains. Also, amino acid sequencing revealed that an altered iso-1-cytochrome *c* normally having a Ac-Ser-Glu-Phe- terminus was unacetylated in a *nat1⁻* strain. Furthermore, Lee *et al.* (1990) have demonstrated N-terminal acetyltransferase activity of *nat1⁻* (denoted *aaa1⁻*) extracts *in vitro* on certain peptides. We have investigated the specificity of the NAT1 acetyltransferase *in vivo* by analyzing the percent acetylation of variant iso-1-cytochromes *c*. Our results clearly distinguished two groups: one containing termini with methionine that are acetylated in both normal *NAT1* and mutant *nat1⁻* strains and the other group, with serine or glycine termini that are not acetylated in *nat1⁻* strains (Table 3). Further information comes from the lack of acetylation of 14 different proteins in *nat1⁻* strains (Table 4). These results suggests that the NAT1 acetyltransferase acts on at least a subset of proteins with serine, glycine, and alanine termini, but not on at least some proteins with methionine N-termini. Furthermore, proteins with a wide range of different sequences adjacent to serine termini are substrates of the NAT1 acetyltransferase (Table 4).

Lee *et al.* (1990) have identified a methionine N-terminal acetyltransferase from yeast that transfers acetyl groups to peptides containing N-terminal methionine but not to serine or alanine. Also, recently Tercero and Wickner (1992) and Tercero *et al.* (1992) described the *MAK3* gene that presumably encodes an N-terminal acetyltransferase which is required for

Table 4. Deficiency of N-terminal acetylation of proteins in *nat1⁻* strains that are normally acetylated in *NAT1⁺* strains.

Protein	N-terminal sequence	Protein	N-terminal sequence
Yeast ribosomal proteins[1]		Yeast non-histone protein 4 (*NHP4*)[2]	
22	Ser-Val-Glu-Pro-Val-		Ser-Asp-Trp-Asp-Thr-
23	Ser-Gln-Pro-Val-Val-	Human ACTH peptide[3]	
30	Ser-Ala-Pro-Gln-Ala-		Ser-Tyr-Ser-Met-Glu-
40	Ser-Ala-Pro-Gln-Ala-	Yeast alcohol dehydogenase I peptide[3]	
39 (L16)	Ser-Ala-Lys-Ala-Gln-		Ser-Ile-Pro-Glu-Thr-
41	Ser-Thr-Glu-Leu-Thr-	Human superoxide dismutase protein[4] & peptide[3]	
50	Ser-Asp-Ala-Val-Thr-		Ala-Thr-Lys-Ala-Val-
61R	Ser-Ala-Val-Pro-Ser-	Human basic fibroblast growth factor[4]	
			Ala-Ala-Gly-Ser- Ile -
		Human acidic fibroblast growth factor[4]	
			Ala-Glu-Gly-Glu-Ile-

[1]*In vivo* (Takakura *et al.*, 1992).
[2]*In vivo* (David Kolodrubetz, unpublished results).
[3]*In vitro* (Lee, Lin and Smith, 1990).
[4]*In vitro* (Lawrence Cousens, unpublished results).

N-terminal acetylation of the viral major coat protein with a Ac-Met-Leu-Arg-Val terminal sequence. The MAK3 N-terminal acetyltransferase is necessary for viral particle assembly and for optimum growth on nonfermentable carbon sources. The relationship between the MAK3 N-acetyl transferase and the methionine acetyltransferase reported by Lee *et al.* (1990) remains to be elucidated.

The NATl N-terminal acetyltransferase has been purified from yeast (Lee *et al.*, 1988) and an N-terminal acetyltransferase that presumably acts co-translationally has been purified from rat liver (Yamada and Bradshaw, 1991a,b). The enzymes from the two species appear to have different structures, as indicated by differences in amino acid compositions and isoelectric points. Also, the action of the rat liver N-terminal acetyltransferase on 65 synthetic peptides *in vitro* revealed specificities that differed from the specificities observed with yeast NATl enzyme *in vivo*. It remains to be seen if this difference is due to inherent differences between the two enzymes or to differences of acetylation occurring *in vivo* and *in vitro*.

BIOLOGICAL IMPORTANCE OF N-TERMINAL ACETYLATION

There are a surprisingly few examples demonstrating the importance of N-terminal acetylation for stability or function of proteins. Unfortunately most of these conclusions are based on comparisons of acetylated and unacetylated forms that had additional differences in amino acid sequence. Alterations at N-termini, including loss of acetylation, decreased thermal stabilities of NADP-specific glutamate dehydrogenase from *Neurospora crassa* (Siddig *et al.*, 1980). Similarly, the unacetylated form of the *Escherichia coli* ribosomal 5S protein is more thermosensitive than the acetylated form (Cumberlidge and Isono, 1979). A decrease in half-life was observed for a form of murine hypoxanthine phosphoribosyl-transferase from mouse, which had an N-terminal replacement of proline for acetylalanine (Johnson *et al.*, 1988). In contrast to early reports that N-terminal acetylation diminished degradation *in vitro* by what was presumed to be a ubiquitin-dependent system (Hershko *et al.*, 1984; Matsuura *et al.*, 1981), Sokolik and Cohen (1992) used $NAT1^{+}$ and $nat1^{-}$ yeast strains to prepare acetylated and unacetylated pairs of rat and yeast cytochrome *c*, respectively (Table 3); they observed equal extents of ubiquitin conjugation within each pair, although both yeast forms were more highly ubiquitinated than both of the rat forms. Thus, the difference in ubiquitination of mammalian and yeast cytochrome *c* is due to differences other than N-terminal acetylation. Furthermore, Mayer *et al.* (1989) observed ubiquitin-dependent degradation of N-terminal acetylated proteins in a crude reticulocyte lysate. The presence of most of the unacetylated derivatives at approximately normal levels in two-dimensional gels demonstrated that acetylation is not generally required for stability (Mullen *et al.*, 1989; Lee *et al.*, 1989c); However, because certain proteins were diminished, enhanced, lost and gained in $nat1^{-}$ strains, some of the alterations on two-dimensional gels could be indirect, perhaps due to the effects of unacetylated regulatory proteins (Lee *et al.*, 1989c).

A few examples of acetylation affecting the function of various proteins have been reported. N-terminal acetylation, occurring post-translationally, causes increased melanotropic effects of α-melanocyte-stimulating hormone, while it reduces the analgesic action of β-endorphin (Symth *et al.*, 1979, 1980; Ramachandran and Li, 1967). Non-acetylated cytoplasmic actin from cultured Drosophila cells is less efficient in the assembly of microfilaments then the acetylated form (Berger *et al.*, 1981). The oxygen affinity of nonacetylated forms of feline and possibly other β-hemoglobin chains becomes insensitive to the modifying effects of organic phosphates than the acetylated forms (Moo-Penn *et al.*, 1977). It should be noted that all of the effects observed *in vivo* and involving proteins modified co-translationally, had differences in amino acid sequences in addition to the lack of acetylation.

The most significant means for assessing the general importance of N-terminal acetylation comes from the defects and, more surprising, the lack of defects in *nat1*⁻ and *ard1*⁻ mutants, which lacks the major N-terminal acetyltransferase, and from *mak3*⁻ strains that presumably lacks a minor N-terminal acetyltransferase. As described above, the silent mating loci, particularly *HMLα*, are partially derepressed in *nat1*⁻ and *ard1*⁻ mutants, leading to a partial mating defect in *MAT*a strains. In addition *nat1*⁻ and *ard1*⁻ mutants exhibit defects of slow growth; inability of homozygous diploid strains to sporulate; and the failure to enter G_o when limited for nutrients (Mullen *et al.*, 1989). Presumably, these multiple defects are due to the lack of N-terminal acetylation of one or more specific proteins requiring acetylation for function. Diminished function by the lack of acetylation of the SIR3 protein, for example, can explain the partial derepression of *HML* (Stone *et al.*, 1991), whereas diminished function of any one of a number of proteins in the cAMP pathway can explain the failure to enter G_o and the inability of homozygous diploids to sporulate (Mullen *et al.*, 1989). Although viability of *nat1*⁻ and *ard1*⁻ mutants certainly was not anticipated, especially in light of the lack of acetylation of many ribosomal proteins in the mutant strains (Table 4), many investigators have been shocked to find viable mutants with few or no obvious detrimental phenotypes after disrupting other "essential" yeast genes.

The requirement of the MAK3 N-terminal acetyltransferase for viral particle assembly clearly illustrates a function for N-terminal acetylation. Also the ill-defined defect of a diminished growth on nonfermentable carbon sources suggests other requirements for MAK3.

CONCLUSIONS

In summary, studies with yeast mutants have indicated that there are at least two acetyltransferases that act co-translationally on many eukaryotic proteins. The major N-terminal acetyltransferase, encoded by *NAT1* and *ARD1*, acts on at least a subset of proteins with serine, glycine, and alanine N-termini, but not on any of the tested methionine termini (Tables 3, 4). A second acetyltransferase may act co-translationally on all or a subset of proteins with methionine termini (Tables 3, 4) (Lee *et al.*, 1990). Each of the acetyltransferases may recognize several groups of sequences at amino termini, and acetylation may be inhibited by amino acid changes within or adjacent to these consensus sequences. Although N-terminal acetylation may be required for stability and function of certain proteins, viability of *nat1*⁻ mutants lacking the major acetyltransferases, as well as viability of *mak3*⁻ mutants, suggests that the role of acetylation may be subtle and not absolute for most proteins. Possibly only a subset of proteins actually requires this modification for activity or stability, whereas the remainder are acetylated only because their termini fortuitously correspond to consensus sequences.

Acknowledgements. This investigation was supported by United States Public Health Service Research Grants R01 GM12702 and GM28220 and by the Ministry of Education, Science, and Culture of Japan Grants-in Aid for International Research Program, Joint Research Grant 63044090.

REFERENCES

Berger, E. M., Cox, G., Weber, L. and Kenney, J. L., 1981 Actin acetylation in *Drosophila* tissue culture cells, *Biochem. Genet.* 19:321.

Chappell, M. C., O'Donohue, T. L., Millington, W. R. and Kempner, E. S., 1986 The size of enzymes acetylating α-melanocyte-stimulating hormone and β-endorphin, *J. Biol. Chem.* 261:1088.

Clements, J. M., O'Connell, L. I., Tsunasawa, S. and Sherman, F., 1989 Expression and activity of a gene encoding rat cytochrome *c* in the yeast *Saccharomyces cerevisiae, Gene* 83:1.

Cook, R. K., Sheff, D. R. and Rubenstein, P. A., 1991 Unusual metabolism of the yeast actin amino terminus, *J. Biol. Chem.* 266:16825.

Cumberlidge, A. G. and Isono, K., 1979 Ribosomal protein modification in *Escherichia coli.* I. A mutant lacking the N-temminal acetylation of proteins S5 exhibits thermosensitivity, *J. Mol. Biol.* 131:169.

Driessen, H. P. C., deJong, W. W., Tesser, G. I. and Bloemendal, H., 1985 The mechanism of N-temminal acetylation of proteins, *CRC Crit. Rev. Biochem.* 18:281-325.

Glembotski, C. C., 1982 Acetylation of oc-melanotropin and α-endorphin in the rat intermediate pituitary, *J. Biol. Chem.* 257:10493.

Glembotski, C. C., 1982 Characterization of the peptide acetyltransferase activity in bovine and rat intemmediate pituitaries responsible for the acetylation of β-endorphin and α-melanotropin, *J. Biol. Chem.* 257:10501.

Hershko, A., Heller, H., Eytan, E., Kaklij, G. and Rose, I. A., 1984 Role of the α-amino group of protein in ubiquitin-medialed protein breakdown, *Proc. Natl. Acad. Sci. USA* 81:7051-7025.

Huang, S., Elliott, R. C., Liu, P. S., Koduri, R. K., Weickmann, J. L., Lee, J. H., Blair, L. C., Gosh-Dastidar, P., Bradshaw, R. A., Bryan, K. M., Einarson, B., Kendall, R. L., Kolacs, K. H. and Saito, K., 1987 Specificity of cotranslational amino-terminal processing of proteins in yeast, *Biochemistry* 26:8242.

Johnson, G. G., Kronert, W. A., Bernstein, S. I., Chapman, V. M. and Smith, K. D., 1988 Altered turnover of allelic variants of hypoxanthine phosphoribosyltransferase is associated with N-terminal amino acid sequence variation, *J. Biol. Chem.* 263:9079.

Kendall, R. L., Yamada, R.and Bradshaw, R. A., 1990 Cotranslational amino-terminal process, *Methods Enzymol.* 185:398.

Lee, F-J. S., Lin, L.-W. and Smith, J. A., 1988 Purification and characterization of an N^{α}-acetyltransferase from *Saccharomyces cerevisiae, J. Biol. Chem.* 263:14948.

Lee, F.-J. S., Lin, L.-W. and Smith, J. A., 1989a, Molecular cloning and sequencing of a cDNA encoding N^{α}-acetyltransferase from *Saccharomyces cerevisiae, J. Biol. Chem.* 264:12339.

Lee, F.-J. S., Lin, L.-W. and Smith, J. A., 1989b N^{α} acetylation is required for normal growth and mating of *Saccharomyces cerevisiae, J. Bacteriol.* 171:5795.

Lee, F.-J. S., Lin, L.-W. and Smith, J. A., 1989c N^{α}-acetyltransferase deficiency alters protein synthesis in *Saccharomyces cerevisiae, FEBS Letters* 256:139.

Lee, F-J. S., Lin, L.-W. and Smith, J. A., 1990 Identification of methionine N^{α}-acetyltransferase from *Saccharomyces cerevisiae, J. Biol. Chem.* 265:3603.

Matsuura, S., Arpin, M., Hannum, C., Margoliash, E., Sabatini, D. D. and Morimoto, T., 1981 *In vitro* synthesis and posttranslational uptake of cytochrome *c* into isolated mitochondria: Role of a specific addressing signal in the apocytochrome, *Proc. Natl. Acad. Sci. USA* 78:4368.

Mayer, A., Siegel, N. R., Schwartz, A. L. and Ciechanover, A., 1989 Degradation of proteins with acetylated amino termini by the ubiquitin system, *Science* 244:1480.

Moerschell, R. P., Tsunasawa, S.,and Sherman, F., 1988 Transformation of yeast with synthetic oligonucleotides, *Proc. Natl. Acad. Sci. USA* 85:524.

Moerschell, R. P., Hosokawa, Y., Tsunasawa, S. and Sherman, F., 1990 The specificities of yeast methionine aminopeptidase and acetylation of amino-terminal methionine *in vivo:* Processing of altered iso-l-cytochromes *c* created by oligonucleotide transformation, *J. Biol. Chem.* 265:19638.

Moerschell, R. P., Das, G. and Sherman, F., 1991 Transformation of yeast directly with synthetic oligonucleotides, *Methods Enzymol.* 194:362.

Moo-Penn, W. F., Bechter, K. C., Schmidt, R. M., Johnson, M. H., Jue, D. L., Schmidt, D. E., Jr., Dunlap, W. M., Opella, S. J., Bonaventura, J. and Bonaventura, C., 1977 Hemoglobin Raleigh (bı Valine Acetylalanine). Structural and functional characterization, *Biochemistry* 16:4872.

Mullen, J. R., Kayne, P. S., Moerschell, R. P., Tsunasawa, S., Gribskov, M., Colavito-Shepanski, M., Grunstein, M., Sherman, F. and Sternglanz, R., 1989 Identification and characterization of genes and mutants for an N-terminal acetyltransferase from yeast, *EMBO J.* 8:2067.

Mullen, J. R., Kayne, P. S., Moerschell, R. P., Tsunasawa, S., Gribskov, M., Colavito-Shepanski, M., Grunstein, M., Sherman, F. and Sternglanz, R., 1989 Identification and characterization of genes and mutants for an N-terminal acetyltransferase from yeast, *EMBO J.* 8:2067.

Park, E.-C. and Szostak, J. W., 1992 ARDl and NATl proteins form a complex that has N-terminal acetyltransferase activity, *EMBO J.* 11:2087.

Persson, B., Flinta, C., von Heijne, G. and Jörnvall, H., 1985 Structures of N-terminally acetylated proteins, *Eur. J. Biochem.* 152:523.

Sheff, D. R. and Rubenstein, P. A., 1989 Identification of N-acetylmethionine as the product released during the NH_2-terminal processing of a pseudo-class I actin, *J. Biol. Chem.* 264:11491.

Siddig, M. A. M., Kinsey, J. A., Fincham, J. R. S. and Keighren, M., 1980 Frameshift mutations affecting the N-terminal sequcnce of *Neurospora* NADP-specific glutamate dehydragenase, *J. Mol. Biol.* 137: 125.

Sokolik, C. W. and Cohen, R. E., 1992 Ubiquitin conjugation to cytochromes *c*. Structure of the yeast iso-l conjugate and possible recognition determinants, *J. Biol. Chem.* 267:10678.

Spencer, N., 1984 Amino terminal acetylation and protein synthesis in eukaryotes, *J. Theor. Biol.* 107:405.

Stone, E. M., Swanson, M. J ., Romeo, A. M., Hicks, J. B. and Sternglanz, R., 1991 The *SIR1* gene of *Saccharomyces cerevisiae* and its role as an extragenic suppressor of several mating-defective mutants, *Mol. Cell Biol.* 11:2253.

Symth, D. G. and Zakarian, S., 1980 Selective processing of β-endorphin in regions of porcine pituitary, *Nature* 288:613.

Symth, D. G., Massey, D. E., Zakarian, S. and Finnie, M. D. A., 1979 Endorphins are stored in biologically active and inactive forms: isolation of α-N-acetyl peptides, *Nature* 279:252.

Takakura, H., Tsunasawa, S., Miyagi, M. and Warner, J. R., 1992 N-terminal acetylation of ribosomal proteins of *Saccharomyces cerevisiae, J. Biol. Chem.* 267:5442.

Tercero, J. C. and Wickner, R. B., 1992 *MAK3* encodes as N-acetyltransferase whose modification of the L-A *gag* NH_2-terminus in necessary for viruse particle assembly, *J. Biol. Chem.* 267:20277.

Tercero, J. C., Riles, L. E. and Wickner, R. B., 1992 Localized mutagensis and evidence for posttranscription regulation of *MAK3,* a putative N-terminal acetyltransferase required for double stranded DNA virus propagation in *Saccharomyces cerevisiae, J. Biol. Chem.* 267:20270.

Tsunasawa, S., Stewart, J. W. and Sherman, F., 1985 Amino-terminal processing of mutant forms of yeast iso-l-cytochrome *c*: The specificities of methionine aminopeptides and acetyltransferase, *J. Biol. Chem.* 260:5382.

Yamada, R. and Bradshaw, R. A., 1991a Rat liver polysome N^{α}-acetyltranferase: Isolation and characterization, *Biochemistry* 30: 1010.

Yamada, R. and Bradshaw, R. A., 1991b Rat liver polysome N^{α}-acetyltranferase: Substrate specificity, *Biochemistry* 30:1017.

Yamamoto, T., Moerschell, R. P., Wakem, L. P., Ferguson, D. and Sherman, F., 1992 Parameters affecting the frequencies of transformation and co-transformation with synthetic oligonucleotides in yeast, *Yeast* 8:935.

POSTTRANSLATIONAL GLUTAMYLATION OF SEVERAL BRAIN TUBULIN ISOTYPES: STRUCTURE OF THE POLYGLUTAMYL SIDE-CHAIN

Virginie Redeker, Jean-Pierre Le Caer
and Jean Rossier

Institut Alfred Fessard, CNRS
91198 Gif-sur-Yvette Cedex, France

INTRODUCTION

Tubulin is the major protein of microtubules which play a major role in flagellar motility, cell division, cell transport and induction of cell morphology. Brain tubulin exhibits a high degree of heterogeneity due to the differential expression of several isotypes which are encoded by different isogenes (Sullivan, 1988) and to several posttranslational modifications. It was reported that α tubulin undergoes removal and possible readdition of the C-terminal tyrosine (Raybin and Flavin, 1977), acetylation of Lys 40 (L'Hernault and Rosenbaum, 1985; Le Dizet and Piperno, 1987; Eddé *et al.*, 1989, 1991), and polyglutamylation of Glu 445 (Eddé *et al.*, 1990). The modifications hitherto identified for β tubulin are phosphorylation (Gard and Kirschner, 1985; Luduena *et al.*, 1988; Eipper, 1974) and polyglutamylation (Alexander *et al.*, 1991).

Glutamylation of brain tubulin has so far been described for the α tubulin of the mouse brain by Eddé *et al.* (1990) and β-III tubulin of the bovine brain by Alexander *et al.* (1991). Recently, we have isolated the glutamylated C-terminal peptides of the major α (α-Ib) and β (β-II) tubulin isotypes from the pig brain. This new posttranslational modification consists of the addition of glutamyl residues into a polyglutamyl side-chain and is located in the tubulin C-terminal acidic domain.

This glutamylation of brain tubulin occurs inside an essential region of the molecule for several inter- and intramolecular interactions. The C-terminal end of tubulin interacts with microtubule-associated proteins (Serrano *et al.*, 1984; Littauer *et al.*, 1986; Maccioni *et al.*, 1988) and modulates tubulin assembly into microtubules (Sackett *et al.*, 1985; Bhattacharrya *et al.*, 1985). This highly charged carboxyl-terminal region is involved in electrostatic as well as conformational effects in the process of tubulin assembly. Microtubule motors, kinesin and dynein, interact also with the COOH-terminal region (Paschal *et al.*, 1989; Rodionov *et al.*, 1990). Moreover, this acidic COOH-terminal

Methods in Protein Sequence Analysis, Edited by K. Imahori
and F. Sakiyama, Plenum Press, New York, 1993

region binds calcium ions and appears to be involved in the inhibitory action of Ca^{2+} on tubulin assembly (Vera *et al.*, 1989). Interestingly, this COOH-terminal end of the protein remains exposed on the outer surface of the polymer and monomer of tubulin (Sackett and Wolff, 1986) and is relatively mobile (Ringel and Sternlicht, 1984). The cellular roles of this polyglutamyl side-chain are still unknown, but the data listed above suggest that polyglutamylation could play a role in regulating the microtubule dynamics and functions.

The aim of our study was to characterize the linkage mode of the glutamyl residues of the polyglutamyl side-chain.

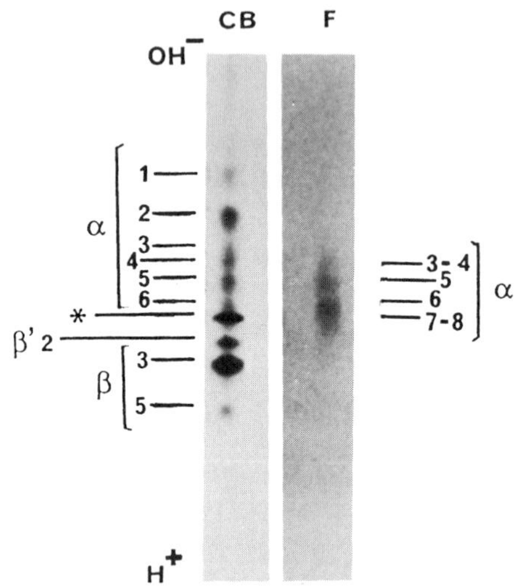

Figure 1. Isoelectric-focusing of purified labeled tubulin.
Cylindrical IEF was performed in the presence of 9.5 M urea as described previously by Wolff *et al.* (1982) except that the pH 5.0-5.5 ampholytes (Serva) were used. The gels were stained with Coomassie Blue (CB) and subjected to fluorography (F). The asterix corresponds to a region of the gels where a comigration of highly acidic α-tubulin isoforms (α7, α8) with a highly basic β-tubulin isoform (β'1) was observed. The nomenclature of Wolff *et al.* (1982) was used.

ISOLATION OF GLUTAMYLATED PEPTIDES

Posttranslational Labeling of Several Acidic Variants of α-Tubulin

Incubation of brain slices from 24 h old mice with [^{3}H]glutamate in the presence of cycloheximide resulted in the posttranslational labeling of several acidic variants of α-tubulin. The posttranslational labeling of tubulin was verified by SDS-PAGE(1) which showed that α tubulin was the only labeled protein. Purified tubulin was subjected to isoelectric-focusing (Figure 1) which revealed the heterogeneity of α-tubulin isoelectric variants by protein staining with Coomassie Blue, while only the 6 most acidic ones designated as α3-α8 were visualized by fluorography.

(1) The abbreviations used are: SDS-PAGE, sodium dodecyl sulfate-polyacrylamide gel electrophoresis; HPLC, high performance liquid chromatography; IEF, isoelectrofocusing. tubulin C-terminal acidic domain.

Isolation of the Peptides Bearing Additional Labeled Glutamyl Units

In order to examine the structure of the glutamylated COOH-terminal peptides, the [^{3}H]glutamylated tubulin was mixed with the same amount of unlabeled tubulin purified from 1 month old mouse brain. The mixture of tubulin was digested with thermolysin, a proteolytic enzyme that yields the glutamylated COOH-terminal peptides of α-tubulin ^{440}VEGEGEEEGEE^{450} according to our previous work (Eddé *et al.*, 1990). The peptides from digested tubulin were purified by anion exchange chromatography. Most of the UV-absorbing material was eluted at the beginning of the chromatogram and corresponded to unlabeled and weakly charged peptides. At the end, 8 radioactive acidic peaks were separated

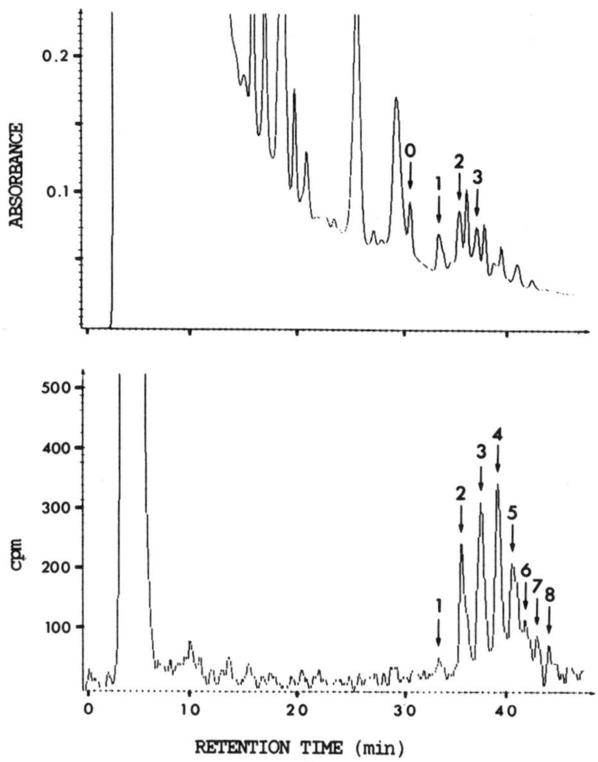

Figure 2. DEAE purification of [^{3}H]-labeled peptides.
Purified [^{3}H]glutamylated tubulin (2 mg/ml) was digested with 1:20 (w/w) thermolysin (Boehringer Mannheim) in 50 mM Tris-HCl buffer, pH 8, containing 0.1 M $CaCl_2$ (5 h, 37°C). The peptides were separated on an anion exchange column (DEAE 5PW, Protein Pack, 7.5 mm x 7.5 cm, Waters) eluted at 1 ml/min with a linear gradient from 0 to 100% B in 60 min (A=20 mM Tris-HCl, pH 8, 50 mM NaCl; B=20 mM Tris-HCl, pH 8, 500 mM NaCl). The HPLC apparatus was equipped with an on-line flow radioactive detector (Flo-one radiomatic) placed after the UV densitometer. Synthetic peptides were used to calibrate the column. In the absorbance panel: numbers 0, 1, 2, and 3 indicate the retention times of the non- (0), mono- (1), bi- (2), and triglutamylated (3) peptides. After digestion, the sample was coinjected with 3 mg of nonglutamylated (0) synthetic peptide used as an internal standard. Most of the radioactivity was eluted between 33 and 44 min, and the radioactive peaks were denoted by 1-6. In that region all the absorbance peaks did not correspond to radioactive glutamylated peptides, in particular the absorbance peak between 2 and 3.

(Figure 2). All the radioactive peaks except the two weakly labeled peaks that were eluted last were collected separately. The separation of [^{3}H]peptides on the DEAE column (Figure 2) indicated that they differed by their increasing acidity.

Each of the peaks denoted by 1 to 6 was purified to homogeneity by chromatography on a C18, 5 μm column, and their amino acid sequences were determined by automatic Edman degradation. They all gave the same sequence VEGEGXEEGXX which corresponded to the sequence of the Mα1 and Mα2 isogene products of the mouse α-tubulin genes (Villasante *et al.*, 1986) from amino acids 440 to 448. The last two amino acids were not identified. When synthetic peptides were sequenced, the last two glutamates could also not be identified. At position 445, instead of obtaining a phenylthiohydantoin glutamyl residue (PTHglu), no amino acids could be detected in that cycle. This gap in sequencing indicates the presence of a modified Glu445 residue as observed previously by Eddé *et al.* (1990). The molecular weight of these peptides was determined by fast atom bombardment-mass spectrometry analysis performed on a ZAB-HS double focusing mass spectrometer. When studied directly, no characteristic molecular ion could be seen. Methylation of free carboxyl groups was performed by reacting the peptides with a hydrochloric acid solution in methanol. After methylation abundant protonated molecular ions (MH^+) were observed with mass/charge ratios (m/z) of 1447.6, 1590.8, 1733.7, 1876.7, 2019.8, 2163.8, respectively. The ion at m/z of 1446.7 corresponded to the mass of the carboxymethylated undecapeptide ^{440}VEGEGEEEGEE^{450} bearing one additional glutamyl residue. The mass of the other more acidic peaks differed by increments of 143. Thus the peaks 1 to 6 separated by the DEAE column corresponded to the undecapeptide ^{440}VEGEGEEEGEE^{450} bearing 1 to 6 additional glutamyl units, respectively

NATURE OF THE LINKAGE CONNECTING THE SECOND AND THIRD ADDITIONAL GLUTAMYL RESIDUES

Obviously, the first additional glutamyl residue must be linked to the γ-carboxyl group of Glu445. Additional glutamyl residues could then be attached through amide bonds to their α- or γ-carboxyl groups. The seven peptides corresponding to each structural possibility of bi- and triglutamylated undecapeptides were synthetized (Figure 3). Under optimized reverse phase HPLC conditions the elution profiles of the two natural peptides were compared with those of the synthetic peptides.

A C18,5 μm reverse phase column was eluted at 15°C with a slow gradient of acetonitrile at pH 3.0. The separations were based on the difference in the pKa of the α-COOH (pKa, 2.1) and γ-COOH (pKa, 4.2) groups. Thus, at pH 3 α-COOH was the only group to be ionized to a large extent. Under reverse phase conditions and at pH 3, the peptides with free α-COOH groups should be eluted before the peptides with free γ-COOH. This assumption was verified by using synthetic peptides, since the γ1γ2 peptide containing two free α-COOH groups was eluted before the γ1α2 peptide containing only one. Additional details on the synthesis and structure of the standard synthetic peptides, and also on the HPLC analysis are given in Redeker *et al.* (1991). Under these conditions, the two biglutamylated peptides (γ1γ2 and γ1α2) were separated by 4.7 min. The monoglutamylated peptide was also separated and eluted between the two biglutamylated peptides. The biglutamylated peptide isolated after tubulin digestion and separation on the DEAE column was eluted in one single peak at 23.8 min, which corresponded to the retention time of the synthetic γ1α2 peptide. No UV absorbing material was detected at 19.1 min, the retention time of the synthetic γ1γ2 peptide. The small peak eluted at 20.7 min has not yet been identified.

Glutamylation degree	Linkage of the glutamyl residues	SEQUENCE	Number of free α COOH on the glutamyl chain
2 Glu	$\gamma_1\alpha_2$	VEGEGEEEGEE I E — E	1
	$\gamma_1\gamma_2$	VEGEGEEEGEE I E I E	2
3 Glu	$\gamma_1\alpha_2\alpha_3$	VEGEGEEEGEE I E — E — E	1
	$\gamma_1\alpha_2\gamma_3$	VEGEGEEEGEE I E — E I E	2
	$\gamma_1\gamma_2\alpha_3$	VEGEGEEEGEE I E I E — E	2
	$\gamma_1\alpha_2\gamma_2$	VEGEGEEEGEE I E — E I E	2
	$\gamma_1\gamma_2\gamma_3$	VEGEGEEEGEE I E I E I E	3

Figure 3. Structure of the synthetic peptides.

For the triglutamylated peptides, five different triglutamylated peptides (γ1γ2γ3, γ1γ2α2, γ1γ2α3, γ1α2γ3 and γ1α2α3) were eluted in four different peaks, while the γ1γ2α3 and γ1α2γ3 peptides were not separated under the current conditions (Figure 4). As expected, the γ1γ2γ3 peptide was eluted first and the γ1α2α3 last. The triglutamylated peptide isolated after tubulin digestion and separation on the DEAE column was eluted in one single peak at 26.0 min, which corresponded to the retention time of the synthetic γ1α2α3 peptide. No UV absorbing material was detected in the other triglutamylated synthetic peptides.

DISCUSSION

By posttranslational labeling of newborn mice brain tubulin with [^{3}H]glutamate and digestion with thermolysin, up to eight modified C-terminal [^{3}H]undecapeptides ^{440}VEGEGEEEGEE^{450} were separated by DEAE chromatography. The six first were isolated. Sequencing and mass spectrometry showed that they corresponded to the C-

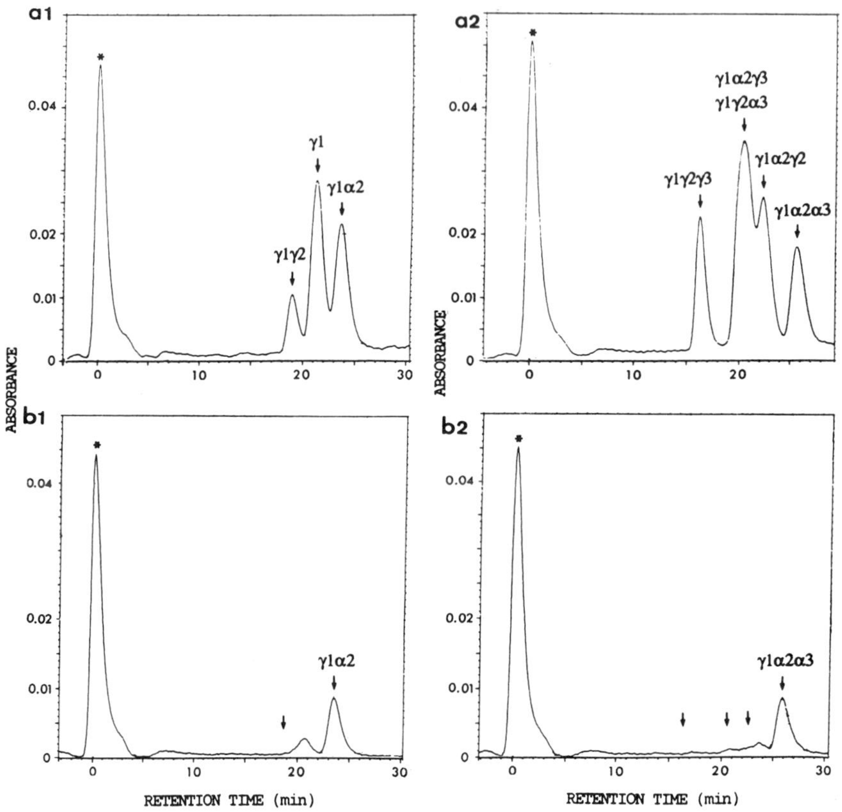

Figure 4. HPLC analysis of the biglutamylated peptides (panel a1,b1) and triglutamylated peptides (panel a2,b2). A reverse phase column (RP18, 220 x 2.1 mm, 5 µm, Brownlee) was eluted at 200 µl/min. After injection of 2% B solvent, 15 min isocratic elution was carried out before the rapid increase of solvent B concentration from 2 to 22% B in 1 min. The gradient was then started: from 22 to 27% B for 40 min at 15°C with A= H_2O, trifluoroacetic acid, pH 3 and B= 35% CH_3CN, trifluoroacetic acid, pH 3. Before injection the sample pH was 3. After calibration with the synthetic peptides, the average retention time (± 0.3 min) of each synthetic peptide was determined relative to the peak (denoted by *) which appeared with the rapid increase of solvent B concentration just before the start of the gradient. a1, elution profile of the synthetic monoglutamylated (γ1) and biglutamylated (γ1α2 and γ1γ2) peptides. a2, elution profile of the synthetic biglutamylated (γ1α2 and γ1γ2) and triglutamylated (γ1α2α3, γ1α2γ2, γ1α2γ3, γ1γ2α3, γ1γ2γ3) peptides. b1 and b2, elution profile of the bi- and triglutamylated tubulin peptides, respectively. *Arrows* indicate the retention time of each standard peptides.

terminal undecapeptide ^{440}VEGEGEEEGEE^{450} bearing one to six glutamyl units, respectively. Optimized reverse phase HPLC and comparison with different synthetic peptides showed that this side-chain corresponded to the γ1α2 and γ1α2α3 chain structure, respectively.

Polyglutamylation plays a major role in the posttranslational heterogeneity of tubulin. Up to 21 variants of adult brain tubulin have been resolved by isoelectric focusing by Wolff *et al.* (1982), and glutamylation which represents a major modification of neuronal α tubulin may account for the existence of several α tubulin isoforms. By DEAE chromatography, we estimated that around 90% of the total α-tubulin present in the brain is glutamylated.

In order to identify consensus sequences which were recognized by the enzymes responsible for this posttranslational modification, isolation of other glutamylated tubulin isotypes from other species and comparison of their primary and tertiary structures are required. Three additional tubulin peptides isolated from other brain species have so far been found to be modified by the addition of a polyglutamic side-chain. Alexander *et al.* (1991) reported that the C-terminal peptide from bovine brain class III-b tubulin which corresponds to the amino acid sequence ^{437}YEDDEEESEAQGPK^{450} was glutamylated at position Glu 438. Recently, we have found that both α (class Ib-α) and β (class II-β) major tubulin isotypes from the pig brain were also glutamylated. These isolated C-terminal peptides corresponding to the sequences ^{434}GEFEEEEGEDEA^{445} (class II-β) and ^{439}SVEGEGEEEGEE(Y)^{451} (class Ib–α) were glutamylated at positions Glu 435 or 445, respectively (Redeker *et al.*, 1992). For each tubulin isotype glutamylation occured on a glutamyl residue located in the C-terminal region, which is the variable region that distinguishes the tubulin isotypes from each other. These peptides are bearing one to at least five additional glutamyl units, but the structure of their polyglutamic side-chains has not yet been studied and compared. No consensus sequence has yet been identified.

Acknowledgements: We thank Serge Plaué from Neosystem for his help in peptide synthesis and Daniel Guénard for the gift of taxol. We are deeply indebted to Jean-Claude Promé for the mass spectrometry analysis. We are grateful to Bernard Eddé and Philippe Denoulet for help. This work was supported by La Ligue Nationale contre le Cancer and by the Association pour la Recherche contre le Cancer (ARC). V. Redeker received a predoctoral fellowship from the Ministère de la Recherche et des Technologies (MRT).

REFERENCES

Alexander, J. E., Hunt, D. F., Lee, M. K., Shabanowitz, J., Michel, H., Berlin, S. C., Macdonald, T. L., Sundberg, R. J., Rebhun, L. I. and Frankfurter, A., 1991 Characterization of posttranslational modifications in neuron-specific class III β-tubulin by mass spectrometry, *Proc. Nat. Acad. Sci. USA* 88:4685.

Bhattacharyya, B., Sackett, D. L. and Wolff, J., 1985 Tubulin, hybrid dimers, and tubulin S, *J. Biol. Chem.* 260:10208.

Eddé, B., Denoulet, P., De Néchaud, B.,Koulakoff, A.,Berwald-Netter, Y. and Gros, F., 1989 Posttranslational modifications of tubulin in cultured mouse brain neurons and astroglia, *Biol. Cell* 65:109.

Eddé, B., Rossier, J., Le Caer, J. P., Desbruyères, E., Gros, F. and Denoulet, P., 1990 Posttranslational glutamylation of α-tubulin, *Science*.247:83.

Eddé, B., Rossier, J ., Le Caer, J. P., Berwald-Netter,Y., Koulakoff, A., Gros, F. and Denoulet, P., 1991 A combination of posttranslational modifications is responsible for the production of neuronal α-tubulin heterogeneity, *J. Cell. Biochem.* 46:134.

Eipper, B. A., 1974 Properties of rat brain tubulin, *J. Biol. Chem.* 249, 1407-1416:746.

Le Dizet, M. and Piperno, G., 1987 Identification of an acetylation site of *Chlamydomonas* α-tubulin, *Proc. Nat. Acad. Sci. USA* 84:5740.

L'Hernault, S. W. and Rosenbaum, J. L., 1985 Chlamydomonas α-tubulin is posttranslationally modified by acetylation on the ε-amino group of a lysine, *Biochemistry* 24:473.

Littauer, U. Z., Giveon, D., Thierauf, M., Ginzburg, I. and Ponstingl, H., 1986 Common and distinct tubulin binding sites for microtubule-associated proteins, *Proc. Natl. Acad. Sci. USA* 83:7162.

Luduena, R. F., Zimmermann, H. and Little, M., 1988 Identification of the phosphorylated β-tubulin isotype in diffentiated neuroblastoma cells, *FEBS Lett.* 230:142.

Maccioni, R.B., Rivas, C.I. and Vera J.C., 1988 Differential interaction of synthetic peptides from the carboxyl terminal regulatory domain of tubulin with microtubule-associated proteins, *EMBO J.* 7:1957.

Paschal, B. M., Obar, R. A. and Vallee, R. B., 1989 Interaction of brain cytoplasmic dynein and MAP2 with a common sequence at the C terminus of tubulin, *Nature (London)* 342:569.

Raybin, D.and Flavin, M., 1975 Modification of tubulin by tyrosylation in cells and extracts and its effect on assembly *in vitro, Biochem. Biophys. Res. Commun.* 65:1088.

Redeker, V., Le Caer, J. P., Rossier, J. and Promé, J. C., 1991 Structure of the polyglutamyl side chain posttranslationally added to α-tubulin, *J. Biol. Chem.* 266:23461.

Redeker, V., Melki, R., Promé, D., Le Caer, J. P. and Rossier, J., 1992 Structure of tubulin C-terminal domain obtained by subtilisin treatment. The major α and β tubulin isotypes from pig brain are glutamylated, *FEBS Lett.* 313:185.

Ringel, I. and Strenlicht, H., 1984 Carbon-13 nuclear magnetic resonance study of microtubule protein: Evidence for a second colchicin site involved in the inhibition of microtubule assembly, *Biochemistry* 28:5644.

Rodionov, V. I., Gyoeva, F. K., Kashina, A. S., Kuznetsov, S. A. and Gelfland, V. I., 1990 Microtubule-associated proteins and microtubule-based translocators have a different binding site on tubulin molecule, *J. Biol. Chem.* 265:5702.

Sackett, D. L., Bhattacharyya, B. and Wolff, J., 1985 Tubulin subunit carboxyl termini determine polymerization efficiency, *J. Biol. Chem.* 260:43.

Sackett, D. L. and Wolff, J., 1986 Proteolysis of tubulin and the substructure of the tubulin dimer, *J. Biol. Chem.* 261:9070.

Serrano, L., Avila, J., and Maccioni, R. B., 1984a Controlled proteolysis of tubulin by subtilisin: Localization of the site for MAP 2 interaction, *Biochemistry* 23:4675.

Serrano, L., De la Torre, J., Maccioni, R. B. and Avila, J., 1984b Involvement of the carboxyl-terminal domain of tubulin in the regulation of its assembly, *Proc. Nat. Acad. Sci. USA* 81:5989.

Sullivan, K. F., 1988 Structure and utilization of tubulin isotypes, *Annu. Rev. Cell. Biol.* 4:687.

Vera, J. C., Rivas, C. I. and Maccioni, R. B., 1989 Biochemical dissection of the role of the one-kilodalton carboxyl-terminal moiety of tubulin in its assembly into microtubules, *Biochemistry* 28:333.

Villasante, A., Wang, D., Dobner, P., Dolph, P., Lewis, S. A. and Cowan, N. J., 1986 Six mouse α-tubulin mRNAs encode five distinct isotypes: Testis-specific expression of two sister genes, *Mol. Cell Biol.* 6:2409.

Wolff, A., Denoulet, P., and Jeantet, C., 1982 High level of tubulin microheterogeneity in the mouse brain, *Neurosci. Lett.* 31:323.

PROTEIN ANCHORING TO MEMBRANE BY GLYCOSYLPHOSPHATIDYLINOSITOL. DETERMINATION OF THE COOH-TERMINAL SIGNAL PEPTIDE SEQUENCE AND GPI-ATTACHMENT SITE IN BOVINE LIVER 5'-NUCLEOTIDASE

Hiroh Ikezawa

Faculty of Pharmaceutical Sciences, Nagoya City University, 3-1 Tanabe-dori, Mizuho-ku, Nagoya, Aichi 467, Japan

POST-TRANSLATIONAL PROTEIN MODIFICATION BY GPI-PRECURSOR(S) AND STRATEGY FOR DETERMINING THE AMINO ACID SEQUENCE THAT WILL BE REPLACED BY GPI

Since 1985, more than seventy proteins in the cells of eukaryotes including protozoa, yeast, insects and vertebrates, have been shown to be anchored to the membranes by glycosylphosphatidylinositol (GPI), a glycolipid consisting mainly of phosphatidylinositol (PI) and oligosaccharide (Ferguson *et al.*, 1985). Attachment of GPI anchor to these proteins is now considered to be a post-translational event. After synthesis of the polypeptide chain of each protein in ER-attached ribosomes, a hydrophobic peptide was cleaved from the COOH-terminus of the protein and replaced by the GPI precursor formed in the ER membrane (Bangs *et al.*, 1986; Ferguson *et al.*, 1986; Low, 1987). The whole scheme of this modification was originally proposed for variant surface glycoprotein (VSG) of *Trypanosoma brucei* and supported by the findings on other GPI-anchored proteins such as Thy- 1 (Low and Kincade, 1985) and decay-accelerating factor, DAF (Caras *et al.*, 1987). The replacement of the COOH-terminal peptide by the GPI precursor inevitably requires a certain recognition process; this hydrophobic peptide in the nascent protein was found to act as a signal for GPI attachment. The COOH-terminal signal peptide is usually composed of 17-31 amino acid residues. Also, the site for attachment of GPI is important.

The strategy for determining the amino acid sequence of this signal peptide and the site for GPI attachment is as follows: At first, cDNA of the target protein must be obtained by cloning, then the amino acid sequence of the nascent protein is deduced from the nucleotide sequence of cDNA. Thereafter, the mature protein isolated from the cells should be subjected to CNBr cleavage and/or limited proteolysis by protease(s), in order to obtain a COOH-terminal peptide linked to the GPI-anchor. After sequencing this peptide fragment, the signal peptide sequence for the attachment of the GPI anchor can be deduced by the difference between the COOH-terminal sequences of nascent and mature proteins. In this study, the sequence of a COOH-terminal signal peptide and the site of attachment to the GPI anchor were determined for bovine liver 5'-nucleotidase as an example.

PROCEDURES FOR CLONING 5'-NUCLEOTIDASE cDNA AND DETERMINING THE SEQUENCES OF NUCLEOTIDES AND AMINO ACIDS

By using the PI-specific phospholipase C (PI-PLC) from *Bacillus thuringiensis*, 5'-nucleotidase was released from minced bovine liver, and isolated by ammonium sulfate precipitation (40-80% saturation) and column chromatographies using ConA-Sepharose, 5'-AMP Sepharose, Toyopearl-HW55, DEAE-Toyopearl and TSK-Gel G4000SW (Suzuki *et al.*, in preparation). The enzyme preparation finally obtained was homogeneous based on SDS-polyacrylamide gel electrophoresis (SDS-PAGE). Through SDS-PAGE and gel filtration analysis, the enzyme was found to occur as a dimer with a molecular weight of 61 kDa per subunit. Purified enzyme (1 mg) was *S*-carboxylmethylated by monoiodoacetate after incubation with DTT in the presence of guanidine chloride and EDTA, and digested with TPCK-trypsin. The tryptic digest was applied to reverse-phase HPLC with ODS. Five of the separated tryptic peptides (T_1-T_5) were sequenced by an Applied Biosystems 477A Protein Sequencer and PTH-amino acids were analyzed by Waters PICO-TAG. From the sequence of T_5 peptide, ETPFLSNPGTNLVF EDEITALQPE(P)VDK(D)L, a mixture of the 19mer deoxyribonucleotides was synthesized on the basis of the underlined heptapeptides, and used as the probe for cloning the 5'-nucleotidase gene. By screening the library from bovine liver, cDNA of 5'-nucleotidase was obtained and sequenced (Suzuki *et al.*, in preparation). From the nucleotide sequence of the cDNA, the whole amino acid sequence of nascent 5'-nucleotidase was deduced as shown in Figure 1.

In order to obtain the COOH-terminal peptide fragment containing the GPI anchor, the isolated 5'-nucleotidase from bovine liver was digested with endoglycosidase F to remove Asn-linked oligosaccharide chains, and reduced with DTT in the presence of guanidine chloride and EDTA. Then, the Cys residues in the molecule of reduced 5'-nucleotidase were *S*-pyridylethylated with 4-vinylpyridine. The modified enzyme protein was cleaved by CNBr or digested with TPCK-trypsin. The peptide fragments obtained were separated by ODS column chromatography or Tricine-SDS-PAGE. From the tryptic peptides, a NH_2-terminal peptide was obtained. Also, five of the CNBr-cleaved peptides were sequenced, and a peptide with the sequence KVLYPAVEGRIQFS, was shown to contain myoinositol by GC analysis with Shimadzu GC12A, after cleavage of the linkage between the oligosaccharide and peptide moieties by HF. Thus, the attachment site for the GPI anchor was found to be Ser523, as indicated in Figure 1. Therefore, a peptide with the sequence AGSHCCG SFSLIFLSVLAVIIILYQ must be cleaved off and replaced by the GPI precursor in the processing of anchor attachment to the protein molecule of 5'-nucleotidase. The site of the GPI attachment (Ser523) and the size of the cleaved peptide (25 amino acid residues) were consistent with the results obtained for the 5'-nucleotidase of rat liver (Ogata *et al.*, 1990) and human placenta (Misumi *et al.*, 1990). The sequence of the cleaved peptide in bovine liver 5'-nucleotidase also shared 60% homology with those in the other two enzymes.

According to Caras (1992), the cleavage / attachment site must be positioned at 10-12 residues from the NH_2-terminal to the hydrophobic domain, for optimal attachment of the GPI anchor. On the other hand, Udenfriend *et al.* (1992) showed that only six amino acids, Asp, Asn, Ser, Gly, Ala and Cys acted as effective GPI-acceptors, and that a small amino

```
        10         20         30         40         50
WELTILHTND VHSRLEQTSE DSSKCVNASR CVGGVARLAT KVHQIRRAEP
        60         70         80         90        100
HVLLLDAGDQ YQGTIWFTVY KGTEVAHFMN ALGYDAMALG NHEFDNGVEG
       110        120        130        140        150
LIDPLLKEVN FPILSANIKA KGPLASKISG LYSPYKILTV GDEVVGIVGY
       160        170        180        190        200
TSKETPFLSN PGTNLVFEDE ITALQPEVDK LKTLNVNKII ALGHSAFEVD
       210        220        230        240        250
KLIAQKVKGV DVVVGGHSNT FLYTGNPPSK EVPAGQYPFI VTSDDGRKVP
       260        270        280        290        300
VVQAYAFGKY LGYLKVEFDE KGNVGTSHGN PILLNSSIPE DPNIKADINK
       310        320        330        340        350
WRVKLDNYST QELGKTIVYL DGTAQSCRFR ECNMGNLICD AMINNNLRHP
       360        370        380        390        400
DEMSWNHVSM CILNGGGIRS PIDERNNGTI TWENLAAVLP FGGTFDLVQL
       410        420        430        440        450
KGSTLKKAFE HTVYRYGQRT GEFLQVGGIH VVYDISRNPG DRVVKLEVLC
       460        470        480        490        500
TQCRVPSYEP LRMDKEYKVI LPSFLVSGGD GFRMIKDEKI KHDSGDQDIN
       510        520        530        540        550
VVSGYISKMK VLYPAVEGRI QFSAGSHCCG SFSLIFLSVL AVIIILYQ*
```

Figure 1. The amino acid sequence of bovine liver 5'-nucleotidase deduced from cDNA. Each of the sequenced peptides is underlined with a single solid line. A stretch of hydrophobic amino acids at the COOH terminal, a possible signal for post-translational modification by the glycophospholipid, is indicated by two lines. Ser 523 becomes the COOH-terminal of the mature protein, i.e. the cleavage/attachment site in the processing of 5'-nucleotidase by the GPI precursor. The amino acid sequences are represented by one-letter code.

acid domain consisting of three amino acid residues occurred at the cleavage / attachment site. The results obtained in bovine liver 5'-nucleotidase also fulfilled the conditions they postulated.

Acknowledgements. The author is greatly indebted to his colleagues and collaborators, Prof. Dr. Yoshiaki Suzuki, National Institute for Basic Biology, Okazaki, Dr. Shin Nakamura, Primate Research Institute, Kyoto University, Dr. Ryo Taguchi, Faculty of Pharmaceutical Sciences, Nagoya City University, and the staffs and several graduate students of his laboratory, for their cooperation .

REFERENCES

Bangs, J. D., Doering, T. L., Englund, P. T. and Hart, G. W., 1988 Biosynthesis of variant surface glycoprotein of *Trypanosoma brucei*, *J. Biol. Chem.* 263: 17697.

Caras, I. W., Weddell, G. N., Davitz, M. A., Nussenzweig, V. and Martin, D. W. Jr., 1987 Signal for attachment of a phospholipid membrane anchor in decay accelerating factor, *Science* 238: 1280.

Caras, I. W., 1992 Probing the signal for glycosylphosphatidylinositol anchor attachment using a decay accelerating factor as a model system, *in*: GPI Membrane Anchors, Cardoso de Almeida, M.L., ed., Academic Press, London.

Ferguson, M. A. J., Low, M. G. and Cross, G. A. M., 1985 Glycosyl-*sn*-1,2-dimyristoylphosphatidylinositol is covalently linked to *Trypanosoma brucei* variant surface glycoprotein, *J. Biol. Chem* . 260:14547.

Ferguson, M. A. J., Duszenko, M., Lamont, G. S., Overath, P. and Cross, G. A. M., 1986 Biosynthesis of *Trypanosoma brucei* variant surface glycoproteins. N-Glycosylation and addition of a phosphatidylinositol membrane anchor, *J. Biol. Chem* . 261:356.

Low, M. G. and Kincade, P. W., 1985 Phosphatidylinositol membrane-anchoring domain of the Thy-1 glycoprotein, *Nature* (London) 318:62.

Low, M. G., 1987 Biochemistry of the glycosyl-phosphatidylinositol membrane protein anchors, *Biochem. J.* 244:1.

Misumi, Y., Ogata, S, Ohkubo, K., Hirose, S. and Ikehara, Y., 1990 Primary structure of human placental 5'-nucleotidase and identification of the glycolipid anchor in the mature form, *Eur. J. Biochem.* 191:563.

Ogata, S., Hayashi, Y., Misumi, Y. and Ikehara, Y., 1990 Membrane-anchoring domain of rat liver 5'-nucleotidase: identification of the COOH-terminal serine-523 covalently attached with a glycolipid, *Biochemistry* 297:7923.

Suzuki, K., Furukawa, Y., Tamura, H., Ejiri, N., Suematsu, H., Taguchi, R., Nakamura, S., Suzuki, Y. and Ikezawa, H., in preparation.

Udenfriend, S., Micanovic, R. and Kodukula, K., 1992 Structural requirements of a nascent protein for processing to a PI-G anchored form: studies in intact cells and cell-free systems, *in*: GPI Membrane Anchors, Cardoso de Almeida, M. L., ed., Academic Press, London.

PYRIDYLAMINATION FOR SENSITIVE ANALYSIS OF SUGAR CHAINS OF GLYCOPROTEINS

Sumihiro Hase

Osaka University College of Science
Toyonaka, Osaka 560, Japan

INTRODUCTION

Biologically active molecules are often present in minute quantity. Much effort and many samples are required to isolate such molecules and to analyze their structures. When sugar chains are analyzed as molecules with biological activity, a method is needed to isolate and analyze sugar chains with a high sensitivity. Therefore, we developed the pyridylamination method (Hase *et al.*, 1978) for sugar chains and improved this method to apply it to sugar chains from glycoproteins (Hase *et al.*, 1984).

The principle of the method consists of tagging the reducing ends of the sugar chains with 2-aminopyridine by reductive amination (Figure 1).

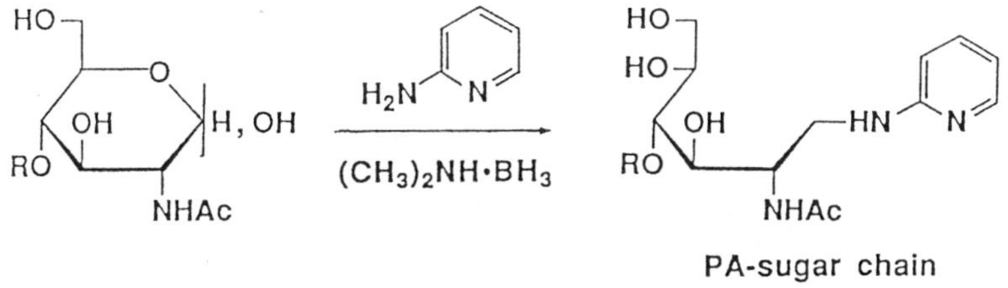

Figure 1. Pyridylamination of sugar chains.

The reaction proceeds quantitatively with minimum side-reactions such as desialylation, epimerization, and β-elimination reactions (Kuraya *et al.*, 1992). Though a labeling procedure is required, the advantages are as follows:
1) About 30 fmol of a PA-sugar chain can be detected by HPLC using a fluorescence spectrophotometer commercially available.
2) Pyridylamino group is chemically stable under reaction conditions such as light, acid hydrolysis, alkaline treatment, hydrazinolysis, methylation analysis, partial acetolysis, Smith periodate oxidation, which are useful methods for determining the sugar structures.

Methods in Protein Sequence Analysis, Edited by K. Imahori
and F. Sakiyama, Plenum Press, New York, 1993

3) Excellent separation was achieved by reverse-phase HPLC.
4) Fluorescence intensities of the PA-sugar chains were almost equal, indicating that the peak area ratios observed represented molar ratios of the sugar chains.

Component Sugar Analysis

Pyridylamination is also used to analyze component sugars. Glycoproteins were hydrolyzed with 4 M trifluoroacetic acid at 100°C for 3 h. After N-acetylation, the sugars released were pyridylaminated. PA-sugars thus obtained were then analyzed by HPLC using a TSKgel Sugar AXI column (Hase *et al.*, 1992). About 50 pmol of sugars usually appearing in glycoconjugates could be quantified.

Analysis of Sugar Chains from Glycoconjugates

Pyridylamination of sugar chains from glycoproteins was performed as follows: sugar chains were released from glycoproteins with anhydrous hydrazine at 100°C for 10 h followed by N-acetylation of free amino groups. The sugar chains without any purification were pyridylaminated. Excess reagents and contaminating materials from polypeptide portions were removed by passing through a TSKgel HW-40F column, and a PA-sugar chain fraction was obtained (Hase *et al.*, 1984).

Separation of PA-Sugar Chains

Separation of the PA-sugar chains was usually carried out using three kinds of HPLC with different separation principles, as follows:
1) A TSKgel DEAE-5PW column or a Mono Q 5/5 EIR column:
PA-sugar chains were separated according to the number of sialic acid residues bound, and the elution times were less affected by the molecular sizes and chemical structures of the PA-sugar chains (Yamamoto *et al.*, 1989).
2) Size-fractionation HPLC:
PA-asialosugar chains could be separated based on their molecular sizes and were less affected by chemical structures (Hase *et al.*, 1986). Under the chromatographic conditions used, the addition of a neutral sugar residue to a PA-sugar chain resulted in the increase of one hexose unit when PA-isomaltooligosaccharides were used as a scale of elution positions. Addition of an N-acetylhexosamine residue caused on increment of 0.6 hexose unit.
3) Reverse-phase HPLC:
Separation of the PA-sugar chains was largely affected by chemical structures such as linkage positions, anomeric configurations, and structures of component sugars. Additivity rule (Hase *et al.*, 1987; 1990) was detected between the structures of the PA-sugar chains and elution times: A sugar residue in a PA-sugar chain affected the elution times under the current chromatographic conditions. The contribution (partial elution time, Ei) was calculated from the elution times of at least two PA-sugar chains. Elution time of a PA-sugar chain (E) can be calculated by the equation; $E = E_1 + E_2 + E_3 + \text{---} + E_i$. Using the rule, elution times of PA-sugar chains can be estimated. A similar rule was also reported (Lee *et al.*, 1990).

Determination of Structure of Sugar Chains

PA-sugar chains thus purified are usually subjected to structure analysis by methods such as component sugar analysis, methylation analysis, Smith periodate degradation, partial acetolysis, mass spectrometry, NMR, and exoglycosidase digestion.

Two-Dimensional HPLC Mapping

When the elution times of the PA-sugar chains obtained by a size-fractionation column are plotted on the ordinate and those by a reverse-phase column on the abscissa, then two-dimensional sugar mapping can be performed. On the map a PA-sugar chain structure is expressed as a dot. When a considerably large number of standard sugar chains are plotted on the map, the PA-sugar structure can be estimated with a high sensitivity. By combining two-dimensional sugar mapping with exoglycosidase digestion and additivity rule, the gross structure of a minute amount of a PA-sugar chains can be estimated.

Application of the Method for Assessment of the Chemical Structures of the Third Component of Human Complement (C3) (Hase *et al.*, 1985).

C3 was hydrazinolyzed followed by N-acetylation, pyridylamination, and purification with TSKgel HW-40F to remove excess reagents. The PA-sugar chain fraction was analyzed by two-dimensional HPLC mapping. Mainly 6 peaks were observed (structures, see below). When the positions on two-dimensional HPLC mapping of these peaks were compared with those of 12 standard PA-sugar chains, which include all the structures considered to be intermediates in the processing of high mannose-type sugar chains, the 6 peaks from C3 showed the same positions as 6 PA-sugar chains with known structures. Next the peak fractions were purified by reverse-phase HPLC. Each PA-sugar chain was digested with α-1,2-mannosidase (*A. oryzae*) and α-mannosidase (jack bean), and the products were analyzed by size-fractionation HPLC and reverse-phase HPLC. A peak corresponding to expected structures, $Man_3GlcNAc_2$ -PA and $Man_1GlcNAc_2$-PA, was observed. The structures of the sugar chains were further confirmed by methylation analysis, partial acetolysis, and Smith periodate oxidation. As C3 is composed of two polypeptide chains, α- and β-chains, the sugar structures on each polypeptide chain were studied. C3 contains two sites for sugar chains; one on the α-chain and the other on the β-chain. C3 was reduced and S-carboxymethylated, and the two chains were separated on a Sepharose 4B column. PA-sugar chains were prepared from α– and β-chains separately and analyzed by reverse-phase HPLC. Sugar structures at the two sites were analyzed by reverse-phase HPLC: $Man_9GlcNAc_2$ and $Man_8GlcNAc_2$ were detected on the α–chain and $Man_6GlcNAc_2$ and $Man_5GlcNAc_2$ on the β-chain.

Proposed structures of sugar chains from C3.

```
Manα1-2Manα1-6\
                Manα1-6\
Manα1-2Manα1-3/         Manβ1-4GlcNAcβ1-4GlcNAc
Manα1-2Manα1-2Manα1-3/

Manα1-2Manα1-6\
                Manα1-6\
          Manα1-3/      Manβ1-4GlcNAcβ1-4GlcNAc
Manα1-2Manα1-2Manα1-3/

Manα1-2Manα1-6\
                Manα1-6\
          Manα1-3/      Manβ1-4GlcNAcβ1-4GlcNAc
          Manα1-2Manα1-3/
```

```
Manα1-6\
         Manα1-6\
Manα1-3/         Manβ1-4GlcNAcβ1-4GlcNAc
Manα1-2Manα1-3/

Manα1-6\
         Manα1-6\
Manα1-3/         Manβ1-4GlcNAcβ1-4GlcNAc
         Manα1-3/
```

REFERENCES

Hase, S. and Ikenaka, T., 1990 Estimation of elution times on reverse-phase high-performance liquid chromatography of pyridylamino derivatives of sugar chains from glycoproteins, *Anal. Biochem.* 184:135.

Hase, S., Hatanaka, K., Ochiai, K. and Schimizu, H., 1992 Improved method for the component sugar analysis of glycoproteins by pyridylamino sugars purified with immobilized boronic acid, *Biosci. Biotech. Biochem.* 56, in press.

Hase, S., Ibuki, T. and Ikenaka, T., 1984 Reexamination of the pyridylamination used for fluorescence labeling of oligosaccharides and its application to glycoproteins, *J. Biochem.* 95:197.

Hase, S., Ikenaka, K., Mikoshiba, K. and Ikenaka, T., 1988 Analysis of tissue glycoprotein sugar chains by two-dimensional high-performance liquid chromatographic mapping, *J. Chromatogr. Biomedical Applications* 434:51.

Hase, S., Ikenaka, T. and Matsushima, Y., 1978 Structure analyses of oligosaccharides by tagging of the reducing end sugars with a fluorescent compound, *Biochem. Biophys. Res. Commun.* 85:257.

Hase, S., Kikuchi, N., Ikenaka, T. and Inoue, K., 1985 Structures of sugar chains of the third component of human complement, *J. Biochem.* 98:863.

Hase, S., Koyama, S., Daiyasu, H., Takemoto, H., Hara, S., Kobayashi, Y., Kyogoku, Y. and Ikenaka, T., 1986 Structure of a sugar chain of a protease inhibitor isolated from Barbados pride (*Caesalpinia pulcherrima* Sw.) seeds, *J. Biochem.* 100:1.

Hase, S., Natsuka, S., Oku, H. and Ikenaka, T., 1987 Identification method for twelve oligomannose-type sugar chains thought to be processing intermediates of glycoproteins, *Anal. Biochem.* 167:321.

Hase, S., Sugimoto, T., Takemoto, H., Ikenaka, T. and Schmid, K., 1986 The structure of sugar chains of Japanese quail ovomucoid. The occurrence of oligosaccharides not expected from the classical biosynthetic pathway for N-glycans; a method for the assessment of the structure of glycans present in picomolar amounts, *J. Biochem.* 99:1725.

Kuraya N. and Hase, S., 1992 Release of O-linked sugar chains from glycoproteins with anhydrous hydrazine and pyridylamination of the sugar chains with improved reaction conditions, *J. Biochem.* 112:122.

Lee, C. Y., Lee, B. I., Tomiya, N. and Takahashi, N., 1990 Parameterization of contribution of sugar units to elution volumes in reverse-phase HPLC of 2-pyridylaminated oligosaccharides, *Anal. Biochem.* 188:259.

Tomiya, N., Awaya, J., Kurono, M., Endo, S., Arata, Y. and Takahashi, N., 1988 Analyses of N-linked oligosaccarides using a two-dimensional mapping technique, *Anal. Biochem.* 171:73.

Yamamoto, S., Hase, S., Fukuda, S., Sano, O. and Ikenaka, T., 1989 Structures of the sugar chains of interferon-γ produced by human myelomonocyte cell line HBL-38, *J. Biochem.* 105: 547.

SEQUENCE ANALYSIS OF PHOSPHOPEPTIDES AND ITS APPLICATION FOR THE DETERMINATION OF PHOSPHORYLATED SITES OF PROTEINS

Tsuneko Uchida, Akira Omori, Koichi Ishiguro
and Kazuki Sato

Mitsubishi Kasei Institute of Life Sciences
Minamiooya 11, Machida
Tokyo 194, Japan

INTRODUCTION

Over the past decade, many investigations on cellular regulation have rapidly developed to find hundreds of regulatory proteins, including protein kinases and phosphatases, their effectors and their targets. At present, it is clear that numerous cellular regulation processes are controlled by a complex network of these regulatory proteins, and that protein phosphorylation plays a crucial role on working of such network. Considering that most of these regulatory proteins are present at extremely low levels in cells, the development of sensitive techniques is essential for analysis of phosphorylated proteins.

Recently, we found a novel protein kinase, tau protein kinase II (TPKII), in microtubules fraction of bovine brain (Ishiguro *et al.*, 1988; 1992). For the study of this kinase, we improved sequence analysis of phosphopeptides in the gas-phase sequencer, and detected the phosphorylation-state of substrate in vivo or in vitro using phosphorylation site-specific antibody (Arioka *et al.*, 1993).

IMPROVEMENT OF SEQUENCE ANALYSIS OF PHOSPHOPEPTIDES IN GAS-PHASE SEQUENCER

Identification of phosphoserine (Ser(P)) or phosphothreonine (Thr(P)) is difficult in the gas-phase sequencer. Treatment with alkali easily splits off phosphate from Ser(P) and Thr(P) by β-elimination. β-Elimination of Ser(P) produces dehydroalanine which is stable when it is surrounded by peptide bond, but easily adds nucleophiles (amines or sulfite) to the double bond upon hydrolysis of the peptide. As the best method to overcome this problem, conversion of Ser(P) to S-ethylcysteine in peptide or protein is recommended (Meyer *et al.*, 1986). This modification, however, is accompanied with a few weak point; (1) The modification occurs also at cysteine residues to give the same product. (2) PTH-S-

Methods in Protein Sequence Analysis, Edited by K. Imahori
and F. Sakiyama, Plenum Press, New York, 1993

ethylcysteine is eluted just before diphenylthiourea which sometimes make the identification confused, particularly in microanalysis. (3) Much more phosphoproteins (0.1-1 nmol) are needed for the modification than those required for usual analysis.

Therefore, we aimed at the development of practical sequence analysis of phosphoprotein at low level as in usual analysis (1-5 pmol)(Omori, 1990) without prior modification in the gas-phase sequencer.

Modifications of the Sequencing Protocol for Analysis Distinguishable Ser(P) from Ser

Since the coupled PTC group increases the reactivity of the α hydrogen, even under the mild alkaline conditions in the gas-phase sequencer, Ser(P) undergoes β-elimination to form dehydroalanine quantitatively, followed by an addition reaction with dithiothreitol (DTT), which is present in the conversion reagent, yielding a DTT adduct of PTH-dehydroalanine as the only end product. While Ser yields both PTH-serine and DTT adduct of PTH-dehydroalanine in the different ratio dependent on the protocol utilized in the gas-phase sequencer (Meyer *et al.*, 1991; Packman, 1990). If we could find the protocol that reduces appearence of DTT adduct of PTH-dehydroalanine in Ser cycle, Ser(P) is clearly distinguishable from Ser and is to be analyzed by a fairly low amount in the gas-phase sequencer without any modification.

As a result, we found that the distribution of PTH-serine and DTT adduct of PTH-dehydroalanine varied greatly with concentration time in a process of concentration of ATZ-amino acid in the conversion vessel containing 25% trifluoroacetic acid and DTT, which is often employed as solvent to receive ATZ-amino acid from the reaction cartridge. Since total yield of PTH-derivatives did not change so greatly, PTH-serine decreased and DTT adduct of PTH-dehydroalanine increased with longer concentration time, as shown in Figure 1A. Considering that the conversion of ATZ-serine to DTT adduct of ATZ-dehydroalanine might occur with higher concentration of H^+ and DTT, we replaced the solvent in the conversion vessel with 20% acetonitrile. In this case, Ser was recovered mainly as PTH-serine (Figure 1B).

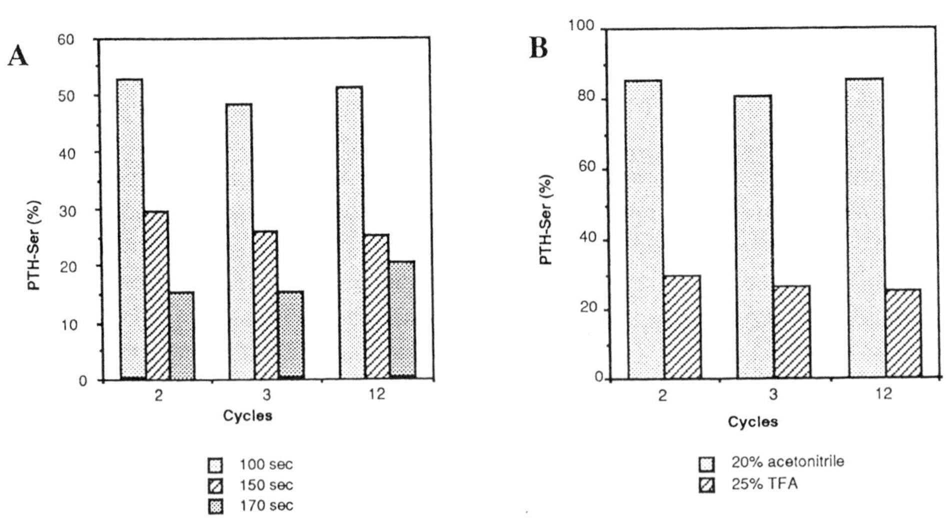

Figure 1. A. Effect of ATZ-amino acid concentration time to the formation of PTH-amino acid. B. Effect of solvent receiving ATZ-amino acid to the formation of PTH-amino acid.

Using this protocol in the gas-phase sequencer, we analyzed a synthetic peptide phosphorylated by a protein kinase. Ser in cycle 2 was clearly detected as PTH-serine. Ser in cycle 9 was detected mainly as DTT adduct of PTH-dehydroalanine with less amount of PTH-serine. We judged that Ser in cycle 9 was mainly phosphorylated by the kinase (Figure 2). Under the same condition, Thr is detected as PTH-threonine, while Thr(P) forms several by-products, which are characteristic of Thr(P) though minor peaks, but not a specific product like DTT adduct of PTH-dehydroalanine in the case of Ser. Therefore, Thr(P) can be also identified to be distinguishable from Thr, though it is difficult to be estimated quantitatively (Figure 2).

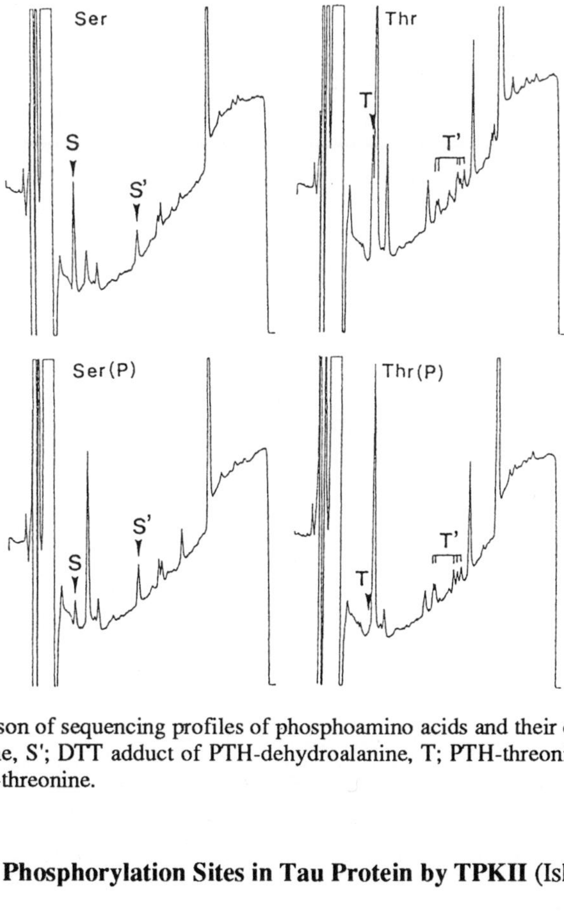

Figure 2 . Comparison of sequencing profiles of phosphoamino acids and their corresponding amino acids. S; PTH-serine, S'; DTT adduct of PTH-dehydroalanine, T; PTH-threonine, T'; characteristic by-products of PTH-threonine.

Determination of Phosphorylation Sites in Tau Protein by TPKII (Ishiguro *et al.*, 1991)

Improved sequence analysis was applied for the identification of phosphorylation sites in tau protein phosphorylated by TPKII. TPKII is one of protein-serine/threonine kinases and seems to be involved in the regulation of microtubule formation (Ishiguro *et al.*, 1992).

Tau protein maximally phosphorylated by TPKII in the presence of [γ-^{32}P]ATP was digested with endoproteinase Lys-C (enzyme/substrate = 1/20) at 37°C for 2 h. Peptides in the digest were separated on C4 column (4.6 x 50 mm) with a linear gradient of acetonitrile/isopropanol (3/7) in 0.1% trifluoroacetic acid (0 - 70%). Three ^{32}P-labeled peptides were detected with scintillation counter, and named K_1, K_2 and K_3 according to elution order. Radioactivity ratio of K_1:K_2:K_3 was 1.6:1.5:1, respectively. For sequence analysis of these phosphopeptides, we used three cold phosphopeptides of K_1, K_2 and K_3 showing retention time corresponding to that of the labeled phosphopeptide, respectively.

The sequences of phosphopeptides were determined using the improved protocol in the gas-phase sequencer. Three phosphopeptides of K_1, K_2 and K_3 were idetified to be the region of 168-182, 133-166 and 307-349 in tau protein, respectively (amino acid numbers in juvenile form of human tau are used).

To confirm these sequences, we chemically synthesized corresponding peptides by the Boc/Bzl mode of solid phase method, and maximally phosphorylated them by TPKII. These synthesized phosphopeptides were confirmed to be eluted at the same positions as those of the phosphopeptides obtained from tau protein.

To determine phosphorylation sites by TPKII, we used the synthesized phosphopeptides instead of the phosphopeptides from tau, in order to avoid ambiguity of data resulted from a small amount of contamination. Phosphorylation sites on each phosphopeptide were determined as follows.

K_3: ^{32}P-Labeled K_3 was digested by endoproteinase Asp-N (enzyme/substrate = 1/20) at 37°C for 2 h, and separated on C_4 column into three non-phosphopeptides and a phosphopeptide (p-K_3-D_7). The corresponding cold phosphopeptide was sequenced. Cycle 3 in p-K_3-D_7 gave only DTT adduct of PTH-dehydroalanine and no PTH-serine, indicating that cycle 3, that is, Ser315 was phosphorylated by TPKII. While cycle 9 in nonphosphorylated K_3, corresponding to Ser315, gave PTH-serine and less amounts of DTT adduct.

K_2: ^{32}P-labeled K_2 was digested by proline-specific endoproteinase, PSP (enzyme/ substrate=1/40) at 30 °C for 16 h, separated on C4 column. Two labeled peptides, p-K_2-$P_{3.5}$ and p-K_2-P_{10} were obtained. Cycle 2 in p-K_2-$P_{3.5}$ gave only DTT-adduct of PTH-dehydroalanine, cycle 5 in p-K_2-$P_{3.5}$ and cycle 2 in p-K_2-P_{10} gave no PTH-threonine, indicating that cycle 2(Ser144) and Cycle 5(Thr147) in p-K_2-$P_{3.5}$ were phosphorylated by TPKII. Furthermore, it is confirmed by sequence analyses that phosphorylaytion by TPKII dose not occur in 10 amino acids of N-terminal part of K_2 and C-terminal peptide (152-166) of K_2 obtained by Arg-C digestion.

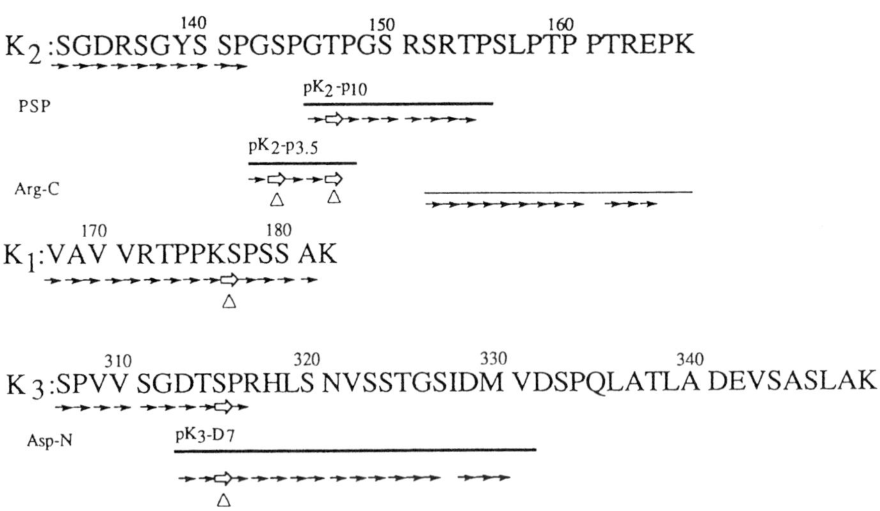

Figure 3. Determination of phosphorylation sites on tau by TPKII. K_1, K_2 and K_3; phosphorylated fragments. PSP, Arg-C and Asp-N; proline-specific, arginine-specific and aspartic acid-specific endoproteinase digests. Thick and thin lines show ^{32}P-labeled peptides and non-labeled peptide, respectively. Arrows show sequenced amino acids and open arrows, phosphoamino acids. Triangles show phosphorylation sites by TPKII.

K_1: By directly sequencing of K_1, Ser177 was determined to be Ser(P), because cycle 9 (Ser177) gave no PTH-serine.

These results are summarized in Figure 3. These phosphorylation sites have a common sequence, SerPro or ThrPro. TPKII phosphorylates the restricted Ser/ThrPro sequences among 14 Ser/ThrPro sequences in the shortest tau. This enzyme may recognize more wide region than only Ser/ThrPro sequence, such as Ser/Thr$X_{2\text{-}3}$ Ser/ThrPro sequence.

PRODUCTION AND APPLICATION OF PHOSPHORYLATION SITE-SPECIFIC ANTIBODY

Since phosphorylation sites of tau by TPKII were determined as described above, we aimed to synthesize a short peptide containing Ser(P) or Thr(P) at a specific phosphorylation site by TPKII, and to produce phosphorylation site-specific antibody using the synthesized peptide as an antigen. Such antibody, if it could be produced, is valuable for the elucidation of the changes in phosphorylation-state of substrate protein by a particular kinase *in vivo* or *in vitro*, which leads us to get the significant information on the functional role(s) of the phosphorylation events.

Synthesis of Phosphopeptides Containing Phosphoamino Acids at Specific Sites (Tsukamoto *et al.*, 1991)

We synthesized the following peptides phosphorylated at specific Ser or Thr for preparation of antisera specific to phosphorylation sites by TPKII (Figure 4).

PS-TF2 (140–150; PO_3H_2 on Ser at position 145)
H-<u>Lys</u>-Ser-Ser-Pro-Gly-Ser-Pro-Gly-Thr-Pro-Gly-Ser-Arg-NH_2

PT-TF2 (PO_3H_2 on Thr)
H-<u>Lys</u>-Ser-Ser-Pro-Gly-Ser-Pro-Gly-Thr-Pro-Gly-Ser-Arg-NH_2

PSPT-TF2 (PO_3H_2 on Ser and PO_3H_2 on Thr)
H-<u>Lys</u>-Ser-Ser-Pro-Gly-Ser-Pro-Gly-Thr-Pro-Gly-Ser-Arg-NH_2

PS-TF3 (307–317; PO_3H_2 on Ser)
H-Lys-Ser-Pro-Val-Val-Ser-Gly-Asp-Thr-Ser-Pro-Arg-NH_2

Figure 4 . Synthetic phosphorylated fragments of tau protein.

In the Boc/Bzl mode of solid phase peptide synthesis (SPPS), phenyl group is widely used as protective group for the phosphate moiety of Ser(P) or Thr(P) and reported to be stable to HF cleavage from resin (Arendt *et al.*, 1989). We observed, however, considerable dephosphorylation of Ser(P) particularly. In search of more stable protective groups for Ser(P), several substituted phenyl groups were examined. The results indicated that 2-alkylphenyl groups were more stable in HF treatment than phenyl group without or with other substituent and could be effectively removed by hydrogenolysis under H_2 with PtO_2 catalyst (Table 1). The larger substituent is used, the more stable it is, resulting that it is more difficult to be removed by hydrogenolysis. 2-Methylphenyl and 2,6-dimethylphenyl groups are recommendable. These protective groups, however, could be inapplicable for synthesis of phosphopeptide containing aromatic amino acids which are reduced by hydrogenolysis.

Table 1. Yields of protected and deprotected peptides by the use of Boc-Ser[$PO(OR)_2$]-OH

Peptide	R	Protected peptide Ratio[a] (P : DP)	Protected peptide Yield (%)	Deprotected peptide Yield (%)
PS-TF2	Phenyl	(60 : 40)	23	68
	2-Methylphenyl	(72 : 28)	46	65
	2,6-Dimethylphenyl	(80 : 20)	52	42
	2-tert-Butylphenyl	(100 : 0)	55	29
PT-TF2	Phenyl	(85 : 15)	57	40
PS-TF3	Phenyl	(40 : 60)	21	61
	2-Methylphenyl	(60 : 40)	40	60
	2,6-Dimethylphenyl	(62 : 38)	39	40

a P(Phosphorylated peptide) and DP (Dephosphorylated peptide) in crude extracts after HF treatment.

Preparation for Antisera Specific to Phosphorylation Sites (Arioka *et al.*, 1993)

For immunization, chemically synthesized phosphopeptides were linked to keyhole limpet hemocyanin in Freund's complete adjuvant and injected intramuscularly into a rabbit at biweekly intervals. Antisera obtained were precipitated by an equal volume of 34% sodium sulfate and passed through Affigel 10 column coupled with the appropriate non-phosphorylated peptide. The antisera were enriched by this procedure to recognize only the phosphorylated forms of TF2 or TF3.

Table 2. Specificities of antisera

Designations	Immunogens	Reactivites to peptides PS-TF2	PT-TF2	PSPT-TF2	PS-TF3
anti-PS144	PS-TF2	++	+	++	-
anti-PS144/PT147	PT-TF2	++	+	+	-
anti-PT147	PSPT-TF2	-	++	++	-
anti-PS315	PS-TF3	-	-	-	++

and showed no cross-reactivity with TF3 nor TF2-(phospho)peptides. This antiserum was thus named anti-PS315. Antiserum to PSPT-TF2 was named anti-PT147, because it reacted with PT-TF2 and PSPT-TF2, but little with PS-TF2. The specificities of antisera to PS-TF2

The binding specificities of these antisera were examined using phospho- or non-phospho-peptides, as shown in Table 2. Antiserum to PS-TF3 reacted only with PS-TF3, and to PT-TF2 were very similar to react with both PS-TF2 and PT-TF2. From antiserum to PS-TF2, however, antibody specific to PS-TF2 was purified by adsorption with PT-TF2 column, and named anti-PS144. In the case of antiserum to PT-TF2, in contrast, reactivities to PS-TF2 and PT-TF2 were inseparable by adsorption with either phosphopeptide. This antiserum was named anti-PS144/PT147.

The consensus Ser(P)/Thr(P)Pro sequences are not the epitopes for these antisera, since the antisera to phosphorylated TF2 peptides showed no cross-reactivities with PS-TF3, another phosphopeptide containing Ser(P)Pro sequence. We noticed that the antibody reactive with PS-TF2 also recognized PT-TF2 in antiserum to PT-TF2, and that antiserum to PS-TF2 was a mixture of the antibody reactive with PS-TF2 and that with PT-TF2. These facts suggest that the immune system may produce antibodies directed to the bulky phosphate group of either Ser(P) or Thr(P) surrounded in a repetitive sequence, SerProGly *** ProGly, where *** means Ser(P) or Thr(P).

We were successful to obtain a set of antisera specific to each of the three phosphorylation sites by TPKII. These antisera strongly reacted with tau protein phosphorylated by TPKII, respectively, confirming that these phosphorylation sites were exactly determined by direct sequence analysis without any modification in the gas-phase sequencer.

Application of Phosphorylation Site-Specific Antibodies (Arioka *et al.*, 1993)

For Phosphorylation-State of Tau Protein. Tau was purified from microtubule proteins of bovine brain, and its phosphorylation state at TPKII sites was examined by immunoblot analyses using these antibodies. Tau from microtubule proteins was immunoreactive to anti-PS144 and anti-PS315, but not to anti-PT147, while tau dephosphorylated by calf intestine phosphatase was no longer recognized by these antibodies. Phosphorylation of dephosphorylated tau by TPKII regained stronger immunoreactivities to these antibodies. These results demonstrate that isoforms of tau from microtubule proteins are already phosphorylated partially at Ser144 and Ser315.

For Developmental Changes in Phosphorylation-State of Tau Protein. Total heat-stable brain extracts from rats of various ages (from embryonic day 18 to postnatal 5 weeks) were prepared and analyzed by immunoblotting with these antibodies. Immunoreactivities were measured quantitatively by scanning the radioactivities of bound [^{125}I]protein A and normalized for the amounts of tau in each sample. At embryonic day 18, juvenile form of tau, the shortest tau, is predominant. The shortest tau gradually disappears to be replaced by adult types, the longer isoforms of tau, till postnatal 5 weeks. Age-related change in immunoreactivity was observed in staining by anti-PS144 and anti-PT147, though staining by anti-PS315 was relatively constant. Phosphorylation of rat brain tau by TPKII occurs evidently in both pre- and post-natal periods, but reaches its highest level at postnatal week 1. In rat brains, formation of neuronal networks proceeds during embryonic and newborn periods. TPKII may be involved in a process, such as neurite extension, by phosphorylating tau and eventually destabilizing preexisting microtubules.

Acknowledgments: This work was supported by the Grant-in-Aid for Scientific Research on Priority Areas No.04268104 from the Ministry of Education, Science and Culture, Japan.

REFERENCES

Arendt, A., Palczewski, K., Moore, W. T., Caprioli, R. M., McDowell, J. H. and Hargrave, P. A., 1989 Synthesis of phosphopeptides containing *O*-phosphoserine or *O*-phosphothreonine, *Int. J. Peptide Protein Res.* 33:468.

Arioka, M., Tsukamoto, M., Ishiguro, K., Kato, R., Sato, K., Imahori K.and Uchida, T., 1993 Tau protein kinase II is involved in regulation of normal phosphorylation state of tau protein, *J. Neurochem.* in press.

Ishiguro, K., Ihara, Y., Uchida, T. and Imahori, K., 1988 A novel tubulin-dependent protein kinase forming a paired helical filament epitope on tau, *J. Biochem. Tokyo* 104:319.

Ishiguro, K., Omori, A.,.Sato, K., Tomizawa, K., Imahori, K. and Uchida, T., 1991 A serine/threonine proline kinase activity is included in the tau protein kinase fraction forming a paired helical filament epitope, *Neurosci. Lett.* 128:195.

Ishiguro, K., Takamatsu, M., Tomizawa, K., Omori, A., Takahashi, M., Arioka, M., Uchida, T. and Imahori, K., 1992 Tau protein kinase I converts normal tau protein into A68-like component of paired helical filaments, *J. Biol. Chem.* 267:10897.

Meyer, H. E., Hoffmann-Posorske, E. and Heilmeyer, Jr. , L. M. G., 1991 Determination and location of phosphoserine in proteins and peptides by conversion to S-ethylcysteine, *in*: Methods in Enzymology Vol.201, pp.169, Hunter, T. and Sefton, B. M. eds., Academic Press, San Diego.

Meyer, H. E., Hoffmann-Posorske, E., Korte, H. and Heilmeyer, Jr., L. M. G., 1986 Sequence analysis of phosphoserine-containing peptides. Modification for picomolar sensitivity, *FEBS Lett.* 204:61.

Omori, A., 1990 Improved microsequencing analysis and its application for the blotted protein from SDS -polyacrylamide gel electrophoresis, *J. Protein Chem.* 9:250.

Packman, L. C., 1990 Identification of phosphoserine in protein sequence analysis, *J. Protein Chem.* 9:307.

Tsukamoto, M., Kato, R., Ishiguro, K., Uchida, T. and Sato, K., 1991 Improved protective groups for phosphate of *O*-phosphoserine useful for the solid-phase peptide synthesis, *Tetrahedron Lett.* 32:7083.

PREDICTION AND DATABASE ANALYSIS

PREDICTION OF PROTEIN STRUCTURE FROM MULTIPLE SEQUENCE ALIGNMENT

Geoffrey J. Barton, Robert B. Russell
and Craig D. Livingstone

Laboratory of Molecular Biophysics
The Rex Richards Building
South Parks Rd., Oxford OX1 3QU, UK

INTRODUCTION

Knowledge of the protein three dimensional structure gives the most accurate guide to the significance of individual residues to protein function. Three dimensional structural information is essential for the rational design and interpretation of site directed mutagenesis studies aimed at enhancing stability or catalytic activity of a protein. Unfortunately, the determination of protein three dimensional structures by X-ray crystallography and NMR techniques is often difficult, and as a consequence, there are only around 1000 protein structures known to atomic resolution. In contrast, the rapid development of nucleic acid and protein sequencing techniques has led current protein sequence databanks to contain over 40,000 unique entries. When a newly determined sequence shows similarity to a protein of known three dimensional structure, then a tertiary model may be built for the new protein with a fair degree of confidence (e.g. see Blundell *et al.*, 1987, for review). However, until recently, attempts to predict protein structure from the amino acid sequence have had little success, with three state (α-helix, β-strand and loop) secondary structure prediction accuracies of 55-65%.

It is now normal for a protein considered important due to its key functional and potential therapeutic role to be sequenced from different tissues and organisms. The multiple alignment of such a family of proteins provides a great deal more information about the structurally and functionally important residues than the sequences considered individually. Exploitation of this additional information can yield a dramatic improvement in the accuracy of secondary and tertiary structure predictions from the amino acid sequences (Barton *et al.*, 1991; Benner and Gerloff, 1990; Crawford *et al.*, 1987; Russell *et al.*, 1992). In addition, a multiple alignment may indicate the residues that confer differing functional specificity to members of the family.

In this article, we first review a method for rapidly generating protein multiple sequence alignments and assessing their reliability. We then describe a recent example of the accurate

Methods in Protein Sequence Analysis, Edited by K. Imahori
and F. Sakiyama, Plenum Press, New York, 1993

prediction of secondary structure and buried residues for a protein family that was of unknown tertiary structure at the time of the predictions. Finally, two new methods for the analysis of protein sequence and structure data are described. AMAS - a tool to identify the most "interesting" features in a multiple sequence alignment, and StAMP - a system for generating reliable multiple structure superpositions.

Multiple Alignment of Protein Sequences

Aligning Two Sequences and Assessing Significance: Methods for optimally aligning two protein sequences were developed in the early 1970s by Needleman and Wunsch (1970) and Sellers (1974). These *dynamic programming* algorithms take a table of weights for matching each possible amino acid pair (e.g. W-W: 17; A-R: -2; etc.) together with a weight for introducing a gap. The algorithm then determines the alignment of the sequences (including gaps) that gives the maximum score. This process takes MN steps, where M and N are the lengths of the two sequences being compared (for further details see:Barton (1990), and refs. therein). Although this method is guaranteed to give the mathematically optimal alignment of the sequences, such an alignment may have no biological significance; in order to be useful, guidelines must be established to assess the likely accuracy of any alignment generated by dynamic programming.

When equivalencing residues in a sequence alignment, one is saying that the residues are performing similar structural roles in the proteins. Comparison of the three dimensional structures for related proteins (e.g the globins) shows that the core secondary structure units (α-helices and β-strands) are more highly conserved than the intervening loop regions. Indeed, whilst the structure of the cores may be very similar for a family, the loops may

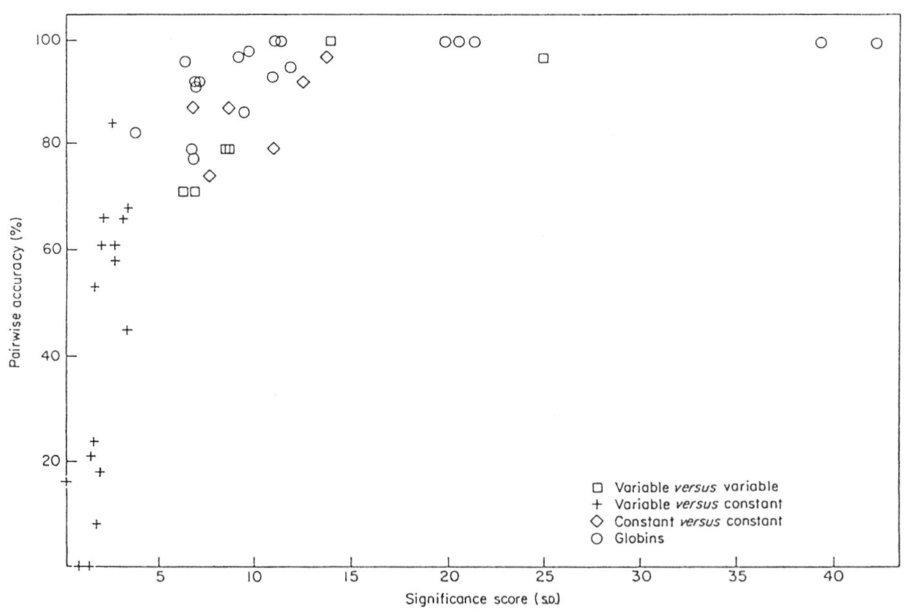

Figure 1. Accuracy of pairwise sequence alignment in the protein core as a function of S.D. score. S.D. scores are obtained by first aligning the pair of sequences using the Needleman and Wunsch algorithm and obtaining the best score V. The sequences are then randomized and re-compared a large number of times (typically 100), the mean S and standard deviation D for the randomized sequence scores is calculated. The S.D. score is given by S.D. = $(V—S)/D$.

show no structural equivalences. Accordingly, we should expect dynamic programming methods at least to equivalence correctly residues in the core secondary structures of the proteins. Figure 1 (from Barton and Sternberg, 1987) illustrates the relationship between accuracy of alignment in the protein core and a standard normalised score for sequence alignment (see legend for details). Clearly, above a score of 6.0 S.D., the accuracy of alignment is very good. Below 6.0 S.D., the quality of alignment is unpredictable. More importantly, there are no bad alignments with scores above 7.0 S.D., and above 15 S.D. the accuracy is near perfect. These threshold values provide guidelines for the belief in a pairwise protein sequence alignment.

Multiple Sequence Alignment: Although the Needleman and Wunsch algorithm can in theory be extended to align optimally any number of sequences, in practice, the memory and CPU requirements become prohibitive for more than three sequences. However, given that we can predict which sequences will align well by pairwise alignment (i.e. those scoring > 6.0 S.D.) then a practical method to generate multiple alignments uses successive applications of the pairwise algorithm (Barton, 1990; Barton and Sternberg, 1987; Feng and Doolittle, 1987). The method for multiple alignment proceeds in three steps. (1) Perform all pairwise sequence comparisons; (2) Cluster the sequences by similarity using the pairwise scores; (3) Generate the multiple alignment. Steps 1 and 2 describe the similarities within the sequence set. Having established which sequences are most similar to which, then step 3

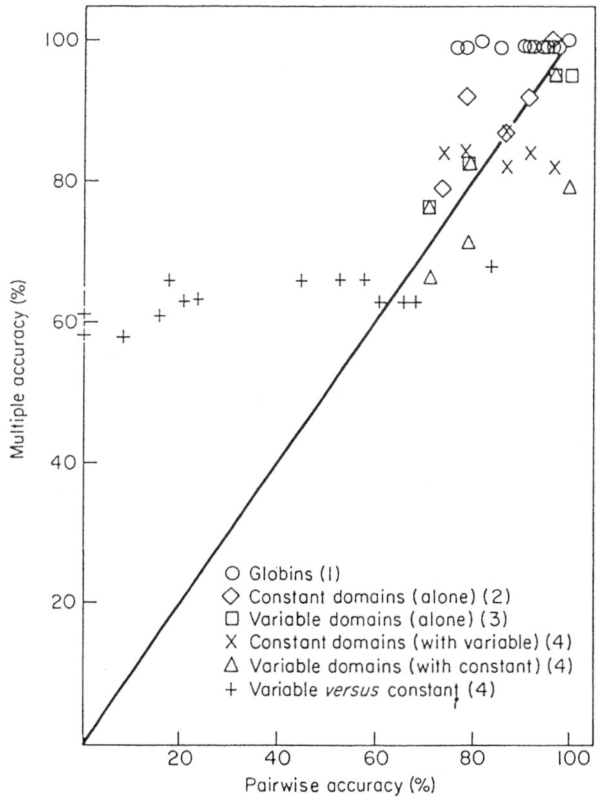

Figure 2. Plot of accuracy for multiple aligment versus pairwise alignment. Points above the line show improvements in accuracy on multiple alignment.

can be performed in two ways. The simplest approach is first to align the most similar pair of sequences, then align the sequence which shows greatest similarity to the pair using average scores at each position in the pre-existing alignment. Further sequences are added one at a time to the growing multiple alignment. A slightly more complex procedure follows the organisation of a tree or dendrogram; this leads to a final alignment that groups the sequences more faithfully by their similarity. Figure 2 illustrates that the alignment accuracy obtained by pairwise and multiple alignment methods is comparable for proteins in the globin and immunoglobulin families (for a more detailed discussion of these methods, see Barton (1990)).

Prediction of Secondary Structure and Buried Residues from Multiple Alignment

Overview: Given the accurately determined multiple sequence alignment of a protein family, positions that have the same amino acid in all members are of clear importance. For example, conserved histidine, or aspartate residues are often indicative of the active site. In addition, the following more subtle inferences may be made. (1) The positions of gaps in the alignment are strong indicators of surface loops in the protein structure. (2) Residues that are distant from gapped regions are likely to be performing similar roles in the folded structure of the proteins. (3) Positions in the alignment that have amino acid residues with conserved hydrophobic character are likely to be buried in the hydrophobic core of the protein and contribute to the stability of the native folded structure. (4) Patterns of residue conservation characteristic of particular secondary structure types may be discerned - e.g. CXXCCXXC for helix and CXCXCXC for strand where C=conserved, usually hydrophobic character. In both examples, one face of the helix or strand is conserved and if hydrophobic, is likely to be facing the core of the protein.

Conventional secondary structure prediction methods may be applied to each sequence in the alignment. Figure 3 illustrates a fragment of a multiple sequence alignment together with predictions on each sequence by two turn prediction methods and three secondary structure prediction algorithms. Although individual predictions on separate sequences show considerable variation, the regions consistently predicted by all methods on all sequences (bottom line) are more reliable. This combined prediction approach provides a good first guess at the secondary structure of the protein. When combined with an analysis of the patterns of conservation, a convincing argument may be built up in favour of a particular structural state. These inferences may *only* be made if the sequence alignment is accurate; attempting to make structure predictions from an alignment that includes proteins of marginal similarity (i.e. < 6.0 S.D.) is usually a fruitless task.

Structure Prediction of the SH2 Domain Family: In meteorology, it is easy to get accurate predictions of yesterday's weather. While such predictions are useful to aid the development of new methods, it is more challenging and useful to predict tomorrow's conditions. Similarly, for protein structure prediction, it is most interesting to predict the structure of proteins that have yet to be solved by X-ray crystallography or NMR

→

Figure 3. Example of multiple secondary structure prediction. A multiple sequence alignment is shown at the top. Turn predictions by the methods of Rose (1978) and Wilmot and Thornton (1988) are shown for each sequence. Helix/Strand predictions by the methods of Robson (1978), Lim (1974) and Chou and Fasman (1978) are also shown. A summary prediction for turn is calculated by summing the results of the two turn prediction methods. Similarly, predictions for helix/strand are obtained by summing the predicted states for all sequences. The summary prediction shown as the last line, is calculated by combining the turn and helix predictions with equal weight and taking the maximum value.

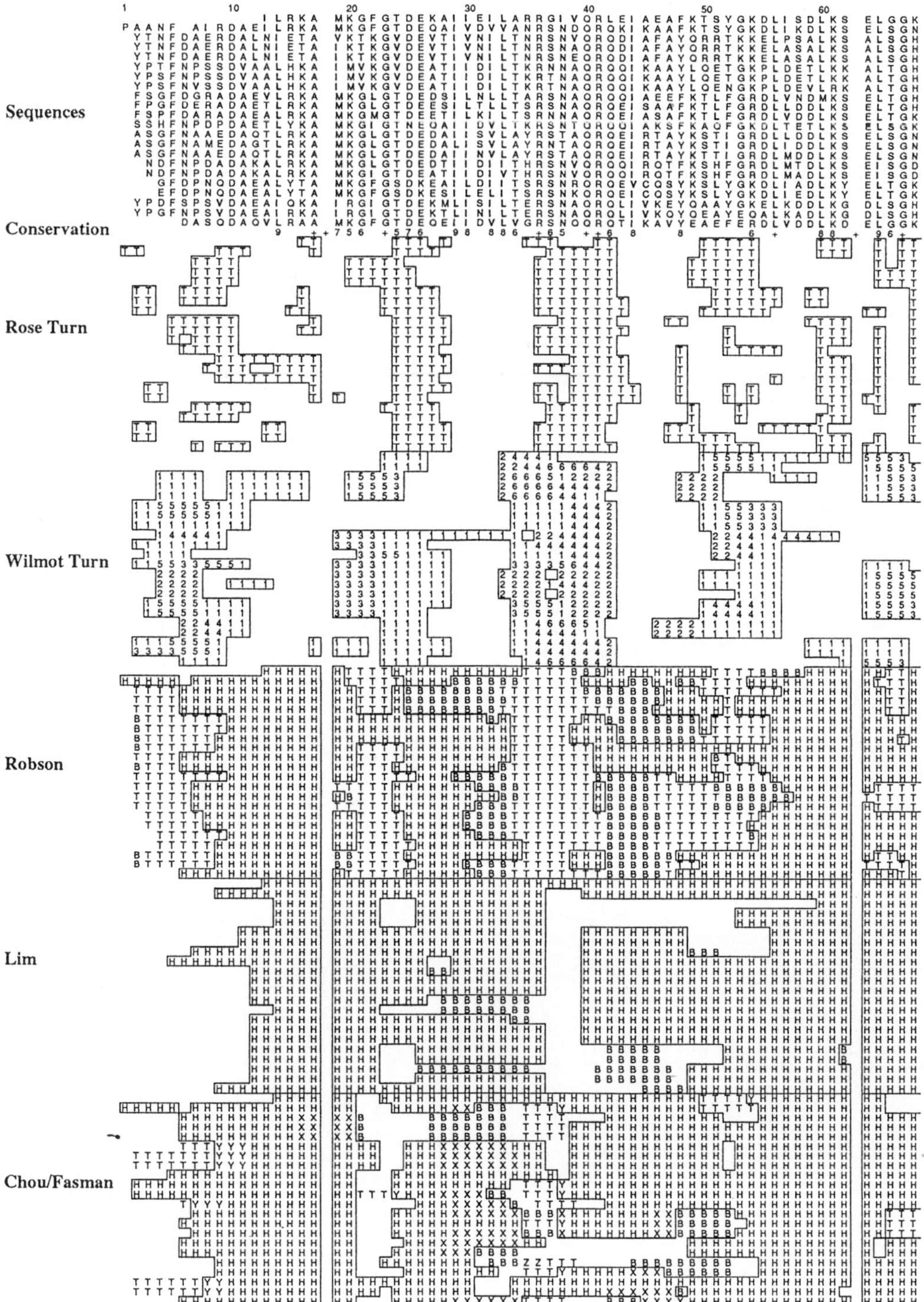
Sequences
Conservation
Rose Turn
Wilmot Turn
Robson
Lim
Chou/Fasman

techniques! Recently, we accurately predicted the secondary structure and position of buried residues for the annexin super-gene family of proteins prior to the crystal structure solution (Barton *et al.*, 1991). This success suggested that other families might be amenable to the technique.

The SH2 (*src* Homology region 2) domains bind to phosphotyrosine residues and are believed to be important in the control of intracellular signalling (see Heldin (1991), Waksman *et al.* (1992) for reviews). In the last quarter of 1991, there were 67 distinct domains available for analysis, multiple alignment of these domains could be performed automatically with little ambiguity and with a high degree of confidence. Figure 4 shows the resulting multiple sequence alignment. There are seven clear blocks of residues (A-F) separated by regions that tolerate insertions/deletions (gaps). Following the points outlined

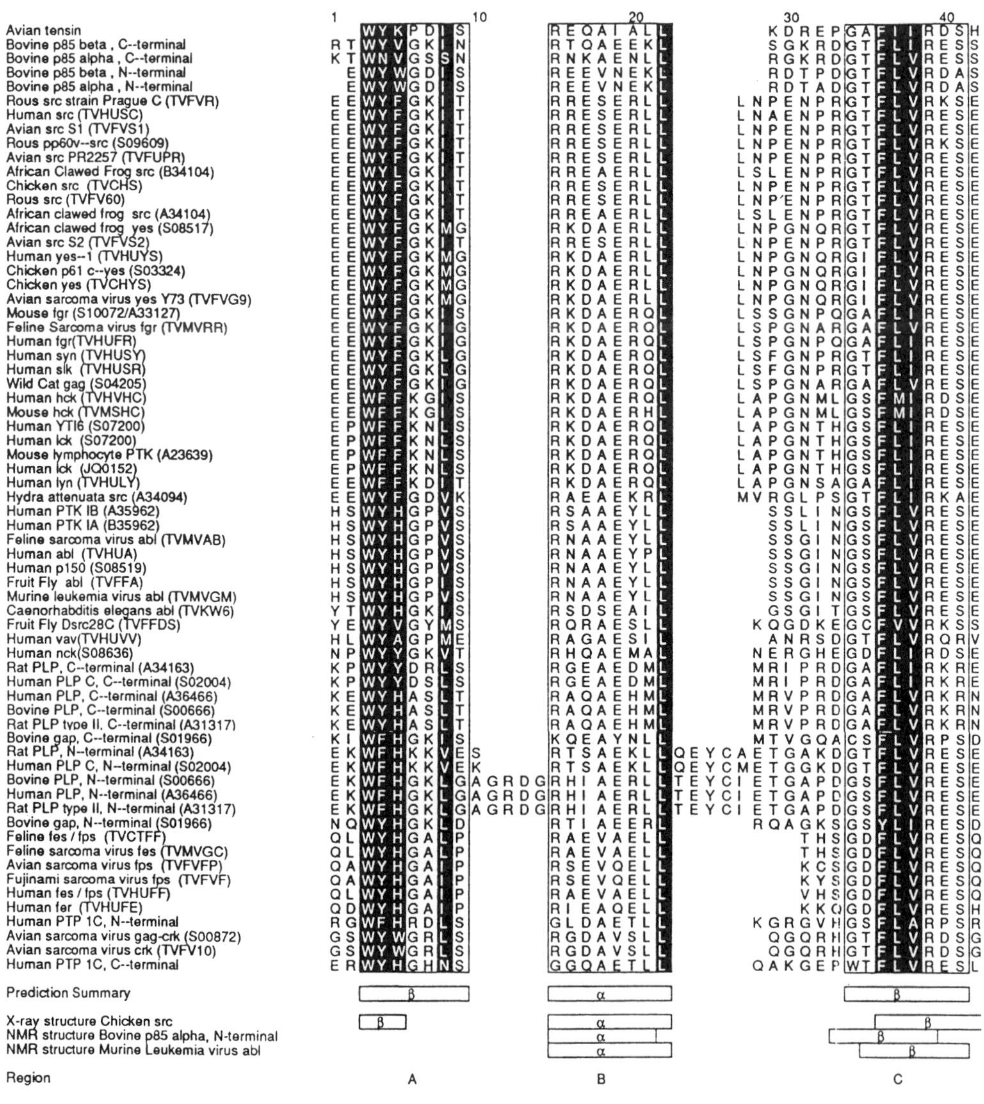

Figure 4. Multiple alignment of the SH2 domains. The summary predicted secondary structure, and the experimentally determined secondary structures are shown below the alignment.

above, the gaps suggest the location of non-core loops that link the core secondary structures of the domain. By inference, the regions without gaps must represent secondary structure units. Combination of secondary structure prediction methods as illustrated in Figure 3 suggested the secondary structures β–α–β–β–β–β–α for these ungapped regions. This prediction is supported by the patterns of conserved hydrophobic amino acids that are highlighted in black. For example, in region D conserved hydrophobics alternate following the pattern CXCXC, suggestive of β-strand, whilst in region G, conserved hydrophobics are consistent with the pattern CXXCCXXC indicating an α-helix. The paper describing this prediction in detail was accepted in March 1992 (Russell *et al.*, 1991), subsequently, the structures of three different SH2 domains have been determined by experimental methods. Chicken *src* by X-ray crystallography (Waksman *et al.*, 1992), c-abl (Overduin *et al.*, 1992), and p85α- subunit of phosphatidylinositol-3-OH kinase (Booker *et al.*, 1992) by NMR

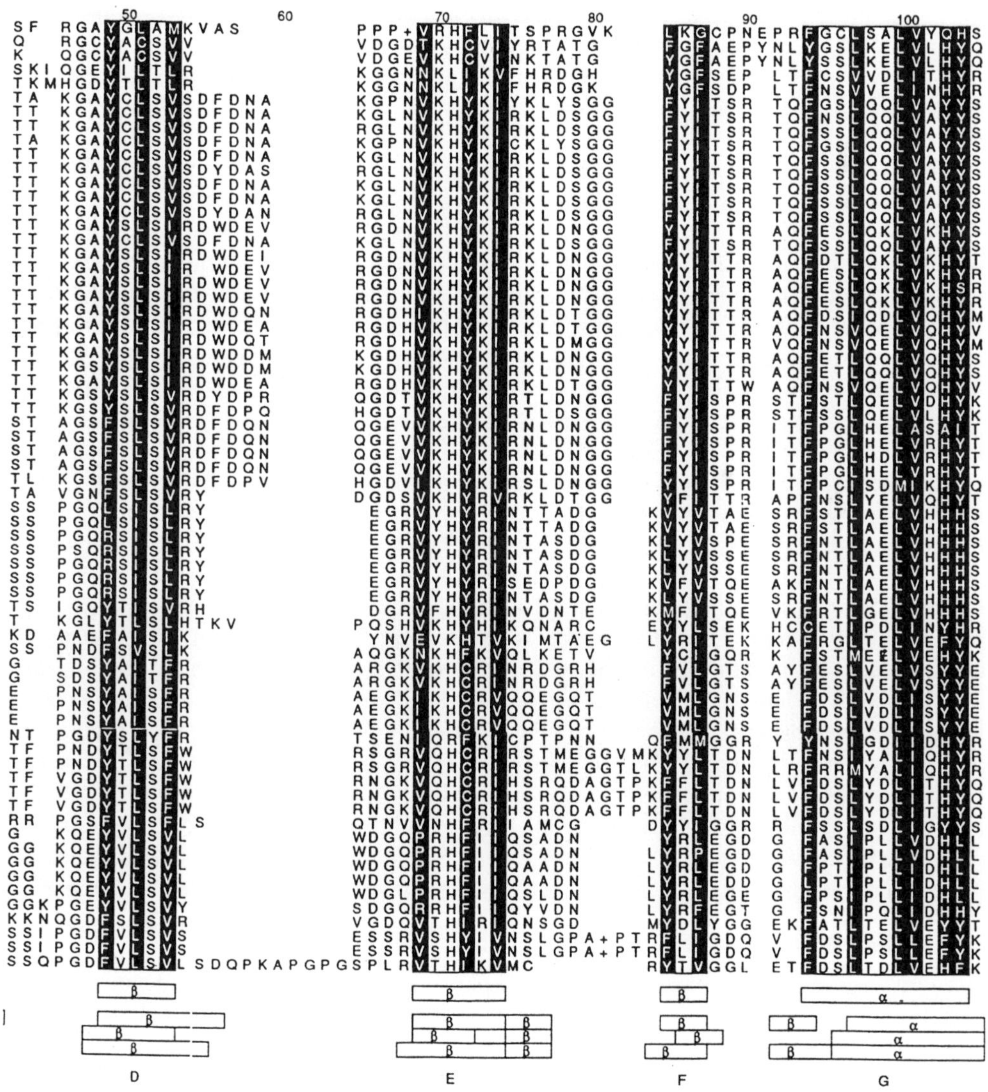

(Figure 4 continued)

techniques. The secondary structure for each of these domains is shown at the bottom of Figure 4. The numbers and types of secondary structure predicted are in good agreement with all three experimental structures. Our consensus prediction achieved 77.9, 76.0 and 75.6 % accuracy when compared to the *src*, abl and p85α- domains. It is interesting to note that the three SH2 domains of known structure only agree with each other at 70% of residue positions. This suggests that our prediction is a good representation of the average secondary structure for the family.

Accurate identification of the secondary structure units for a protein is one step towards predicting the native fold. In addition, it is important to locate those positions that pack in the protein core and stabilise the structure. Following from the work on annexins (Barton *et al.*, 1991), we considered those positions that were highly conserved and of general hydrophobic character to be buried. Careful scrutiny of the *src* SH2 domain coordinates, kindly supplied by Dr. J. Kuriyan, suggests that all 22 predicted buried positions are indeed inaccessible to solvent.

In summary, although, three short, β-strands in a variable surface sheet were not correctly predicted, the principal conserved SH2-domain core was accurately located. This degree of accuracy coupled with distance constraints from three residues predicted to interact with the phosphotyrosine substrate (Arg-15, Arg-39 and His-71) was sufficient to speculate correctly the most likely fold for the SH2 domain.

Automatic Analysis of Large Multiple Sequence Alignments - AMAS

The dendrogram, or tree generated during multiple sequence alignment provides a guide to the overall similarity shared by sub-groups of the protein sequences aligned. For example, the tree for the SH2 domain alignment in Figure 4 clearly delineates the *src* like SH2 domains from those of p85 (data not shown). However, we are often interested in the individual residues that confer the functional specificity of a protein, perhaps as a guide for site-directed mutagenesis experiments. This can often be determined by scanning the sequence alignment by eye for similiarities or differences in amino acid type or property. For example, in the detailed analysis of the residue conservation seen across 88 repeat units in the annexin super-gene family, conservation of Arg in repeats 4 and Glu in repeats 2, but unconserved amino acids in repeats 1 and 3 suggested repeats 2 and 4 may interact *via* a buried salt-bridge (Barton *et al.*, 1991). This observation was subsequently confirmed by X-ray crystallographic analysis (Huber *et al.*, 1990). When many long sequences are to be scrutinized, the task becomes unmanageable and the risk of missing interesting residue substitutions is great. Accordingly, we have developed techniques for the analysis of multiply aligned protein sequences that allow residue and physico-chemical property differences to be identified rapidly in large sequence alignments. These methods are encoded in the computer package AMAS (Analysis of Multiply Aligned Sequences) (Livingstone and Barton, in preparation). Briefly, each amino acid is assigned one or more physico-chemical properties, developed from the work of Taylor (1986) and Zvelebil *et al.* (1987). For example, Lysine might be assigned the properties of charge, positive and hydrophobic (by virtue of its long aliphatic side chain), Alanine the properties of small and hydrophobic. Residue conservation at any position in an alignment may then be quantified by comparing the properties represented at the position in the alignment. Given the property table, a multiple sequence alignment, and pre-determined clustering of the sequences (perhaps into functional sub-families), AMAS compares all pairs of grouped sequences and annotates the positions with shading, bolding and font changes. More detailed information is also provided in a textual output.

Figure 5 illustrates part of the AMAS output for the analysis of the SH2 domain alignment shown in Figure 4. Eight groups have been chosen, inverted characters (e.g. at

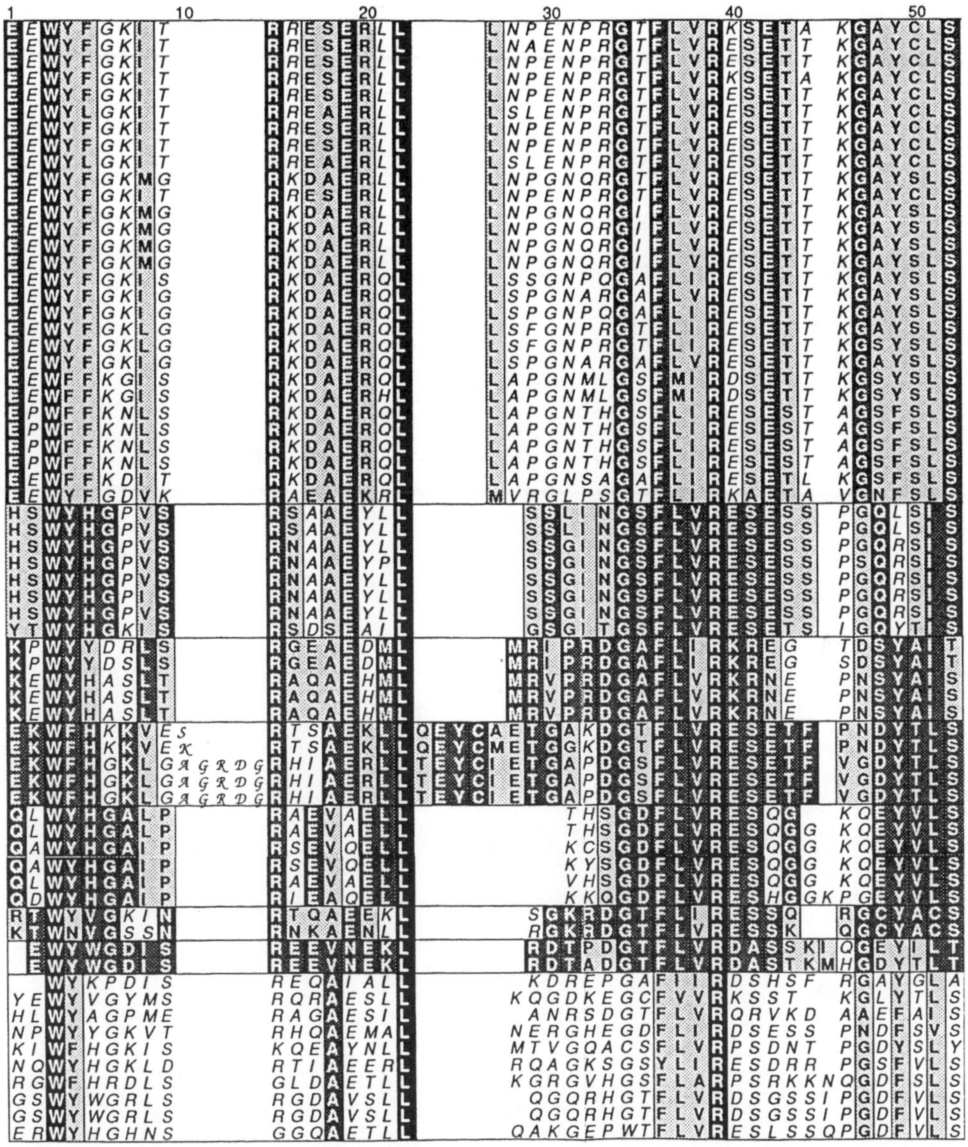

Figure 5. AMAS analysis of the SH2 domains. The shading illustrates similarities in physicochemical propertied within each of 8 predetermined groups. Note that the order of sequences is different to Figure 4.

position 39) illustrate identities across all groups. Heavy shading highlights identities within groups, whilst lighter shading shows regions sharing common physico-chemical properties. Italics show positions unconserved within a group, whilst Chancery script positions are unconserved across the entire alignment. One feature that is apparent from Figure 5 is the greater preponderance of conserved Lys and Arg residues within the *src*-like

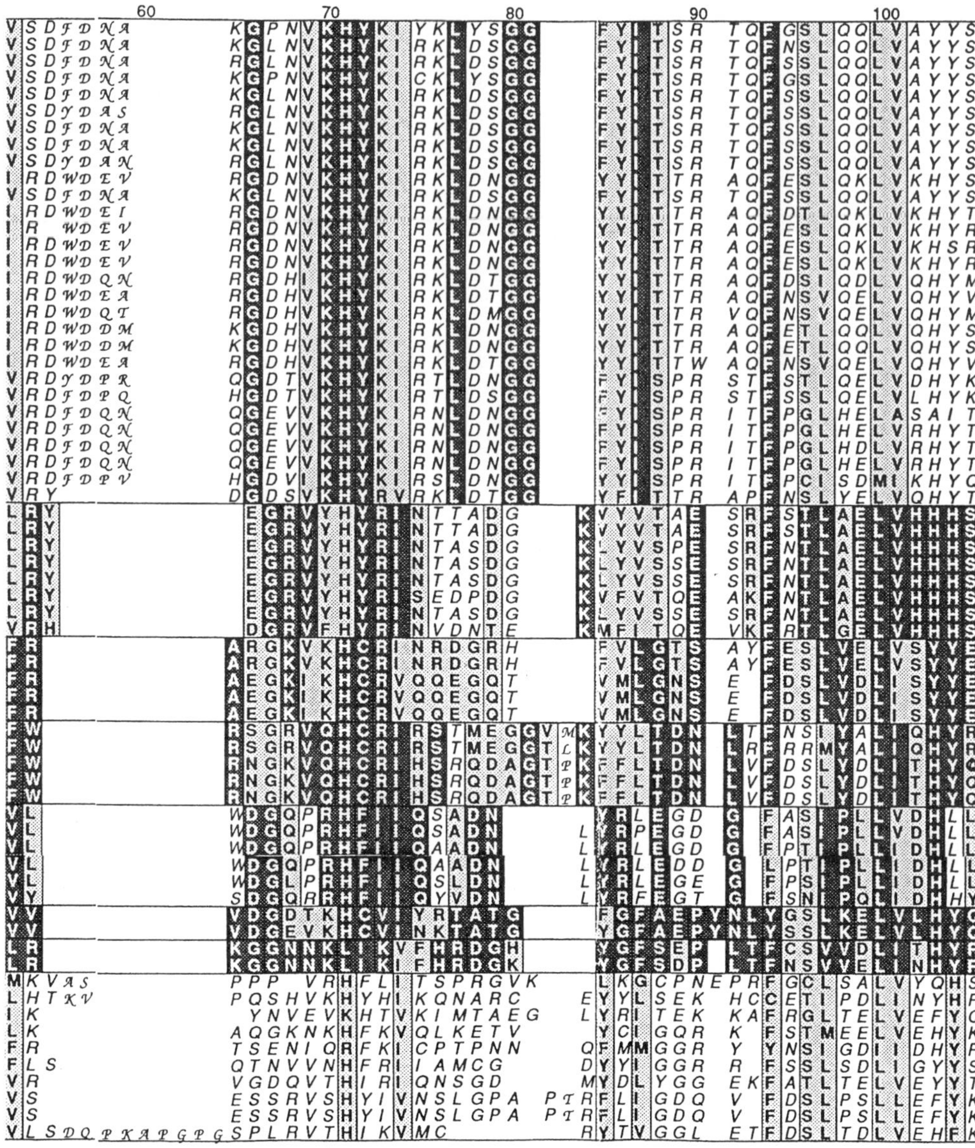

(Figure 5 continued)

sub-family. Examination of the crystal structure of the chicken *src* SH2 domain (Waksman *et al.*, 1992) shows these Lys and Arg residues to be on the side of the molecule away from the phosphotyrosine binding site, though no clear function for the residues has yet been determined.

Alignment of Multiple Protein Three-Dimensional Structures

Although the database of protein tertiary structures is growing at a slower rate than the sequence database, there are now a number of protein families for which more than one homologous structure is available. Analysis of these families provides valuable insights into the rules governing the stability of individual folds (e.g. see Bashford (1987)).

However, if this information is to be exploited effectively, reproducible structure-based multiple alignments must be generated. The method that we have developed (Russell and Barton, 1992) follows a similar strategy to the techniques outlined above for multiple sequence alignment. (1) An initial structure superposition is generated; (2) All structures are compared pairwise using dynamic programming; the score being optimised is a function of the conformational similarity and distance between each residue pair; (3) Cluster analysis is performed and a dendrogram drawn; (4) Multiple structure alignments are calculated by progressing from the leaves to the root of the dendrogram, structurally aligning the proteins at each node. The program StAMP (Structural Alignment of Multiple Proteins) that implements this algorithm, provides transformation matrices to superimpose the protein structures and multiple sequence alignments for each node in the dendrogram. The program also assigns an overall similarity value to each pair of structures or aligned structures (S_c) and a normalised quality value to each aligned position ($P'i,j$). These values allow the most structurally similar regions of the protein family to be identified automatically. StAMP has been used to derive a library of multiple structure alignments for all protein families for which at least two three-dimensional structures have been solved.

Acknowledgements: We thank Prof. L. N. Johnson for her continued support; Drs M. J. E. Sternberg, P. Freemont, R. Newman, M. Crumpton and Mr. J. Breed for their contributions to the work described here.

REFERENCES

Barton, G. J.,1990 Protein multiple sequence alignment and flexible pattern matching, *Methods Enzymol.* 183:403.

Barton, G. J., Newman, R. H., Freemont, P. F. and Crumpton, M. J., 1991 Amino acid sequence analysis of the annexin super-gene family of proteins, *Eur. J. Biochem.* 198:749.

Barton, G. J. and Sternberg, M. J. E., 1987 A strategy for the rapid multiple alignment of protein sequences: Confidence levels from tertiary structure comparison, *J. Mol. Biol.* 198:327.

Bashford, D., Chothia, C. and Lesk, A. M., 1987 Determinants of a protein fold: Unique features of the globin amino acid sequences, *J. Mol. Biol.* 196:199.

Benner, S. A. and Gerloff, D., 1990 Patterns of divergence in homologous proteins as indicators of secondary and tertiary structure: A prediction of the structure of the catalytic domain of protein kinases, *Adv. Enz. Reg.* 31:121.

Blundell, T. L., Sibanda, B. L., Sternberg, M. J. E. and Thornton, J. , 1987 Knowledge-based prediction of protein structures and the design of novel molecules, *Nature (London)* 326:347.

Booker, G. W., Breeze, A. L., Downing, A. K., Panayotou, G., Gout, I., Waterfield, M. D. and Campbell, I. D., 1992 Structure of an sh2 domain of the p85alpha subunit of phosphatidylinositol-3-oh kinase, *Nature (London)* 358:684.

Chou, P. Y. and Fasman, G. D., 1978 Prediction of the secondary structure of protein from their amino acid sequence, *Adv. Enzymol.* 47:45.

Crawford, I. P., Niermann, T. and Kirchner, K., 1987 Prediction of secondary structure by evolutionary comparison: Application to the alpha subunit of tryptophan synthase, *Proteins: Struct. Funct. Genet.* 2:118.

Feng, D. F. and Doolittle, R. F., 1987 Progressive sequence alignment as a prerequisite top correct phylogenetic trees, *J. Mol. Evol.* 25:351.

Garnier, J., Osguthorpe, D. J. and Robson, B., 1978 Analysis of the accuracy and implications of simple methods for predicting the secondary structure of globular proteins, *J. Mol. Biol.* 120:97.

Heldin, C., 1991 Sh2 domains: Elements that control protein interactions during signal transduction, *Trends Biochem. Sci.* 16:450.

Huber, R., Romsich, J. and Paques, E-P., 1990 The crystal and molecular structure of human annexin v, an anticoagulant protein that binds to calcium and membranes, *EMBO J.* 9:3867.

Koch, C. A., Anderson, D., Moran, M. F., Ellis, C. and Pawson, T., 1992 Sh2 and sh3 domains: Elements that control interactions of cytoplasmic signaling proteins, *Science* 252:66-673.

Lim, V. I. 1974, Algorithms for prediction of α helices and β structural regions in globular proteins, *J. Mol. Biol.* 88:73.

Needleman, S. B. and Wunsch, C. D., 1970 A general method applicable to the search for smilarities in the amino acid sequence of two proteins, *J. Mol. Biol.* 48:443.

Overduin, M., Rios, C. B., Mayer, B. J., Baltimore, D. and Cowburn, D., 1992 Three dimensional solution structure of the src homology 2 domain of c-*abl*, *Cell* 70:697.

Rose, G.D., 1978 Prediction of chain turns in globular proteins on a hydrophobic basis, *Nature (London)* 272:586.

Russell, R. B. and Barton, G. J.,1992 Multiple protein sequence alignment from tertiary structure comparison: assinment of global and residue confidence levels, *Proteins: Struct. Funct. Genet.* 14:309.

Russell, R. B., Breed, J. and Barton, G. J., 1992 Conservation analysis and secondary structure prediction of the sh2 family of phosphotyrosine binding domains, *FEBS Lett.* 304:5.

Sellers, P. H., 1974 On the theory and computation of evolutionary distances, *SIAM J. Appl. Math.* 26:787.

Taylor, W. R., 1986 Classification of amino acid sonservation, *J. Theor. Biol.* 119:205.

Waksman, G., Kominos, D., Robertson, S. C., Pant, N., Baltimore, D., Birge, R. B., Cowburn, D., Hanafusa, H., Mayer, B. J., Overduin, M., Resh, M. D., Rios, C. B., Silverman, L. and Kuriyan, J., 1992 Crystal structure of the phosphotyrosine recognition domain of sh2 of v-*src* complexed with tyrosine-phosphorylated peptides, *Nature (London)* 358:646.

Wilmot, A. C. M. and Thornton, J. M., 1988 Analysis and prediction of the different types of beta-turn in proteins, *J. Mol. Biol.* 203:221.

Zvelebil, M. J . J. M., Barton, G. J., Taylor, W. R. and Sternberg, M. J. E., 1987 Prediction of protein secondary structure and active sites using the alignment of homologous sequences, *J. Mol. Biol.* 195:957.

SEQUENCE PATTERNS THAT CHARACTERIZE PROTEIN FAMILIES WITH A COMMON FOLD

Mark S. Johnson and Thomas L. Blundell

Imperial Cancer Research Fund Unit of Structural Molecular Biology
Department of Crystallography
Birkbeck College
Malet Street, London WClE1HX, United Kingdom

INTRODUCTION

Divergent evolution has led to families of proteins with similar three-dimensional structures (Bajaj and Blundell, 1984). The identification of sequences that can produce a given fold is one approach that is likely to yield solutions to the protein folding problem, since the three-dimensional structures of proteins place severe constraints on the sequences that can adopt a particular fold. For example, within a family, equivalent amino acids of the solvent inaccessible core are more conserved in identity and position than those on the protein surface (Chothia and Lesk, 1982a,b; Lesk and Chothia, 1986; Hubbard and Blundell, 1987), except where functional restraints such as catalytic activity or ligand binding require the presence of a particular amino acid.

Our approach has been the comparison of three-dimensional structures and the analysis of the constraints that act upon them (Overington *et al.*, 1990a,b; 1992). Hence we correlate the local environment at a position in the fold with the allowed residues at that site (Johnson *et al.*, 1992). Arising from this work is the flexible treatment of both structure prediction and alignment using information or rules from known folds. Since the number of basic folding modules is likely to be limited, the problem of protein folding then becomes one of classification and identification of these folding units.

From a classification of tertiary structures we have derived rules that can be used directly in a finely-tuned comparison between structure and sequence. The local environments in known homologous structures (i.e. residue accessibility to solvent, secondary structure and patterns of hydrogen bonding) are used to select position-specific amino acid substitution scores and to produce a rule-base suitable for alignment of sequences or searching data banks. These amino acid substitution scores are themselves specific to the location within a protein: they have been accumulated from the large pool of aligned families of protein structures in which the observed substitutions have been classified according to features of the local structure.

Methods in Protein Sequence Analysis, Edited by K. Imahori
and F. Sakiyama, Plenum Press, New York, 1993

Hence, in the case of structure-sequence alignments, the value attributed to a particular amino acid interchange in the comparison is not constant, but is dependent upon the environmental-context. The penalties for the appearance of insertions/deletions within the alignment also vary with position and environment within the known protein structure: gaps are encouraged to appear at solvent accessible positions, within loops, and at sites of existing gaps as defined by the structural alignment of two or more structures.

This approach has been adapted to search for likely folds given the patterns of amino acid substitutions in multiply-aligned sequences, the alignment of structures and sequences (also see Sali *et al.*, 1990; Lüthy *et al.*, 1992; Bowie *et al.*, 1991; Jones *et al.*, 1992), the search of sequence databases, structure prediction and modelling, to name but a few applications of our methodology.

STRUCTURAL ALIGNMENTS

A computer program COMPARER was employed to generate multiple alignments among families of homologous structures (Sali and Blundell, 1990; Zhu *et al.*, 1992). COMPARER uses a dynamic programming approach to trace the full-length alignment of three-dimensional structures. Instead of matching the proteins on the basis of sequence, this procedure considers many features of protein structure. These features include properties of individual amino acids or portions of the three-dimensional structure: main chain torsion angles, chain direction, solvent accessibility, secondary structure designations, to name a few. Relationships, which involve more than one position within the protein fold, can also be included: for instance, hydrogen bonding patterns and van der Waals interactions.

We used the structural alignment procedure to align up to 65 homologous families, for a total of 235 proteins. The individual pairwise alignments vary from as low as 8 percent sequence identity to over 90 percent. However, the majority of comparisons have between 15 and 35 percent sequence identity and the representation of the amino acids are very similar to those found in a sequence data bank of over 25,000 sequences (Johnson *et al.*, 1992).

ENVIRONMENTAL SUBSTITUTION DATA

A program JOY (Overington *et al.*, 1990a; 1992) was used to tabulate amino acid substitutions from the structural alignments of families of proteins and to determine the local structural environment for each residue in the alignment. Nearly 250,000 amino acid replacements were tabulated and segregated into 64 20-by-20 matrices on the basis of the following local environmental features:

1. The amino acid substitution.
2. The residue solvent accessibility.
3. The secondary structure class of the residue (α-helical, β-strand, coil or positive-ϕ torsion angle).
4. The hydrogen bonding potential of the side chain: side-chain to-side chain hydrogen bonding, side-chain to main-chain NH or side-chain to main-chain CO hydrogen bonds.

The 64 matrices arise as a consequence of their being 2 accessibility classes, 4 secondary structure classes and 3 hydrogen bonding states each with 2 classes each: $2 \times 4 \times 2^3 = 64$. Not all residues can participate in hydrogen bonding or have polar atoms that allow participation in all three types of hydrogen bonds considered here. Therefore, some of the tables will not be fully populated.

OVERVIEW OF THE SUBSTITUTION MATRICES

The substitution frequency matrices for each of the 64 environments are unique. If one integrates over all the features to reveal a single 20-by-20 matrix, this matrix is analogous to amino acid scoring matrices used in sequence comparisons (i.e. the Dayhoff matrix; Dayhoff *et al.*, 1978). Indeed, we have explored the usefulness of this structure-based amino acid scoring matrix in sequence alignments and sequence data bank searches (Johnson and Overington, 1992).

One can also integrate over fewer features, for instance on the basis of solvent accessibility. The resulting two matrices are very different from one another and it is quite evident that main diagonal exchanges (i.e. conservation of residue type) are more likely to be conserved within the solvent-inaccessible protein core than for residues lying on the protein surface.

As an example of the differences in the pattern of substitutions that are observed for the same amino acid in two different environments (solvent accessible *versus* solvent-inaccessible), we compare leucine and aspartate in Figure 1.

The major conclusions one can extrapolate from an examination of the trends within the substitution data are as follows:

1. Residues within, β-strands are most conserved, followed by those in α-helices and then

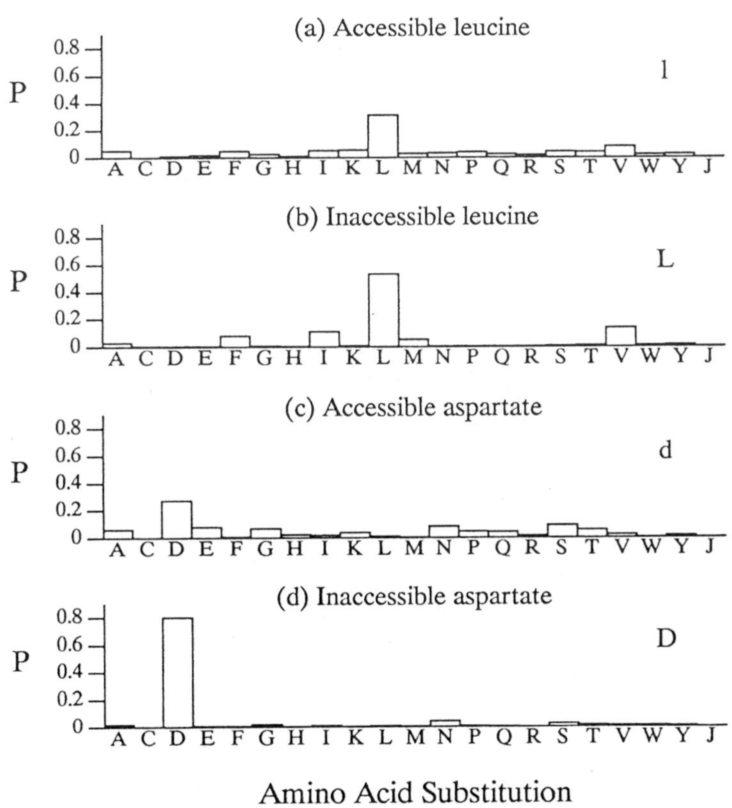

Figure 1. Probability (P) of replacement for accessible and inaccessible environments (from Johnson *et al.*, 1992).

coil or loop positions. This is as expected. Strands are generally completely buried within the protein core; helices many times have a solvent-exposed face, and coil residues are mostly on the surface of proteins where substitutions, insertions and deletions are more easily accommodated.

2. Buried residues are more conserved than solvent-exposed residues.

3. Residues involved in side-chain to main-chain NH hydrogen bonding are more conserved than those participating in main chain CO hydrogen bonds or side chain-side chain hydrogen bonds.

4. Polar residues buried within the protein core are more conserved than surface polar residues.

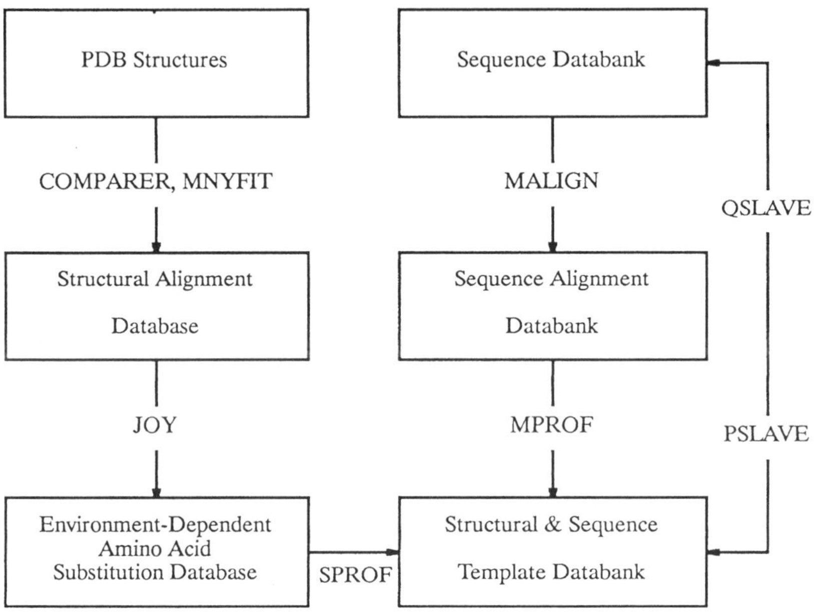

Figure 2. Unification of structure and sequence comparisons (from Johnson *et al.*, 1992)

ID-3D MAPPING

In Figure 2 we show the unification of protein structure and sequence comparisons. Homologous protein structures from the Brookhaven Protein Data Base (PDB; Bernstein *et al.*, 1977) can be aligned with structural comparison procedures such as COMPARER (Sali and Blundell, 1990) and MNYFIT (Sutcliffe *et al.*, 1987). The alignments form the structural alignment database. The program JOY (Overington *et al.*, 1990a; 1992) can then be used both to create an environment-dependent amino acid substitution data base, as well as encode all families and individual structures according to environmental features. Template descriptions of the family alignments or individual structures can then be constructed using the amino acid substitution database to describe protein folds (Johnson *et al.*, 1992).

On the other hand, sequences from a sequence data bank can be multiply aligned with the program MALIGN (Johnson *et al.*, 1992) and sequence-based template descriptions produced (MPROF; Johnson *et al.*, 1992).

A structural (or sequence template) can then be compared against all known amino acid sequences with a search/alignment program QSLAVE; in turn, a sequence or alignment of sequences can be searched/aligned against all known three-dimensional folds (PSLAVE; Johnson *et al.*, 1992).

The structures, singly or as a family alignment, are encoded with the patterns of residue substitution as specified by the individual local environments of each residue and the corresponding rules obtained from the environment-dependent substitution database. For instance, a serine in the familiar Dayhoff matrix has twenty scores associated with matching each of the twenty amino acids and these scores are the same regardless of the location within the protein. In contrast, our substitution tables predict up to 64 different scoring distributions for each of the amino acids. After accounting for those amino acids that cannot participate in each of the classes of hydrogen bonding we considered, there are over 750 distinct amino acid types instead of the more standard 20 types considered in sequence-based comparisons (Johnson *et al.*, 1992).

The alignments are made using dynamic programming procedures. As a consequence, a second aspect is critical to the resulting alignments: namely, the gap penalty. Like the scores for amino acid matches themselves, gap penalties are also tailored to the environment along the known structure. The number of Cα-carbons lying within a 14Å radius of each residue position is a very good indicator of the likely position where gaps occur. This number will be low (< 20) for loop regions at the protein surface and where packing density is low. In addition, the positions of standard secondary structure elements are used to moderate the penalty. The overall goal is to make the penalty more severe at positions corresponding to the protein core while reducing the penalty for likely positions for gaps.

Table 1. Globins located prior to the first non-globin in a search of a sequence data bank (> 25,000 sequences) containing 624 globins (from Johnson *et al.*, 1992).

Comparisont+	Structure-based template	Sequence-based template+
1. 11 globins	622	602
2. lecd, 1lh1	615	107(538)
3. 4mbn	602	547
4. 2lhb	577	532
5. lecd	490	244

+ - lecd, erythrocruorin; 1lh1, leghemoglobin; 4mbn, myoglobin; 2lhb, lamprey globin; the first two comparisons, 11 globins or 2 globins, are based on a composite template created from 11 and 2 aligned structures/sequences respectively. + - based on the Dayhoff matrix (Dayhoff *et al.*, 1978). § - the second non-globin appeared at position 538.

We have examined the success of this approach in making data bank searches (Table 1; Johnson *et al.*, 1992; Overington *et al.*, 1992). It is clear from our analysis that the procedure is superior to using the typical static 20-by-20 matrix generally considered in sequence comparisons. For instance, a search of the structural-template for the erythrocruorin (lecd) structure against a sequence data bank containing 624 globins locates 490 globins prior to the first non-globin. For a sequence-based template, 244 globins are found (Table 1).

We have also considered the search of a sequence against the templates for all available three-dimensional structures (Johnson *et al.*, 1992). The procedure is successful in locating homologues with very low sequence similarity: between 8 and 20 percent sequence identity. For example, a search of a bacterial serine proteinase sequence against the structural templates finds all of the mammalian serine proteinases although the similarity is extremely low.

CONCLUDING REMARKS

We have examined the amino acid replacements within a data base of accurately aligned three-dimensional structures and the local protein environments in which these replacements take place. From this tabulation of substitutions in known three-dimensional structures, procedures have been developed for the comparison of sequence and structure leading to very powerful approaches to data bank searching and alignment (Johnson *et al.*, 1992). In addition such procedures have proved useful in assessing the accuracy of modelled proteins (Overington *et al.*, 1990a), search and selection of loop regions in modelling exercises (Topham *et al.*, in press), and prediction of structure from multiply aligned sequences (Overington *et al.*, 1992).

Acknowledgements: We thank our former colleagues who have contributed towards this work especially A. Sali and J. P. Overington. Current funding is provided by the Imperial Cancer Research Fund.

REFERENCES

Bajaj, M. and Blundell T. L, 1984 Evolution and tertiary structure of proteins, *Ann. Rev. Biophys. Bioeng.* 13:453.

Bernstein, F. *et al.*, 1977 The Protein Data Bank: a computer based archival file for macromolecular structures, *J. Mol. Biol.* 112:535.

Bowie, J. U., Lüthy, R. and Eisenberg, D., A method to identify protein sequences that fold into known three-dimensional structure, 1991 *Science* 253:164.

Chothia, C. and Lesk A. M., 1982a Evolution of proteins formed by β-sheets: I. plastocyanin and azurin, *J. Mol. Biol.* 160:309.

Chothia, C. and Lesk A. M., 1982b Evolution of proteins formed by β-sheets: II. the core of the immunoglobulins, *J. Mol. Biol.* 160:325.

Dayhoff, M. O., Schwartz, R. M. and Orcutt, B. C., 1978 A model for evolutionary change, *in*: Atlas of Protein Sequence and Structure, Dayhoff, M.O., ed., Vol. 5, Suppl. 3, National Biomedical Research Foundation, Washington, D.C., 345.

Hubbard, T. J. P. and Blundell, T. L., 1987 Compariosn of solvent-inaccessible core of homologous proteins: definitions useful for protein modelling, *Prot. Engin.* 1:159.

Johnson, M. S., Overington, J. P and Blundell, T. L., Alignment and searching for common protein folds using a data bank of structural templates, submitted.

Johnson, M. S. and Overington, J. P., A structural basis for sequence comparisons: a critical evaluation of scoring methodologies, submitted.

Jones, D. T., Taylor, W. R. and Thornton, J. M., 1992 A new approach to protein fold recognition, *Nature* 358:86.

Lesk A. M. and Chothia, C., 1986 The response of protein structure to amino acid changes, *Phil. Trans. Roy. Soc. Lond.* A317:345.

Lüthy, R., McLachlan, A. D. and Eisenberg, D. Secondary structure-based profiles, *Proteins* 10:229.

Overington, J. P., Johnson, M. S., Sali, A. and Blundell, T. L., 1990a Tertiary structural constraints on protein evolutionary diversity: templates, key residues and structure prediciton, *Proc. Roy. Soc.* B241:132.

Overington, J. P., Johnson, M. S., Topham, C., McLeod, A., Sali, A., Zhu, Z.-Y., Sibanda, L. and Blundell, T. L., 1990b Applications of environment specific amino acid substitution tables to identify key residues in protein tertiary structure, *Current Science (India)* 59:867.

Overington, J. P., Donnelly, D., Johnson, M. S., Sali, A. and Blundell, T. L., 1992 Environment-specific amino acid substitution tables: tertiary templates and prediction of protein folds, *Prot. Sci.* 1:216.

Sali, A. and Blundell, T. L. J., 1990 Definition of general topological equivalence in protein structures, *Mol. Biol.* 212: 403.

Sali, A., Overington, J. P., Johnson, M. S. and Blundell, T. L., 1990 From comparisons of protein sequences and structures to protein modelling and design, *Trends in Biol. Sci.* 15:235.

Topham, C. M., McLeod, A.,. Eisenmenger, F., Overington, J. P., Johnson, M. S. and Blundell, T. L. J., Fragment ranking in modelling of protein structure: conformationally-constrained environmental amino acid substitution tables, *Mol. Biol.*, in press.

Zhu, Z.-Y., Sali, A. and Blundell, T. L., 1992 A variable gap penalty function and feature weights for protein 3-D structure comparisons, *Prot. Engin.* 5: 43.

DOMAINS AND MODULES OF PROTEINS

Mitiko Gô[1], Michiko Nosaka[2] and Shirou Tomoda[1]

[1]*Department of Biology, Faculty of Science, Nagoya University, Furo-cho, Chikusa-ku, Nagoya 464-01 and* [2]*Department of Biology, Faculty of Science, Kyushu University, Hakozaki, Higashi-ku, Fukuoka 812, Japan*

INTRODUCTION

Investigations on the three-dimensional structure of water soluble proteins showed that they consist of one to several globular domains (Richardson, 1981). Module, introduced as a compact structure in globular domains, revealed that it consists of contiguous 10 to 40 amino acids, and domains and modules have a close correlation with exon-intron organization of the DNA sequence encoding proteins (Gô, 1985).

Introns are often found at domain boundaries. For example they are located between domains of immunoglobulin light and heavy chains (Williams and Barclay, 1988), between those of alcohol dehydrogenase (Dennis *et al.*, 1984) and between domains of ovomucoid (Stein *et al.*, 1980), etc.

However, introns are also located within the coding sequence of a globular domain; such as in alcohol dehydrogenase, glycelaldehyde-3-phosphate dehydrogenase and pyruvate kinase. Four modules of leghemoglobin from plants have a one-to-one correspondence with four exons of the leghemoglobin gene; i.e. the module boundaries are located close to the positions of three introns (Gô, 1981). Module boundaries in other proteins such as lysozyme (Gô, 1983), triose phosphate isomerase (Gô, 1987), ovomucoid 3rd domain (Gô, 1985), epidermal growth factor (Mizushima *et al.*, unpublished results), etc. show a close correlation with intron positions.

These observations imply that proteins have evolved by combining modules first, followed by the fusion of domains. The remnant of such shuffling is observed in the exon-intron structures of eukaryotic genes.

SIZE DISTRIBUTION OF MODULES

Using a computerized algorithm to identify module boundaries, we have analyzed 89 non-homologous proteins, the three-dimensional structures of which are known. The average

Methods in Protein Sequence Analysis, Edited by K. Imahori and F. Sakiyama, Plenum Press, New York, 1993

length of identified modules is around 15 residues. From a comparison between distribution of module size and exon length, it was estimated that as many as three times the present introns located in ancient genes of eukaryotes (Gô and Nosaka, 1987). It was also observed that introns were lost non-randomly (Nosaka *et al.*, in preparation), therefore joining of two or three modules into a segment encoded by one exon occurred frequently and this size of segment was favored as a shuffling unit.

Contemporary prokaryotic genes encoding proteins are not split by introns; they seem to have been lost during evolution, as well as some introns presumably lost in eukaryotic genes. However, module organization of prokaryotic and eukaryotic proteins provide information not only on how to locate positions of introns in ancestral genes but also to understand how small parts were recruited into various globular domains.

PREDICTION OF MODULE BOUNDARIES

Identification of module structures of proteins was carried out by partitioning the polypeptide chain into the segments containing the residue pairs located closely from each other, space-wise. The assignment of module was originally carried out using a map where the distance of α carbons is drawn and by searching module boundaries as residues present not far from the center of the chain (Gô, 1981).

However, we proposed an alternative algorithm to assign module boundaries by locating a center of local segment with $2k+1$ residues along the polypeptide chain (Gô and Nosaka, 1987). To do so, we used a plot called centripetal profile, defined for the ith residue by,

$$F_i = \sum_{i-k \leqq j \leqq i+k} r^2_{ij} \,/(2k+1) \qquad (1)$$

where r_{ij} is the distance between the ith and jth residues and $2k+1$ is the window size including the ith residue. Module boundaries are assigned as the stable local minima of the profile, under several different values of the window size.

Recently, we developed an empirical method to predict module boundaries on the basis of only amino acid sequences (Tomoda *et al.*, in preparation). The prediction is based on distance information between amino acid residues of proteins whose three-dimensional structures are known. We used 93 non-homologous proteins listed in the Protein Data Bank (Bernstein *et al.*, 1977; Abola *et al.*, 1987) from which the average distance between amino acids separated up to 60 residues in sequence-wise was calculated. We used the average distance between amino acid residues to calculate the predicted centripetal profile according to equation (1). Module boundaries were predicted as the stable local minima of the predicted centripetal profiles.

Predicted module boundaries from the amino acid sequence of hen egg white lysozyme are shown as well as module boundaries identified from the three-dimensional structure of the lysozyme (Figure 1). Ten modules are identified from the tertiary structure of the lysozyme. We assigned at least five modules, using a distance map (Gô, 1983). The present refinement in identification of modules shows that each of the former modules is split into two. On the other hand, seven modules are predicted from the sequence of the lysozyme; three modules are under-predicted. All the predicted module boundaries correspond to the assigned module boundaries from the atomic coordinates allowing for three residue deviation. However 67%

of the assigned module boundaries were predicted. This result shows that the module boundaries in chicken lysozyme were predicted with about a 80% accuracy. The correlation of the intron positions with the predicted module boundaries as well as identified ones using atomic coordinates is shown in Table 1.

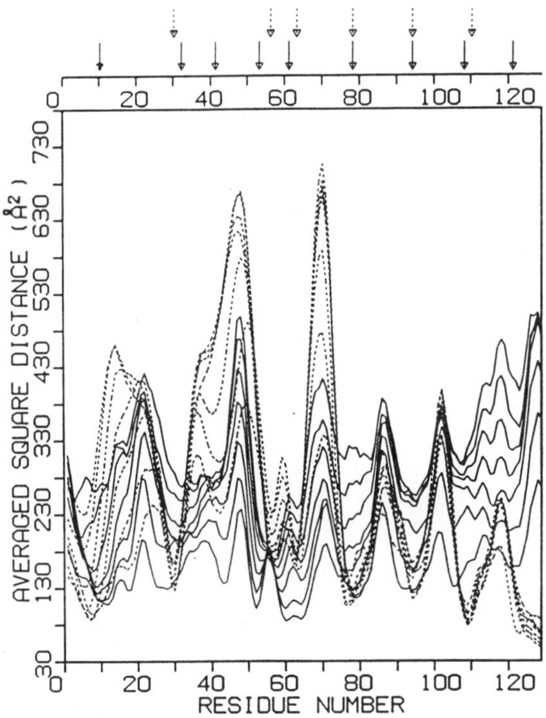

Figure 1. Identified and predicted module boundaries of lysozyme. Observed centripetal profiles (solid curves) and the predicted ones (dotted curves) for the window lengths $(2k+1)$, 31,41,51,61,71,81,91 are shown. The module boundaries and the predicted ones are shown by solid arrows (lower) and dotted arrows (upper), respectively.

We tested the reliability of the prediction method on 93 proteins used in data base construction; we compared the predicted module boundaries from the sequence information and the assigned ones on the basis of the atomic coordinates. The module boundaries were predicted, allowing for a three residue deviation, with the accuracy of about 70%. This quality is similar to the joint-method for secondary structure prediction (Nishikawa and Ooi, 1986), though a different evaluation was applied. We found this method for prediction of modules to be useful, particularly when a protein sequence under consideration has homologs, since the prediction is improved by taking consensus module boundaries. Predicted module boundaries in some proteins show a close correlation with intron positions of their genes (Nakashima *et al.*, submitted). Predicted module boundaries provide insight into the structural organization and evolution of proteins.

Table 1. Module boundaries of lysozyme and introns of its genes.

	Module boundary		Corresponding intron[1]
	Assigned	Predicted	
Ml - M2	10		
M2 - M3	32	30	28
M3 - M4	41		
M4 - M5	53		
M5 - M6	61	59	
M6 - M7	78	78	82
M7 - M8	94	94	
M8 - M9	108	109	108
M9 - M10	121		

[1]Jung *et al.*, 1980

Acknowledgements: We thank M. Ohara for critical comments on the manuscript. Part of this research work was supported by Grants-in-Aid on Priority Area and for Scientific Research from the Ministry of Education, Science and Culture of Japan and by a grant from the Ogasawara Foundation for the Promotion of Science & Engineering.

REFERENCES

Abola, E., Bernstein, F. C., Bryant, S. H., Koetzle, T. F. and Weng, J., 1987 Protein data bank, *in:* Crystallographic Databases - Information Content, Software Systems, Scientific Applications, Allen, F.H., Bergerhoff, G. and Sievers, R., eds., Data Commission of the Intemational Union of Crystallography, Bonn/Cambridge/Chester.

Bernstein, F. C., Koetzle, T. F., Williams, G. J. B., Meyer, E. F. Jr., Brice, M. D., Rodgers, J. R., Kennard, O., Shimanouchi, T. and Tasumi, M., 1977 The protein data bank: a computer-based archival file for macromolecular structures, *J. Mol. Biol.* 112:535.

Dennis, E. S., Gerlach, W. L., Pryor, A. J., Bennetzen, J. L., Inglis, A., Llewellyn, D., Sachs, M. M., Ferl, R. J. and Peacock,W. J., 1984 Molecular analysis of the alcohol dehydrogenase (*Adhl*) gene of maize, *Nucleic Acids Res.* 12:3983.

Gô, M., 1981 Correlation of DNA exonic regions with protein structural units in haemoglobin, *Nature* 291:90.

Gô, M., 1983 Modular structural units, exons, and function in chicken lysozyme, *Proc. Natl. Acad. Sci. USA* 80:1964.

Gô, M., 1985 Protein structures and split genes. *Adv. Biophys.* 19:91.

Gô, M. and Nosaka, M., 1987 Protein architecture and the origin of introns, *Cold Spring Harbor Symp. Quaut. Biol.* 52:915.

Jung, A., Sippel, A. E., Grez, M. and Schütz, G., 1980 Exons encode functional and structural units of chicken lysozyme, *Proc. Natl. Acad. Sci. USA* 77:5759.

Nishikawa, K. and Ooi, T., 1986 Amino acid sequence homology applied to the prediction of protein secondary structures, and joint prediction with existing methods, *Biochem. Biophys. Acta.* 871:45.

Richardson, J. S., 1981 The anatomy and taxonomy of protein structure, *Adv. Protein Chem.* 34:167.

Stein, J. P., Caterall, J. F., Kristo, P., Means, A. R. and O'Malley, B. W., 1980 Ovomucoid intervening sequences specify functional domains and generate protein polymorphism, *Cell* 21:681.

Williams, A. F. and Barclay, A. N., 1988 The Immunoglobulin Superfamily, *Ann. Rev. Immunol.* 6:381.

HOMOLOGY SEARCH AND PREDICTION OF BIOLOGICAL FUNCTION OF PROTEIN FROM AMINO ACID SEQUENCES

Hiroyuki Toh

Protein Engineering Research Institute
6-2-3 Furuedai, Suita 565, Japan

INTRODUCTION

Amino acid sequence data is a rich source of knowledge about structure, function, and molecular evolution of proteins. However, it is still difficult to extract such information from a single amino acid sequence. Sequence comparison of homologous proteins is regarded as a practical approach to investigate sequence data, and is widely utilized in the research of molecular biology and molecular evolution. In this paper, we will review the general procedure of sequence analysis and use the firefly luciferase family to describe how protein function can be investigated by sequence analysis. The sequence data used in this paper were obtained from NBRF and PRF protein primary structure databases.

HOMOLOGY SEARCH

The accumulation of amino acid sequence data of proteins has enabled us to inquire the function and structure of proteins by sequence comparison. When an investigator determines a new sequence, it is one of the first things to compare it with all available sequences. If he finds resemblance between the query sequence and some sequences whose structures or functions are already known, the detected similarity may guide him to characterize his sequence. However, the investigator should keep in mind that there are two types of sequence similarity, 'homology' and 'analogy'. The former is similarity which originates from a common ancestry, while the latter is accidental similarity. It is important to distinguish them, because structural or functional information of proteins is obtained from sequence comparison of homologous proteins. The judgment, homology or analogy, is required, especially when the detected sequence similarity is very low. We explain the problem with an example of our analysis.

Microorganisms produce a wide variety of peptide antibiotics such as gramicidin S, tyrocidine, cyclosporine and so forth. These peptide antibiotics, which contain non-protein

Methods in Protein Sequence Analysis, Edited by K. Imahori and F. Sakiyama, Plenum Press, New York, 1993

amino acids such as D-amino acids, hydroxy acids, and other unusual constituents, are synthesized by a non-ribosomal system called the thiotemplate mechanism (Kleinkauf and von Dohren, 1990). The thiotemplate mechanism is a multi-enzymatic system including the activation of the constituent residues as adenylates on the enzymic template under the presence of Mg^{2+}, and the acylation of specific template thiol groups. Gramicidin S synthetase 1 (GS1) from *B. brevis* is a constituent of a multienzyme complex which catalyzes the formation of gramicidin S. Tyrocidine synthetase 1 (TS1) from *B. brevis* is a constituent of other multienzyme complex to catalyze the formation of tyrocidine. They are about 1100 amino acid residues in length, and are homologous one another in all regions.

In order to identify the evolutionary origin of the non-ribosomal peptide synthesis, and to get clues about the catalytic mechanism, the amino acid sequences of these two enzymes were subjected to homology search (Toh, 1990). A computer program with a local alignment algorithm (Nishikawa *et al.*, 1987) was used for the searching. This search revealed that the N-terminal halves of the two peptide antibiotic synthetases resemble the entire regions of insect luciferases and plant 4-coumarate:CoA ligases.

As described above, it should be checked whether the detected sequence similarity is homology or analogy. A jumbling test is a popular method to judge the similarity of a pair of amino acid sequences (Schwartz and Dayhoff, 1978). First, a multiple alignment of the enzymes was constructed. The length of the alignment was about 550 amino acid residues. The sequence identity between the peptide antibiotic synthetases and the other eukaryotic enzymes was found to range from 18 to 21 %. Then, each aligned pair was subjected to the jumbling test with 100 randomized sequences. The normalized alignment scores for sequence similarity between TS1 and six insect luciferases were 3.4 - 5.7 SD. However, the scores for the similarity between TS1 and 4-coumarate:CoA ligases were 1.9SD, and those between GS1 and other eukaryotic enzymes were 1.4 - 3.5 SD. As the N-terminal regions of the alignment were highly diverged, the procedure for the jumbling test was repeated using the C-terminal 150 residues of the multiple alignment. In this case, the normalized alignment scores were much higher (3.8 - 5.5 SD) (Toh, 1990).

Other information such as tertiary structure and biological activity may assist us in the judgment of the sequence similarity. The tertiary structures of the above enzymes have not been identified, but their catalytic activities have been extensively investigated. Insect luciferases are involved in bioluminescence, and most of the knowledge about the catalytic activities of the enzymes has been obtained from the research of firefly luciferase. Firefly luciferase catalyzes the adenylation of luciferin under the presence of Mg^{2+}. Then, the luciferyl adenylate is subjected to oxidative decarboxylation to produce light. However, the enzyme can also catalyze the formation of dehydroluciferyl adenylate under the presence of Mg^{2+}, when dehydroluciferin is given as a substrate. Furthermore, the enzyme can catalyze the formation of luciferyl-CoA from dehydroluciferyl adenylate under the presence of CoA. On the other hand, 4-coumarate:CoA ligase is involved in the general phenylpropanoid metabolism, which catalyzes the formation of 4-coumarate:CoA from cinnamate or 4-coumarate via adenylation of the substrate. That is, the reactions catalyzed by luciferases and 4-coumarate:CoA ligases are equivalent to those by GS1 and TS1, which are schematically expressed as follows.

X-COOH + ATP <---------> X-COO-AMP + PPi

X-COO-AMP + Y-SH <-----------> X-COS-Y + AMP

The results of jumbling tests, graphic representation by dot matrix, and resemblance in catalytic reactions suggest that the detected sequence similarity is homology, and that the N-terminal halves of the prokaryotic peptide antibiotic synthetases, insect luciferases, and plant 4-coumarate:CoA ligases constitute a protein family, which are here referred to as 'firefly luciferase family' (Toh, 1990).

REVERSE SEARCHING

Reverse searching (RS) is a method to search through sequence databases with information obtained from a multiple alignment, which shows greater sensitivity than ordinary searching with a single query (Doolittle, 1990). One of the simple ways for RS is the consensus sequence searching. The consensus sequence obtained from a multiple alignment is called a motif, which often includes the active sites and their neighboring residues. Thus, the consensus method is regarded as an effective tool to predict protein function. We have applied this method to several proteins to predict their functions. We will here show two examples of the consensus methods.

A multiple alignment of retroviral proteases was constructed (Toh *et al.*, 1985a), and a conservative region was detected. RS with the consensus sequence of the conservative region was carried out. It was found that the motif sequence of the retroviral proteases is quite similar to the conservative amino acid stretch of a wide variety of aspartic acid proteases (Figure 1) (Toh *et al.*, 1985a, b), although global sequence smilarity between retroviral proteases and aspartic acid proteases was not detected. '*' in Figure 1 indicates the active site for aspartic acid proteases. The result of RS suggested that the catalytic mechanism of the retroviral proteases may be the same as that of aspartic acid proteases. Recently, the crystal structures of the retroviral proteases were determined, and found to be quite similar to those of aspartic acid proteases in spite of high sequence divergence between them (Navia *et al.*, 1989; Miller *et al.*, 1989).

Leukotriene A (LTA) hydrolase catalyzes the formation of leukotriene B. LTA hydrolase shows sequence similarity to prokaryotic and eukaryotic aminopeptidases-N. A multiple alignment of the LTA hydrolase and aminopeptidases-N was constructed, and a conservative amino acid stretch was obtained. It was found by RS that the conservative amino acid stretch corresponds with a ubiquitous sequence motif for zinc-metalloproteases (Figure 2), which include amino acid residues acting as zinc-ligands ('Z' in Figure 2) and an active site for peptidase activity ('*' in Figure 2) (Toh *et al.*, 1990; Vallee and Auld, 1990). Led by the observation, experimental investigation revealed that LTA hydrolase

```
Viral protease
  CaMV ORF5                       LHCFV-DTGASLCIAS
  17.6 ORF2                       LKCLI-DTGSTVNMTS
  M-MuLV pol                      VTFLV-DTGAQHSVLT
  RSV gag p15                     ITALL-DSGADITIIS
  HTLV (gag-pol)                  IEALL-DTGADMTVLP
Acid protease (C-terminal)              *
  pepsinogen (human)              CQAIV-DTGTSLLTGP
             (porcine)            CQAIV-DTGTSLLTGP
  prochymosin (bovine)            CQAIL-DTGTSKLVGP
  penicillopepsin                 FSGIA-DTGTTLLLLB
  renin (mouse)                   CEVVV-DTGSSFISAP
        (human)                   CLALV-DTGASYISGS
Acid protease (N-terminal)              *
  pepsinogen (human)              DFTVVFDTGSSNLWVP
             (bovine)             DFTVIFDTGSSNLWVP
             (porcine)            DFTVIFDTGSSNLWVP
  prochymosin (bovine)            EFTVLFDTGSSDFWVP
  penicillopepsin                 TLNLNFDTGSADLWVF
  renin (mouse)                   TFKVMFDTGSANLWVP
        (human)                   TFKVVFDTGSSNVWVP
```

Figure 1. Alignment of motifs for retroviral proteases and aspartic acid proteases. Abbreviation is as follows. CaMV, cauliflower mosaic virus; 17.6, copia-like element 17.6; M-MuLV, Moloney murine leukemia virus; RSV, Rous sarcoma virus; HTLV, human T-cell leukaemia virus.

```
                                           Z*  Z                          Z
leukotriene A4 hydrolase                   VIAHEISHSW   (14 residues)   LNEGH

aminopeptidase N human                     VIAHELAHQW   (14 residues)   LNEGF
aminopeptidase N rabbit                    VVAHELAHQW   (14 residues)   LNEGF
aminopeltidase M rat                       VIAHELAHQW   (14 residues)   LNEGF
aminopeptidase N E.coli                    V1GHEYFHNW   (14 residues)   LKEGL

thermolysin B.stearothermophilus           VVAHELTHAV   (15 residues)   INEAI
thermolysin B.thermoproteolyticus          VVAHELTHAV   (15 resldues)   INEAI
thermolysin B.subtilis                     VTAHEMTHGV   (15 resldues)   LNESF
thermolysin B.amyloliquefaciens            VTAHEMTHGV   (15 residues)   LNESF
neutral proteinasc B.cercus                VIGHELTHAV   (15 residues)   LNEAI
elastase P.aerguirlosa                     VAAHEVSHGF   (15 residues)   MNEAF

endopeptidase rabbit                       VIGHEITHGF
endopeptidase rat                          VIGHEITHGF

collagenase rabbit                         VAAHELGHSL
collagenase human                          VAAHELGHSL
collagenase IV human                       VAAHEFGHAM
transin 2 rat                              VAAHELGHSL
stromelysin human                          VAAHEIGHSL
stromelysin 2 human                        VAAHELGHSL
stromelysin rat                            VAAHELGHSL
proactivator rabbit                        VAAHELGHSL
```

Figure 2. Alignment of motifs of zinc metalloproteases and LTA hydrolase

contains a zinc ion per molecule, and shows peptidase activity. Furthermore, it was found that a zinc ion is required both for LTA hydrolase activity and for peptidase activity (Minami et al., 1990; Haeggström *et al.*, 1990a, b).

Thus, RS with a motif is a simple, but effective method to predict protein function. We extracted a conservative sequence motif from the multiple alignment of the firefly luciferase family, and tried the consensus sequence searching. Then, fungal acetyl-CoA synthetases and *V. anguillarum* AngR were found to contain a motif sequence of the firefly luciferase family (Toh, 1991).

Acetyl-CoA synthetase is an enzyme of about 650 amino acid residues in length, which is involved in the formation of acetyl-CoA from acetate under the presence of Mg^{2+} and ATP. The primary structures of the enzymes from *A. nidulans* and *N. crassa* were determined, which are homologous not only one another but also to the members of the firefly luciferase family in their entire regions. Two models have been proposed for the catalytic mechanism of this enzyme. One model contains a step of acetyl-adenylate formation, and another does not. The members of the firefly luciferase family catalyze two step reaction as described in the previous section, which includes a step of the adenylation of their substrates. Therefore, the detected sequence homology supports the former model for two fungal acetyl-CoA synthetases.

On the other hand, AngR is a protein of about 1300 amino acid residues in length, which acts as a transcriptional regulator for anguibactin synthesis by *V. anguillarum*. Anguibactin is a kind of siderophores, soluble iron-binding bioactive peptides which contain unusual constituents in them like bacterial peptide antibiotics. The mechanism for the biosynthesis of siderophores has not been known. Sequence comparison revealed that the C-terminal half of AngR is homologous to the members of the firefly luciferase family in their entire regions. The sequence homology suggests that AngR not only is involved in the regulation of anguibactin synthesis but also catalyzes the formation of anguibactin by the thiotemplate mechanism used as the synthetases of peptide antibiotics. We will further discuss this problem in the next section.

Matrix searching is an improved way for RS, and uses a position-specific scoring matrix derived from a multiple alignment instead of the consensus sequence (Staden, 1990; Storomo, 1990). Figure 3 shows the results of matrix searching with the motif of the firefly luciferase family. NBRF protein primary structure database (rel.32.0) was used for the RS. All the members of this family stored in the database were detected by the RS. An amino acid stretch detected by the RS, the statistical significance of the similarity, the accession number in the database, and the name of the sequence including the stretch are written in a line.

TSGTTGNPKGTM	5.655	YGBSG1	Gramicidin S synthetase component
TSGTTGKPKGTM	5.476	YGBSTB	Tyrocidine synthetase component I
TSGTTGFPKGIF	5.704	YGPLV8	alpha-Aminoadipyl-cysteinyl-valin
TSGTTGRPKGVT	5.201	YGPLV8	alpha-Aminoadipyl-cysteinyl-valin
TSGTSGKPKGVL	5.760	YGPLV8	alpha-Aminoadipyl-cysteinyl-valin
TSGTTGFPKGIF	5.704	YGPLV3	alpha-Aminoadipyl-cysteinyl-valin
TSGTTGRPKGVT	5.201	YGPLV3	alpha-Aminoadipyl-cysteinyl-valin
TSGTSGKPKGVL	5.760	YGPLV3	alpha-Aminoadipyl-cysteinyl-valin
TSGTTGFPKGIY	5.331	YGCEVC	alpha-Aminoadipyl-cysteinyl-valin
TSGTTGNPKGVM	5.765	YGCEVC	alpha-Aminoadipyl-cysteinyl-valin
TSGTTGKPKGVL	5.592	YGCEVC	alpha-Aminoadipyl-cysteinyl-valin
TSGTTGKPKGVM	5.535	YGBSG2	Gramicidin S synthetase component
TSGSTGTPKGVV	5.312	YGVCAR	angR protein - Vibrio anguillarum
TSGSTGRPKGVM	5.495	YGECEF	Enterochelin synthetase component
TSGSEGIPKGVL	5.391	YGBYAD	alpha-Aminoadipate reductase - Ye
SGGTTGTPKLIP	5.454	S08076	Enterobactin synthetase component
SSGTTGLPKGVM	5.854	S01667	4-Coumarate--CoA ligase - Parsley
SSGSTGLPKGVQ	5.629	JS0181	Photinus-luciferin 4-monooxygenas
SSGSTGLPKGVA	5.533	A26772	Photinus-luciferin 4-monooxygenas
TSGTTGNPKGAM	5.699	JX0202	Long-chain-acid--CoA ligase - Hum
TSGTTGVPKGVP	4.988	S18268	*alpha-Aminoadipyl-L-cysteinyl-D-
TSGTTGKPKAVL	5.403	S18268	*alpha-Aminoadipyl-L-cysteinyl-D-
TSGTTGKPKAVL	5.403	S18268	*alpha-Aminoadipyl-L-cysteinyl-D-
TSGTTGKPKGIE	5.286	A41043	*Acetate--CoA ligase - Methanothr
SSGTTGLPKGVM	5.854	A39827	*4-Coumarate--CoA ligase 1 - Pota
SSGTTGLPKGVM	5.854	B39827	*4-Coumarate--CoA ligase 2a - Pot
SSGTTGLPKGVM	5.854	C39827	*4-Coumarate--CoA ligase 2b - Pot
SSGTTGLPKGVM	5.854	JU0311	*4-Coumarate--CoA ligase 1 - Rice
TSGSTGKPKGVM	5.270	S09245	*Acetate--CoA ligase - Emericella
TSGTTGFPKGIL	5.778	A40889	*delta-(L-alpha-aminoadipyl)-L-cy
TSGTTGKPKGVM	5.535	A40889	*delta-(L-alpha-aminoadipyl)-L-cy
TSGTSGKPKGVL	5.760	A40889	*delta-(L-alpha-aminoadipyl)-L-cy
TSGSTGKPKGVA	5.125	S09244	*Acetate--CoA ligase - Neurospora
TSGTTGNPKGAM	5.699	JX0205	*Long-chain-acid--CoA ligase - Ra
TSGTTGNPKGAM	5.699	A36275	*Long-chain-fatty-acid-CoA ligase

Figure 3. The result of matrix searching with the motif of the firefly luciferase family.

PHYLOGENETIC TREE

One of the goals of sequence comparison analysis is the reconstruction of evolutionary history of species or genes. In addition, the tree construction is regarded as a useful method to inquire protein function, because the operation for tree construction is equivalent to the cluster analysis.

Figure 4 shows an unrooted phylogenetic tree of the α-type DNA polymerase family. DNA polymerases require primer to initiate new DNA chains. In many cases, RNA is utilized as primer. But, several polymerases utilize proteins as primers. As shown in the

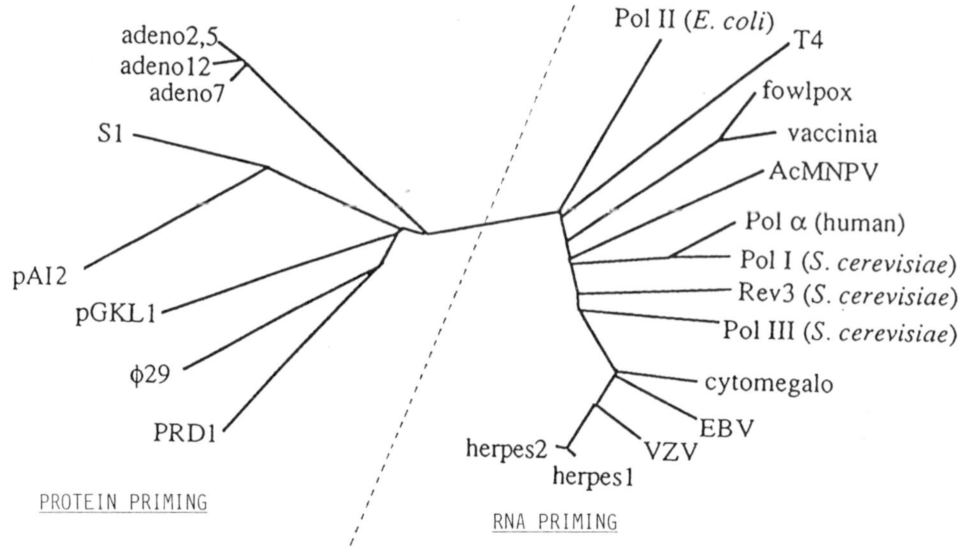

Figure 4. An unrooted phylogenetic tree of α–type DNA polymerases

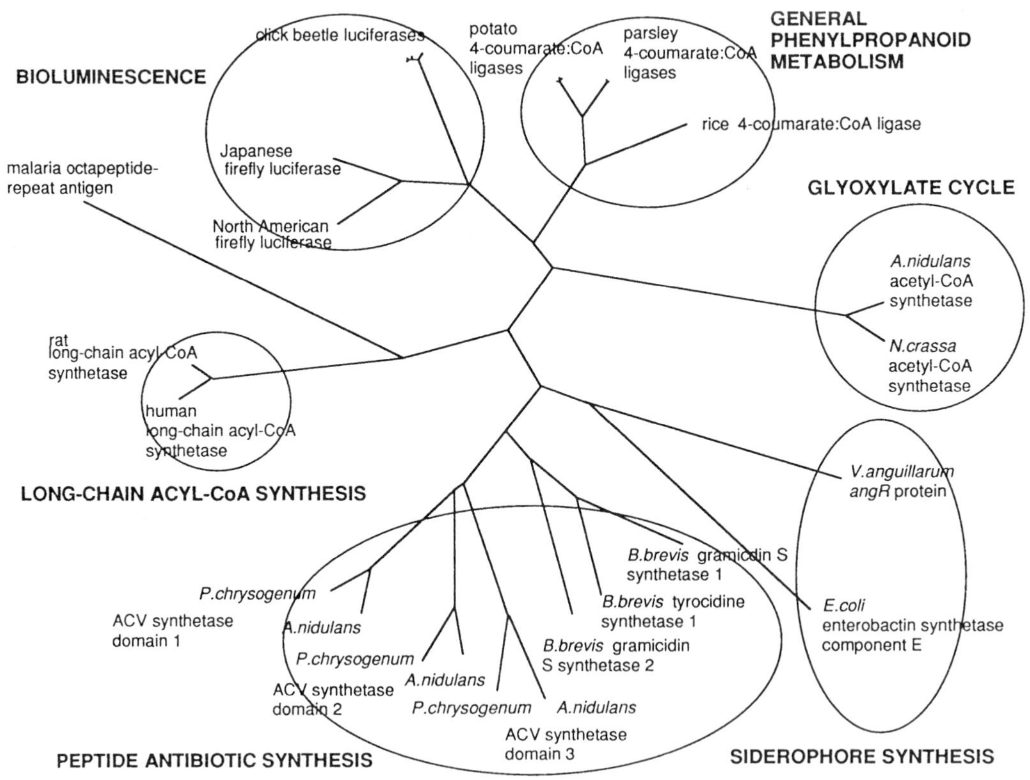

Figure 5 . An unrooted phylogenetic tree of the firefly luciferase family.

tree, the family is clearly divided into two subfamilies. One of them is a group of DNA polymerases which utilize RNA primer, while another is a group whose members utilize protein primer (Iwasaki *et al.*, 1991). Suppose that we have sequence data of a new member of α-type DNA polymerase. Which one does the enzyme utilize as its primer, RNA or protein? Phylogenetic tree may answer this question. At first, a phylogenetic tree including the new data is reconstructed. If the new sequence occupies a position within or adjacent to the subfamily utilizing RNA (protein) primer, the enzyme is considered to utilize the RNA (protein) primer.

We applied the tree method to investigate the function of AngR (Toh, 1991). Figure 5 is an unrooted phylogenetic tree of the firefly luciferase family. The tree has several clusters and isolated nodes, which correspond with their biological functions. AngR occupies a position evolutionarily close to enterobactin synthetase component E from *E. coli*. Enterobactin is a kind of bacterial siderophores. The proteins involved in the siderophore synthesis constitute a cluster which neighbors the cluster of the peptide antibiotics synthetases. This observation supports the hypothesis described in the previous section that AngR may act as the synthetase for anguibactin, and siderophores may be synthesized by the thiotemplate mechanism.

CONCLUDING REMARKS

We have reviewed the prediction of biological function from primary structure. Sequence comparisons of homologous proteins often give clues about structure and function, and various tools for sequence comparison have been elaborated. However, the important information about structures and/or functions of proteins often resides in weak homology, or so-called 'twilight zone' (Doolittle, 1990). The similarity in the twilight zone is often beyond the limit for application of the tools available now. Additional information such as tertiary structure and biological activity enables sequence analysis to probe sequence analysis in the twilight zone, although such approach is rather arbitrary. Therefore, it is vital not only for molecular evolution but also for prediction of structure and function to establish systematic approaches to inquire about residents in the twilight zone.

REFERENCES

Doolittle, R. F., 1990 Searching through sequence databases. *in*: Methods in Enzymology, Vol. 183, Doolittle, R. F., ed., Academic Press, Inc., San Diego, New York, Boston, London, Sydney, Tokyo, Toronto.

Haeggström, J. Z., Wetterholm, A., Shapiro, R., Vallee, B. L. and Samuelsson, B., 1990a Leukotriene A4 hydrolase: A zinc metalloenzyme, *Biochem. Biophys. Res. Commun.* 172:965.

Haeggström, J. ., Wetterholm, A., Vallee, B. L. and Samuelsson, B., 1990b Leukotriene A4 hydrolase: An epoxide hydrolase with peptidase activity, *Biochem. Biophys. Res. Commun.* 173:431.

Iwasaki, H., Ishino, Y., Toh, H., Nakata, A. and Shinagawa, H., 1991 *Esherichia coli* DNA polymerase II is homologous to α-like DNA polymerases, *Mol. Gen. Genet.* 226:24.

Kleinkauf, H. and von Döhren, H., 1990 Nonribosomal biosynthesis of peptide antibiotics, *Eur. J. Biochem.* 192:1.

Miller, M., Jaskolski, M., Rao, J. K. M., Leis, J. and Wlodawer, A., 1989 Crystal structure of a retroviral protease proves relationship to aspartic protease family, *Nature* 337:576.

Minami, M., Ohishi, N., Mutoh, H., Izumi, T., Bito, H., Wada, H., Seyama, Y., Toh, H. and Shimizu, T., 1990 Leukotriene A4 hydrolase is a zinc-containing aminopeptidase, *Biochem. Biophys. Res. Commun.* 173:620.

Navia, M. A., Fitzgerald, P. M. D., McKeever, B. M., Leu, C.-T., Heibach, J. C., Herber, W. K., Sigal, I. S., Darke, P. L. and Springer, J. P., 1989 Three-dimensional structure of aspartyl protease from human immunodeficiency virus HIV-1, *Nature* 337:615.

Nishikawa, K., Nakashima, H., Kanehisa, M. and Ooi, T., 1987 Detection of weak sequence homology of proteins for tertiary structure prediction. *Prot. Seq. Data Anal.* 1:107.

Schwartz, R. M. and Dayhoff, M. O., 1978 Metrices for detecting distant relationships. *in*: Atlas of Protein Sequence and Structure, Vol. 5, Dayhoff, M. O., eds., National Biomedical Research Foundation, Washington DC.

Staden, R., 1990 Searching for patterns in proteins and nucleic acid sequences, *in*: Methods in Enzymology, Vol. 183, Doolittle, R. F., eds., Academic Press, Inc., San Diego, New York, Boston, London, Sydney, Tokyo,Toronto.

Storomo, G. D., 1990 Consensus Patterns in DNA, *in*: Methods in Enzymology, Vol. 183, Doolittle, R. F., eds., Academic Press, Inc., San Diego,New York, Boston, London, Sydney, Tokyo, Toronto.

Toh, H., 1990 N-terminal halves of gramicidin S synthetase 1 and tyrocidine synthetase 1 as novel members of firefly luciferase family, *Prot. Seq. Data Anal.* 3:517-521.

Toh, H., 1991 Sequence analysis of firefly luciferase family reveals a conservative sequence motif, *Prot. Seq. Data Anal.* 4:111.

Toh, H., Kikuno, R., Hayashida, H., Miyata, T., Kugimiya, W., Inouye, S., Yuki, S. and Saigo, K., 1985a Close structural resemblance between putative polymerase of a *Dorosophila* transposable genetic element 17.6 and *pol* gene product of Moloney murine leukaemia virus, *EMBO J.* 4:1267.

Toh, H., Minami, M. and Shimizu, T., 1990 Molecular evolution and zinc ion binding motif of leukotriene A4 hydrolase, *Biochem. Biophys. Res. Commun.* 171:216.

Toh, H., Ono, M., Saigo, K. and Miyata, T., 1985b Retroviral protease-like sequence in the yeast transposon, *Ty*1. *Nature* 315:691.

Vallee, B. L. and Auld, D. S., 1990 Zinc coordination, function, and structure of zinc enzymes and other proteins, *Biochemistry* 29:5647.

AMINO ACID SEQUENCE COMPARISON AS AN AID TO DETERMINING EVOLUTIONARY ORIGINS

Russell F. Doolittle

Center for Molecular Genetics
University of California
San Diego, La Jolla, CA 92093-0634 USA

INTRODUCTION

The development of rapid methods for determining amino acid and nucleic acid sequences has stimulated a great awakening among biochemists about matters of evolution. These endeavors have depended on the use of computers for manipulating and storing the vast amounts of accruing sequence data. Together these two advances, the collection of sequences and computer-assisted analysis, are allowing a remarkable reconstruction of the history of life on Earth. Just as there have always been technical problems associated with sequence determinations, however, so are there problems and limitations in the computer analysis field. Although in principle it is possible to reconstruct past events with present-day sequences, there are a number of confounding aspects that must be confronted.

Various computer algorithms have been devised that can align sequences "properly" and cluster them into the dendrograms we call "trees" (Fitch and Margoliash, 1967; Needleman and Wunsch, 1970; Feng and Doolittle, 1987, *inter alia*). Generally speaking, the more distantly related the organisms from which the sequences for a particular protein are drawn, the more different are the sequences themselves. There are a number of complicating factors, however, and occasionally such trees are not an accurate reflection of the biological relationship of the organisms. The confounding phenomena include unequal rates of change, gene conversions, overly distant relationships, gene mosaicism, unwitting comparison of paralogous genes, incorrect sequences, and occasional horizontal gene transfers. Some of these considerations are obvious and can be accommodated by common sense and putting the appropriate data in the right context. Thus, rapidly changing sequences are most useful in cataloguing closely related organisms, and slower changing ones better suited to more distant relationships. Some other problems are more troublesome, however. The matter of "unequal rates" is especially vexing, and it is this aspect that I will focus on here.

Methods in Protein Sequence Analysis, Edited by K. Imahori and F. Sakiyama, Plenum Press, New York, 1993

RATES OF CHANGE

From the outset, it was clear that proteins change at different but characteristic rates. Moreover, different parts of proteins were observed to change at different rates. For example, surface residues change much more rapidly than interior ones, and connector regions faster than catalytic units. There are also genetic factors that bear on the overall rate of amino acid replacement, including both matters of genetic occurrence (mutation) and survival (evolutionary fixation). Given all the many variables, it is remarkable how steady the rate of change is for a given protein (Figure 1). It is also noteworthy how limited is the "envelope" of rates of change for all proteins taken together. Thus, the fastest and slowest changing animal proteins only differ in rates by two orders of magnitude. These limits are the natural consequence of some proteins, like some histones, hardly changing at all, and others, like albumin, changing at a pace restricted mostly by the bounds of protein folding.

It has long been recognized that, if proteins have changed at constant rates, then they can be used as "clocks" to date the great biological divergences. For example, many efforts have been made in the past to date the time of divergence of prokaryotes and eukaryotes (reviewed in Doolittle *et al*, 1989b). In the earliest studies, the small number of sequences available limited the accuracy of the estimates, and even today there are not enough data to pin the answer down with any precision. At the moment, the best consensus data put the divergence at about 1.8 billion years ago. That average number belies a large scatter in the data, however. Furthermore, there are some curious anomalies that make the picture even more confusing.

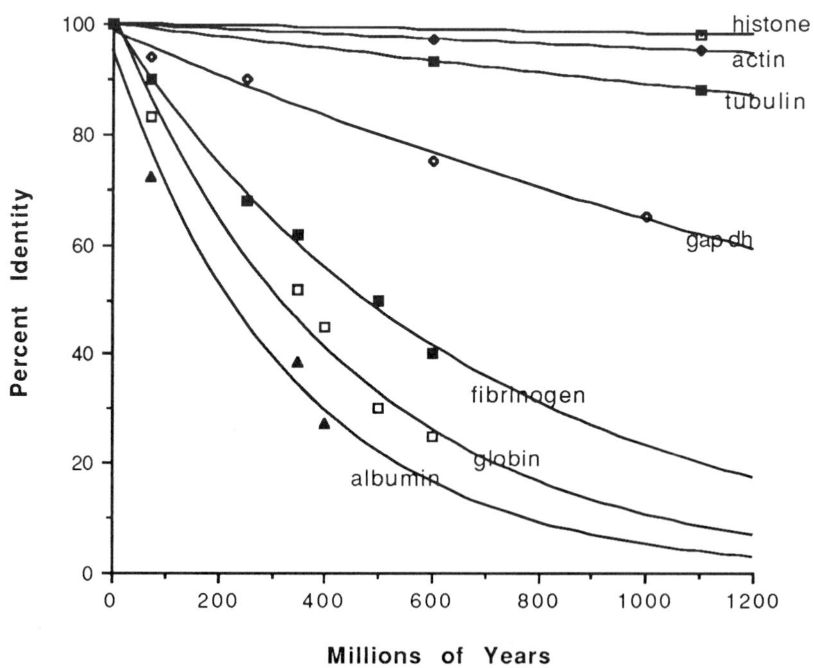

Figure 1. Different rates of change for different proteins. The divergence times out to 600 million years are based on the fossil record. The points at 1,000 and 1,100 million years denote divergence of plants and fungi from animals (from Doolittle, 1992).

I will try to illustrate the problem by comparing a set of protein sequences from three well studied organisms: humans, yeast and *E. coli* . This should eliminate the problem of varying genetic distances between organisms. Thus, human and yeast are both eukaryotes and are obviously genetically closer than either is to *E. coli.* (Figure 2). The last common ancestor of humans and yeast existed about a billion years ago. As noted above, the last common ancestor of prokaryotes and eukaryotes is thought to have existed about 1.8 billion years ago, and therefore one expects prokaryotic sequences to be significantly more different from human sequences than those from yeast, assuming anywhere near constant rates of change (Figure 2a). To my surprise, however, even though the majority of sequences meet the expectation, there are a number of *E. coli* sequences that are as similar (or more!) to one or both of the eukaryotes as the latter are to each other (Table 1). How can this be so?

There are several possible explanations. Consider the case of ornithine carbamoyl transferase. The human and yeast sequences are 36.4% identical, but the human and *E. coli* sequences are 37.4% identical. This could be explained if, for some unknown reason, the lineage leading to the yeast sequence has been changing at a significantly faster rate than that going to human (Figure 2b). That such might have been the case is suggested by the even lower similarity of the yeast and *E. coli* sequences (29.6%).

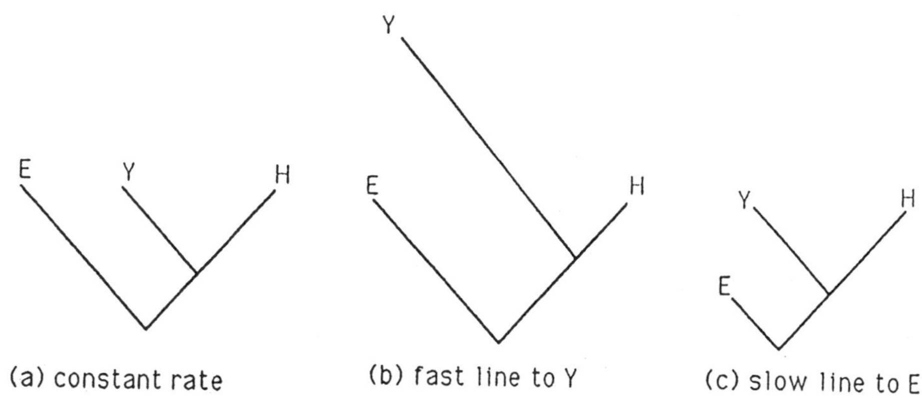

Figure 2. Simple trees showing evolutionary distances for *E. coli* (E), yeast (Y) and human (H). Lengths of lines intended to be proportional to amount of difference in particular protein sequences.

In the case of fructose-bis-phosphatase, all three sequences differ to approximately the same extent, a condition that can be explained if the lineage leading to *E. coli* has witnessed a very slow rate of change for this protein (Figure 2c).

Why should a protein slow its rate of change? Are there limits beyond which two diverging sequences cannot further change? Certainly in most proteins there are a few irreplaceable residues without which the protein cannot function. Generally, however, the fraction of such residues is quite low and within the natural variance of random comparisons. We can see this "limitless" phenomenon and the decay into virtual oblivion when we compare sequences from RNA viruses, organisms whose mutation rates are very fast and whose proteins quickly change to the point of nonrecognition at the sequence level (Doolittle *et al.*, 1989a).

In any event, it is not immediately obvious why the rate of change of a given protein should vary along different lineages when other proteins are maintaining a relatively constant pace. This forces us to consider some alternative explanations. Accordingly, for our next example, I'd like to discuss glyceraldehyde-3-phosphate dehydrogenase (GAP-DH).

Table 1. Pairwise similarities for 18 human, yeast and *E. coli* proteins

protein	hum/yst	hum/col	yst/col
gapdh A*	**64.5**	**65.4**	**67.6**
heat shock, large	59.1	47.9	45.5
SOD, Mn	47.7	44.9	41.0
triose-P-isomerase	49.6	44.8	45.2
fructose-bis P'ase	**44.1**	**43.7**	**44.6**
P-fructokinase	54.5	41.9	37.8
dihydrolipoamide dh	54.9	41.9	40.0
P-glycerate kinase	62.4	38.2	42.4
porph'gen syn	52.7	37.9	34.3
orn. carb. tf	**36.4**	**37.4**	**29.6**
gapdh B	64.5	35.6	37.6
adenylate kinase	**30.4**	**35.1**	**47.2**
gln tRNA syn	**30.9**	**32.9**	**41.3**
dihydrofolate red.	**34.8**	**32.7**	**34.0**
alkaline P'ase	**27.7**	**28.0**	**26.5**
his tRNA syn	44.3	21.2	26.2
UDP-GLU P-U tf	30.5	15.1	12.5
catalase	39.1	12.5	11.6

*Values are pairwise percent identities. Entries in bold denote anomalous similarities for eukaryote/prokaryote relationship.

HORIZONTAL TRANSFER

As it happens, *E. coli* (and some other bacteria) have two GAP-DH enzymes. In one case, the sequences from human, yeast and *E. coli* (GAP-DH-A) are all about 65% identical to each other (Table 1). In contrast, the human and yeast enzymes are only 36-37% identical with the *E. coli* GAP-DH-B enzyme. Does this mean the *E. coli* B sequence has been changing twice as fast as its A? It would be surprising, because the enzymes presumably fold the same way and are subject to the same structure-function restraints.

It would seem more reasonable to me to presume that the GAP-DH-A sequence has been acquired from some eukaryote by a horizontal gene transfer (Doolittle *et al.*, 1990). In this regard, its sequence is simply too similar to those of eukaryotes. Indeed, of 30 pairs of human and *E. coli* proteins I was able to extract from the Protein Identification Resource data bank, the GAP-DH-A comparison is the most similar (Figure 3).

There are two other entries in Table 1 that seem even more anomalous. In the cases of adenylate kinase and glutamine aminoacyl synthetase, the yeast and *E. coli* enzymes are much more similar to each other than either is to human. Whether these discrepancies are due to rate inequalities, horizontal gene transfers, or other factors, remains to be seen. One possibility is that they were brought into the eukaryotic line during the endosymbiotic enslavement of mitochondria. For a while eukaryotes might have maintained two versions of those enzymes which were common to both host and symbiont, discarding one or the other at some later time. In this case it may have been after the divergence of fungi and animals, the line leading to yeast discarding its own gene and the line leading to animals discarding the bacterial one.

OTHER RATE CONSIDERATIONS

How far back into the past can we look? If we examine very slowly changing proteins, shouldn't we be able to look back right to their beginnings? The problem with such logic is that it assumes a constant rate of change throughout history. Such an assumption must be incorrect, as shown by the following example.

Actin is a protein found only in eukaryotic cells. Remarkably, when its X-ray structure was determined, it turned out to resemble the structure of a portion of a heat shock protein (Kabsch *et al.*, 1990). Until that discovery, no one had ever detected any sequence

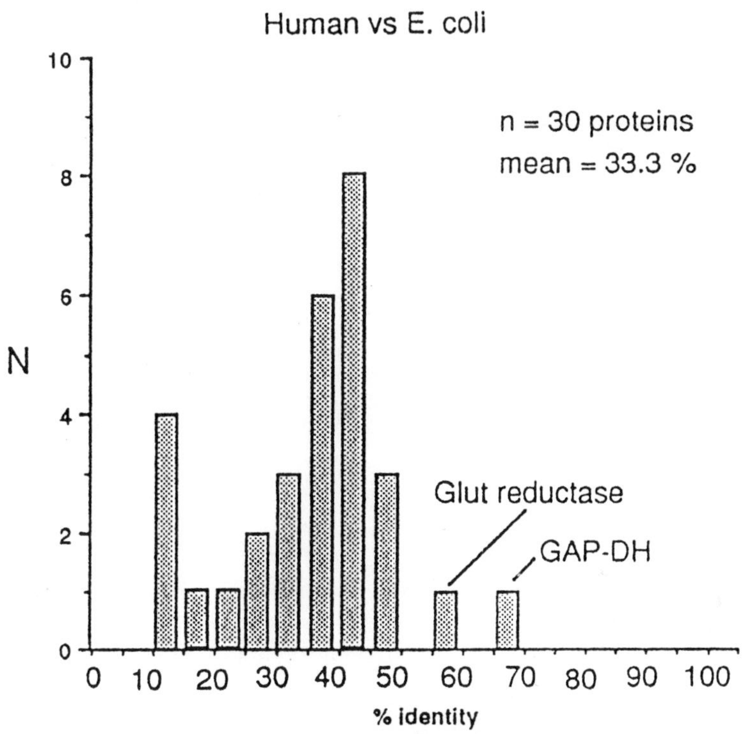

Figure 3. Distribution of similarities for 30 sequence pairs from humans and *E. coli*. 28 of the 30 are enzymes (identical EC numbers). The two non-enzymes are heat shock 70 and thioredoxin.

similarity between actin and any heat shock protein. With the aid of the X-ray structure, however, it was possible to show some faint resemblance (Doolittle, 1992).

The paradox is that actin is among the slowest changing of eukaryotic proteins, and heat shock proteins themselves are also very conserved. The inescapable conclusion is that actin, and maybe heat shock proteins too, changed at a much faster rate in the past. If we assume the X-ray structures demonstrate common ancestry, when did all the sequence change occur? Certainly we can assume that the earliest proteins must have changed at a very fast rate for much the same reasons that the proteins of RNA viruses still do; namely, that in an RNA-based world the mutation rate must have been considerably heightened. More than that, the opportunities for structural improvement by natural selection must have been much greater.

Beyond that, other paradoxes remain. The observed rates of change of proteins, slow as they are, imply that the heritage of most existing proteins ought to be readily apparent. Yet the data banks contain large numbers of "orphan sequences" with no apparent relatives. Homologues of some of these "orphans" have been sequenced from two or more species and found to be changing at ordinary rates; where did they come from? Why haven't we seen their ancestors? Clearly, there is much to be disentangled in our goal of reconstructing the past with amino acid sequences.

REFERENCES

Doolittle, R. F., 1992 Reconstructing history with amino acid sequences, *Prot. Sci.*, 1:191.

Doolittle, R. F., Feng, D. F., Anderson, K. L. and Alberro, M. R., 1990 A naturally occurring horizontal gene transfer from a eukaryote to a prokaryote, *J. Mol. Evol.* 31:383.

Doolittle, R. F., Feng, D. F., Johnson, M. S. and McClure, M. A., 1989a Origins and evolutionary relationships of retroviruses, *Quart. Rev. Biol.* 64:1.

Doolittle, R. F., Anderson, K. L. and Feng, D. F., 1989b Estimating the prokaryote-eukaryote divergence time from protein sequences, *in*: The Hierarchy of Life. Molecules and Morphology in Phylogenetic Analysis, Chapter 6, pp. 73-85, Fernholm, Kare and Jornvall, eds., Elsevier Sci. Publ. V.V., Amsterdam.

Feng, D. F. and Doolittle, R. F., 1987 Progressive sequence alignment as a prerequisite to correct phylogenetic trees, *J. Mol. Evol.* 25:351.

Fitch, W. M. and Margoliash, E., 1967 Construction of phylogenetic trees, *Science*, 155:279.

Kabsch, W., Mannherz, H. G., Suck, D., Pai, E. F. and Holmes, K. C., 1990 Atomic structure of the actin: DNase I complex, *Nature*, 347:37.

Needleman, S. B. and Wunsch, C. D., 1970 A general method applicable to the search for similarities in the amino acid sequence of two proteins, *J. Mol. Biol.* 48:443.

PROTEIN DATABASES CONSTRUCTED BY QUANTITATIVE TWO-DIMENSIONAL GEL ELECTROPHORESIS

James I. Garrels[1], B. Robert Franza, Jr.[1], Scott D. Patterson[1], Keith Latham[2], Davor Solter[3], Cecile Chang[1], and Gerald Latter[1]

[1]*Cold Spring Harbor Laboratory*
Cold Spring Harbor, NY 11724
[2]*Fels Institute for Cancer Research*
Temple University School of Medicine
Philadelphia, PA 19140
[3]*Max Planck Institute for Immunology*
Freiburg-Zahringen, Germany

INTRODUCTION

Two-dimensional gels can provide quantitative information about the levels of expression and degree of modification for thousands of proteins from biological samples. However, obtaining that information is a difficult task involving 1) preparation of reproducible two-dimensional gels, 2) detection and quantitation of protein spots from complex gel images, 3) comparison of multiple gel images through matching techniques, 4) identification of the protein spots, and 5) characterization of protein modifications. At the QUEST Protein Database Center we have developed and refined methods for all aspects of 2D gel production and analysis, and we have used the QUEST system to build protein databases for rat and mouse cells. These databases have been explored quantitatively using cluster analysis to find proteins with similar expression patterns. These databases, and a potentially powerful new method for identification of protein spots in two-dimensional gels, are presented below.

RAT PROTEIN DATABASE

The rat database is the first major application of the QUEST system (Garrels, 1989; Garrels and Franza, 1989a; Garrels and Franza, 1989b). This database contains 10 major groups of gels (matchsets) analyzed for REF52 fibroblast cells and virally-transformed derivative cell lines. A set of approximately 1600 high-quality spots were analyzed from each of 80 gels. The matchset for normal REF52 cells is illustrated by the data in Figure 1.

Methods in Protein Sequence Analysis, Edited by K. Imahori
and F. Sakiyama, Plenum Press, New York, 1993

Expression profiles for two proteins, a nuclear protein of 85 kD and non-muscle caldesmon, are shown in a quantitative comparison of 8 samples. The samples represent REF52 cells labeled on days 2, 3, 4, 6, and 12 after plating. Cells at day 6 are confluent, and cells at day 12 are totally quiescent with respect to DNA synthesis. Cells at days 6 and 12 were also labeled at 5 or 15 hr after refeeding with fresh serum. The expression profiles seen in Figure 1 illustrate two of the characteristic patterns of growth regulation seen among the REF52 proteins.

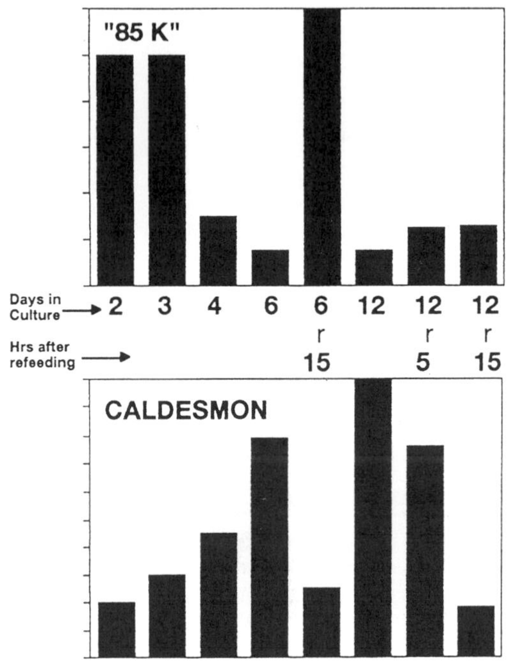

Figure 1. Expression profiles from the REF52 database. Non-transformed REF52 cells were labeled at 2, 3, 4, 6, and 12 days after plating. One culture of Day 6 cells was refed ("r") and labeled 15 hr later. Cultures of Day 12 cells were labeled at 5 and 15 h after refeeding.

To find other proteins that share these, or other patterns of growth regulation, we performed cluster analysis on the matchset described above (Garrels *et al.*, 1990). Cluster analysis is a statistical technique for finding groups of data points (in this case, spots) that are related by some measure of distance (in this case, a score for similarity of the expression patterns). The 6 major clusters for REF52 growth regulation are shown in Figure 2. The first cluster, called PCNA-like because it includes the DNA-replication protein PCNA, is related to the "85K" pattern seen in Figure 1. About 10% of the proteins analyzed share this pattern, and of these, nearly 25% are nuclear. These proteins decrease at confluence and quiescence, and appear to have roles in growth, DNA synthesis, and the cell cycle.

The second cluster is the set of proteins that are not regulated by growth. Many proteins, including β-F1-ATPase, tropomyosin 5, and calpactin II, are synthesized at constant rates regardless of the rate of cell growth and division. The third cluster also contains proteins that are not affected by the rate of growth or cell division, as seen by their constant rate of synthesis up to confluence, but members of this set are highly induced in confluent cells. Such proteins are likely to include differentiated products that appear only after cells have entered a G_0 state. The fourth cluster is the inverse of cluster 3. These proteins do not show significant growth regulation, but they are strongly repressed in Go cells. Refeeding quiescent (day 12) cells does not affect the rate of synthesis of proteins in clusters 3 and 4, at least for a period of 15 h.

Cluster 5 contains proteins whose expression profiles are related to caldesmon (Figure 1). Such proteins increase in rate of synthesis as cells reach confluence, and are strongly repressed after refeeding. This set includes many of the nuclear proteins and many

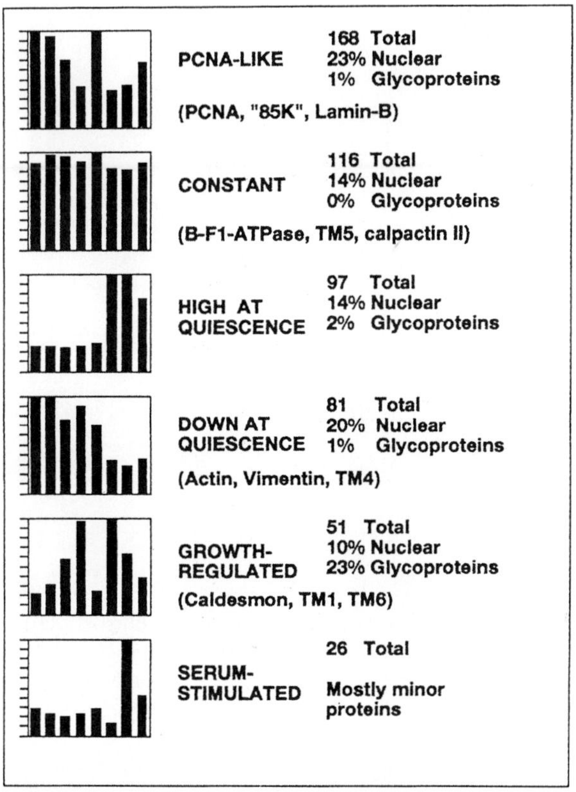

Figure 2. Cluster analysis of protein expression profiles. The eight-sample experiment described in Figure 1 was analyzed by cluster analysis. Each cluster is a group of proteins with similar expression profiles, and each graph shows the average expression profile for the members of the cluster.

glycoproteins. The final cluster contains proteins that are specifically stimulated by serum early, but not late, after addition to quiescent cells. These proteins tend to be minor components and none have been identified, however this set could include transcription factors such as Fos and Jun that are known to be part of the early serum response.

Cluster analysis of the REF52 database has brought into focus the types of regulation that occur during growth control of non-transformed cells in culture, and it has given us lists of proteins for each cluster that are of particular interest for identification and further characterization.

Major patterns of synthesis from the 2-cell stage mouse embryo.

description	fraction of analyzed (%)	profile
increases	37	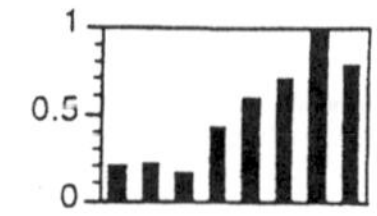
decreases	20	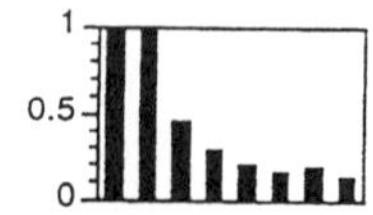
transient decreases, high early	16	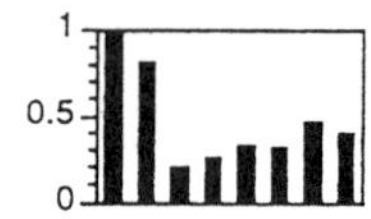
transient increases	6	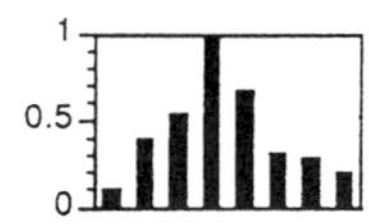
transient decreases, low early	4	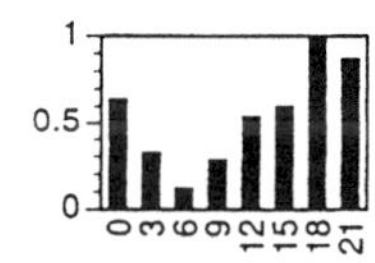
Total	77	

Figure 3. Cluster analysis of expression profiles from the mouse embryo database. Cells at the 2-cell stage were analyzed by labeling at 3 h intervals, and cluster analysis was performed as in Figure 2. This data represents the analysis of approximately 800 proteins.

MOUSE PROTEIN DATABASE

Protein database techniques have been used to analyze the proteins of mouse embryos during normal early development (Latham *et al.*, 1991; Latham *et al.*, 1992). This database has provided an important baseline for subsequent studies of experimental situation, such as nuclear transplantation, developmental blocks with protein kinase inhibitors, etc. The mouse database contains data from embryos labeled every 3 hours from fertilization through blastocyst. Cluster analysis was applied to find the important patterns of change at the 1-cell, 2-cell, and 4-cell stages. Most changes in protein synthesis occur at the 2-cell stage.

The major expression profiles of the 2-cell stage are shown in Figure 3. A large fraction (37%) of the proteins increase markedly in rate of synthesis or appear for the first time early in the 2-cell stage. Two other clusters (containing 36% of the proteins) represent proteins that decrease or disappear during the 2-cell stage. These clusters represent proteins that are either turned on with the appearance of embryonic mRNA or cease to be synthesized with the disappearance of maternal mRNA. These patterns of change were expected, but the number and magnitude of the changes was a surprise. More than one fourth of the proteins either disappear or appears for the first time during the 2-cell stage. Protein patterns from the beginning and end of the 2-cell stage are so different that matching them is difficult. The amount of change over a 3 hr interval of the 2-cell stage is greater than the amount of change found over the entire growth curve of REF52 cells.

The two remaining clusters in Figure 3 were unexpected. In the 2-cell stage, precisely at the time when the switch from maternal to embryonic mRNA utilization is most rapid, one group of proteins is induced transiently and another group is repressed transiently. None of the members of these clusters have been identified, and their role in development remains to be discovered.

A POTENTIAL NEW METHOD OF PROTEIN IDENTIFICATION

The availability of quantitative imaging technology makes possible a new approach to protein identification. Samples labeled with two amino acids can be analyzed quantitatively to measure amino acid ratios. If the ratios are accurate enough, it is likely that enough information can be gained from several such determinations to uniquely identify proteins of known sequence. Our approach is to label cells simultaneously with one ^{35}S-labeled amino

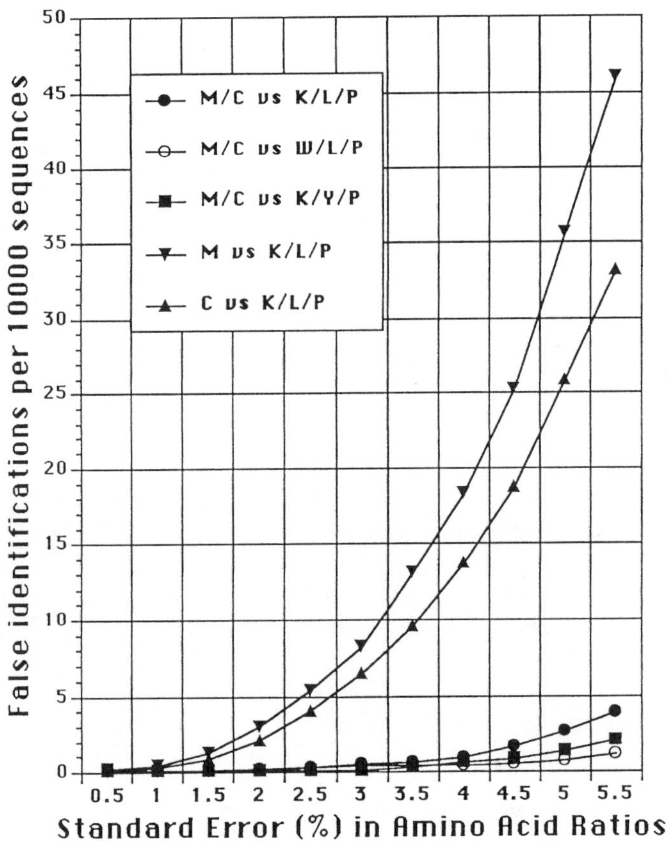

Figure 4. Relationship between accuracy of amino acid ratios and number of false identifications expected per 10000 protein sequences. This simulation was carried out using every non-redundant sequence in the yeast protein database (PIR release 30) to give sample amino acid ratios followed by searching the rest of the database for other sequences that match these ratios within a 2 standard deviation window of error.

acid and one ^{14}C-labeled amino acid. By comparing images of each gel obtained over a period of 3-4 months, as much of the ^{35}S signal decays, it is possible to determine the relative amount of the two isotopes pixel by pixel throughout the gel.

Computer simulations have been run, using the database of yeast protein sequences, to determine the number of ratios and the accuracy of the ratios that are needed to reduce the number of false identification to below 1 per 10000 sequences, which is enough to uniquely identify a protein coded by the yeast genome. Figure 4 shows the number of "hits" among yeast sequences at various levels of accuracy for combinations of 4 amino acids (one ^{35}S-labeled) and for 5 amino acids (two ^{35}S-labeled). The result is that if the ratios are accurate to within 4%, then only 4 ratios relating 5 amino acids are necessary.

This approach is currently being applied to the identification of yeast proteins in two-dimensional gels. The necessary degree of accuracy is being achieved by phosphorimager technology, which provides excellent linearity of detection over a wide latitude of intensities. Software has been developed to align the images from successive exposures to within a fraction of a pixel, and to calculate the rate of decay for each pixel of the image. From the decay curve for each pixel, we can calculate the proportion of the signal generated by each isotope, and by normalization to proteins of known composition, we can calculate actual amino acid ratios for each major protein. The ratio data for each pixel within a spot are averaged to increase the accuracy. Preliminary data suggest that periodic (biweekly) exposures over 3-4 month interval will give the necessary accuracy. The amino acid composition data can then be used, in conjunction with predicted molecular weight and isoelectric point, to correlate spots on the 2D gel image with proteins from the yeast protein sequence database.

The new method should allow us to identify many spots that represent proteins of known sequence. It will also be extremely useful for helping us to recognize spots (identified or not) that are related by post-translational modification. The identification of the spots detected on two-dimensional gels, when combined with construction of quantitative protein databases as described above for rat and mouse cells, should provide a much clearer picture of the pathways of protein synthesis, degradation, and modification that occur during complex events of signal transduction and gene regulation.

Acknowledgements: We thank the QUEST 2D gel laboratory for production of all 2D gels discussed here, Phyllis Myers for analysis of rat and mouse 2D gels using the QUEST protein database software, and Phil Renna for photographic assistance. This work was supported by grant P41 RR02188 from the National Institutes of Health Biomedical Research Technology Program.

REFERENCES

Garrels, J. I., 1989 The QUEST system for quantitative analysis of two-dimensional gels, *J. Biol. Chem.* 264:5269.

Garrels, J. I. and Franza, B. R. Jr., 1989a The REF52 protein database: Methods of database construction and analysis using the QUEST system, and characterizations of protein patterns for proliferating and quiescent REF52 cell cultures, *J. Biol. Chem.* 264:5283.

Garrels, J. I. and Franza, B. R. Jr., 1989b Transformation-sensitive and growth-related changes of protein synthesis in REF52 cells, *J. Biol. Chem.* 264:5299.

Garrels, J. I., Franza, B. R. Jr., Chang, C. and Latter, G. I., 1990 Quantitative exploration of the REF52 protein database: Cluster analysis reveals the major protein expression profiles in responses to growth regulation, serum stimulation, and viral transformation, *Electrophoresis*, 11:1114.

Latham, K. E., Garrels, J. I., Chang, C. and Solter, D., 1991 Quantitative analysis of protein synthesis in mouse embryos I: Extensive reprogramming at the one- and two-cell stages, *Development* 112:921.

Latham, K. E., Garrels, J. I., Chang, C. and Solter, D., 1992 Analysis of embryonic mouse development: construction of a high-resolution, two-dimensional gel protein database, *Appl. Theor. Electrophoresis* 2:163.

PROTEIN STRUCTURES OF SPECIAL INTERESTS

BIOCHEMISTRY OF A NATRIURETIC PEPTIDE FAMILY

Hisayuki Matsuo

National Cardiovascular Center Research Institute
Suita, Osaka 565, Japan

It was 8 years ago, when ANP (atrial/A-type natriuretic peptide) was first identified in mammalian heart. The peptide elicits vasorelaxant, diuretic and natriuretic activities. Furthermore, ANP is known to be secreted from the heart, circulates in the blood and participates in the regulation of salts-water balance and blood pressure. These facts revealed that the heart is not only a mechanical pump impelling blood, but is also an endocrine organ regulating the circulation system. Thus, the peptide represents an exciting and hopeful new direction in the circulation research. Identification of ANP induced an explosive series of investigations in the related fields. However, there is much more to be learned for understanding the still puzzling significance of the peptide. Subsequent to the discovery of ANP, another natriuretic peptide designated BNP (brain/B-type natriuretic peptide), was isolated from porcine brain (Sudoh *et al.*, 1988). BNP is very similar to ANP structurally as well as pharmacologically.

Identification of ANP and BNP disclosed the possibility that a family comprising similar natriuretic peptides participates in regulating homeostatic balance of body fluid volume and blood pressure. ANP and BNP, though they are derived from distinct genes, share a highly homologous ring-structure composed of 17 amino acid residues formed by a pair of cysteine residues, which is known to be essential for exerting biological activities. Moreover, we have very recently discovered in porcine brain a third member of the natriuretic peptide family and designated it C-type natriuretic peptide (CNP) (Sudoh *et al.*, 1990). Porcine CNP of 22 amino acid residues (CNP-22) which is an endogenous form in porcine brain, shows remarkable sequence homology to mammalian ANP and BNP so far identified, within a 17- residue ring portion. Surprisingly, however, porcine CNP terminates at the second cysteine residue participating in the ring formation, and completely lacks further C-terminal extension, which occurrs both in ANP and BNP. Porcine CNP-22 potently stimulates guanylate cyclase activity in the rat cultured vascular smooth muscle cell (Furuya *et al.*, 1990), which is not sensitive to ANP and BNP, implying that CNP may function in a manner different from ANP and BNP. It is now essential to characterize the specified physiological roles of CNP and to define its functional difference from two other natriuretic peptides.

Methods in Protein Sequence Analysis, Edited by K. Imahori
and F. Sakiyama, Plenum Press, New York, 1993

Brain/B-type Natriuretic Peptide (BNP)

Mammalian ANP and BNP share a 17-amino acid ring-structure formed by an intramolecular disulfide linkage. It is quite evident that the ring portion is highly conserved between ANP and BNP. Between human ANP and BNP, however, five amino acid replacements are observed in the ring system and remarkable substitutions occur both in the N-and C-terminal tails. Accordingly, it is clear that BNP is definitely distinct from, but very similar to ANP. As was suggested by the structural resemblances, BNP shows a pharmacological spectrum very similar to that of ANP. Natriuretic and diuretic activities as well as hypotensive effect are quite comparable to those induced by ANP at the same dosage (Sudoh *et al.*, 1988). BNP is thus qualified as a member of the natriuretic peptide family.

It is well known that the amino acid sequence of ANP are fully conserved among all mammals. In contrast with ANP, immunological observations using the antiserum for porcine BNP suggested that human and rat BNP may be more or less different from porcine BNP. We have cloned and sequenced cDNAs encoding the BNP precursors, and deduced the amino acid sequences of pig, human and rat BNP-precursors (Maekawa *et al.*, 1988; Kojima *et al.*, 1990; Sudoh *et al.*, 1989), showing the structural similarity, on the whole, to pig, human and rat ANP-precursors. They carry in common the bioactive unit including the ring-structure flanked by two cysteine residues at their C-terminal regions. In spite of such structural similarity, the chain-length of BNP-precursors varies from species to species. Furthermore, in the BNP series, homologous sequences localize only in the N- and C-terminal regions of the precursors. Even in the bioactive unit, sequence homology among three species is not so high compared with that observed in mammalian ANPs. Although BNP was first identified in porcine brain, the highest concentration of BNP-immunoreactivity was found in mammalian heart, though not more than 5% of ANP (Aburaya *et al.*, 1989). BNP was also found in the circulation, both in secretory level and in plasma level (Saito *et al.*, 1989). These facts clearly indicate that BNP is also a cardiac hormone, secreted from the heart and circulating in the blood. Furthermore, it should be emphasized that synthesis of BNP in the heart and its secretion were markedly enhanced in patients with severe heart failure (Mukoyama *et al.*, 1990). These facts suggest the possibility that expression of BNP is regulated in a manner different from that of ANP, and plays a role as an emergency aid for ANP. In fact, the possibility may be implicated in the presence of the AT-rich sequence (clustered ATTTA motifs) occurring in the 3'-untranslated region of BNP-cDNA (Maekawa *et al.*, 1988; Kojima *et al.*, 1990; Sudoh *et al.*, 1989), since such an AT-rich nucleotide sequence, which has never been found in ANP-cDNA, is now known to destabilize mRNA in the cell, as seen in several lymphokines and proto-oncogenes.

C-Type Natriuretic Peptide (CNP)

The amino acid sequence of CNP shows remarkable similarity to ANP and BNP, especially in the 17-residue ring system. However, CNP terminates in the second cysteine residue, lacking a further C-terminal extension. Our recent cDNA analyses for mammalian CNP-precursors demonstrated that a termination codon TAG directly follows a C-terminal cysteine residue (Tawaragi *et al.*, 1991; Kojima *et al.*, 1990). This fact confirmed that the C-terminal truncated form is an intrinsic structure of CNP. Furthermore, it should be mentioned that a pharmacological spectrum of CNP is very similar to those of ANP and BNP, though much weaker than ANP and BNP (Sudoh *et al.*, 1990). These results clearly indicate that CNP is structurally and pharmacologically a third member of the natriuretic peptide family. Thus, natriuretic peptides so far identified are classified into three subclasses, ANP, BNP and CNP, which genetically distinct from each other. The occurrence of BNP and CNP along with ANP strongly suggests that a complex control mechanism through the family comprising ANP, BNP and CNP exists in mammals for regulating the blood pressure. In addition, it should be mentioned that CNP is localized in brain and spinal cord . Thus, it is most likely that ANP and BNP are synthesized in heart, secreted into blood stream and function as circulating hormones, while CNP is expressed exclusively in brain and presumably functions as a neuropeptide.

Receptors for the Natriuretic Peptide Family

Biological activities of the natriuretic peptides are now thought to be mediated by intracellular accumulation of cGMP through the activation of particulate guanylate cyclases (Brenner *et al.*, 1990). Two types of receptors, designated A- and B-receptor, beside a so-called clearance receptor, have been cloned as those specific for ANP (Chinkers *et al.*, 1989; Schulz *et al.*, 1989). Both A- and B- receptors include a guanylate cyclase domain within the molecules. It has been well known that ANP and, to a lesser extent, BNP are efficient activators of A-receptor. However, neither peptide can significantly stimulate B-receptor, even though B-receptor has reportedly higher affinity for BNP. Thus, a question as to whether BNP is really a specific ligand or not still remains to be solved. Recent isolation of CNP gives rise to another question; "Is there a receptor specific for CNP?" In this context, it should be noted that natriuretic-diuretic and hypotensive activities of CNP are about 100 times less potent than those induced by ANP and BNP, while the rectum-relaxant effect of CNP is 3~4 times more potent than that of ANP and comparable to BNP (Sudoh *et al.*, 1990). In addition, CNP potently increased cGMP level in cultured rat vascular smooth muscle cells (VSMC) and stimulated guanylate cyclase activity in the particulate franction of the cells (Furuya *et al.*, 1990). The extent of maximum activation of the enzyme induced by CNP was 4 times higher than that by hANP. In contrast, CNP caused only a slight elevation of cGMP in rat renal glomeruli. These results indicate that CNP is a potent stimulator of cGMP formation in VSMC, but not in glomeruli, suggesting that CNP is pharmacologically distinct from that of ANP. Such pharmacological dissociation of CNP from ANP and BNP suggests a possibility that CNP has its own receptor. In this context, interesting results should be mentioned as follows, which have recently been obtained by our collaborative work with Genentech group (Klooer *et al.*, 1991). CNP, a new member of the family, potently and selectively activated the human B-receptor. In contrast, CNP does not increase cGMP accumulation in cells expressing human A-receptor. In addition, the affinity of CNP for B-receptor is 50- or 500-fold higher than ANP or BNP, respectively. Considering that CNP is localized in brain, this ligand-receptor pair may be involved in the regulation of body fluid homeostasis by the central nervous system.

Occurrence of three related but distinct natriuretic peptides, ANP, BNP and CNP, suggests that they may function in a complex manner to coordinate their physiological roles, not only as the hormones, but also as the neuropeptides. The presence of two subtypes of the receptor guanylate cyclases may support this hypothesis discussed above.

REFERENCES

Aburaya, M., Minamino, N., Kangawa, K., Tanaka, K. and Matsuo, H., 1989 Distribution and molecular forms of brain natriuretic peptide in porcine heart and blood, *Biochem. Biophys. Res. Commun.* 165:872.

Brenner, B. M., 1990 Diverse biological actions of atrial natriuretic peptide, *Physiol. Rev.* 70:665.

Chinkers, M., Garbers, D. L., Chang, M. S., Lowe, D. G., Chin, H., Goeddel, D. V. and Schulz, S., 1989 A membrane form of guanylate cyclase is an atrial natriuretic peptide receptor, *Nature* 338:78.

Furuya, M., Takehisa, M., Minamitake, Y., Kitajima, Y., Hayashi, Y., Ohnuma, N., Ishihara, T., Minamino, N., Kangawa, K. and Matsuo, H., 1990 Novel natriuretic peptide, CNP, potently stimulates cyclic GMP production in rat cultured vascular smooth muscle cells, *Biochem. Biophys. Res. Commun.* 170:201.

Kojima, M., Minamino, N., Kangawa, K. and Matsuo, H., 1989 Cloning and sequence anlysis of cDNA encoding a precursor for rat brain natriuretic peptide, *Biochem. Biophys. Res. Commun.* 159:1420.

Kojima, M., Minamino, N., Kangawa, K. and Matsuo, H., 1990 Cloning and sequence analysis of cDNA encoding a precursor for rat C-type natriuretic peptide (CNP), *FEBS Lett.* 276:209.

Koller, K. J., Lowe, D. G., Bennett, G. L., Minamino, N., Kangawa, K., Matsuo, H. and Goeddel, D. V. 1991 Selective activation of the B natriuretic peptide receptor by C-type natriuretic peptide (CNP), *Science* 252:120.

Maekawa, K., Sudoh, T., Furusawa, M., Minamino, N., Kangawa, K., Ohkubo, H., Nakanishi, S. and Matsuo, H., 1988 Clonig and sequence analysis of cDNA encoding a precursor for porcine brain natriuretic peptide, *Biochem. Biophys. Res. Commun.* 157:410.

Mukoyama, M., Nakao, K., Saito, Y., Ogawa, Y., Hosoda, K., Suga, S., Shirakami, G., Jougasaki, M. and Imura, H., 1990 Increased human brain natriuretic peptide in congestive heart failure, *New Engl. J. Med.* 323:757.

Saito, Y., Nakao, K., Itoh, H., Yamada, T., Mukoyama, M., Arai, H., Hosoda, K., Shirakami, G., Suga, S., Minamino, N., Kangawa, K., Matsuo, H. and Imura, H., 1989 Brain natriuretic peptide is a novel cardiac hormone, *Biochem. Biophys. Res. Commun.* 158:360.

Schulz, S., Singh, S., Bellet, R. A., Singh, G., Tubb, D. J., Chin, H. and Garbers, D. L., 1989 The primary structure of a plasma membrane guanylate cyclase demonstrates diversity within this new receptor family, *Cell* 158:1155.

Sudoh, T., Kangawa, K., Minamino, N. and Matsuo, H., 1988 A new natriuretic peptide in porcine brain, *Nature* 332:78.

Sudoh, T., Maekawa, K., Kojima, M., Minamino, N., Kangawa, K. and Matsuo, H., 1989 Cloning and sequence analysis of cDNA encoding a precursor for human brain natriuretic peptide, *Biochem. Biophys. Res. Commun.* 159:1427.

Sudoh, T., Minamino, N., Kangawa, K. and Matsuo, H., 1990 C-type natriuretic peptide (CNP) : A new member of natriuretic peptide family identified in porcine brain, *Biochem. Biophys. Res. Commun.* 168:863.

Tawaragi, Y., Fuchimura, K., Tanaka, S., Nakazato, H., Minamino, N., Kangawa, K. and Matsuo, H., 1991 Gene and precursor of human C-type natriuretic peptide, *Biochem. Biophys. Res. Commun.* 175:645.

ACTIVIN, ACTIVIN-BINDING PROTEIN (FOLLISTATIN) AND ACTIVIN RECEPTOR

Hiromu Sugino[1], Takanori Nakamura[1], Kishiko Sugino[1] and Koiti Titani[2]

[1]*Institute for Enzyme Research, University of Tokushima Tokushima 770, Japan*
[2]*Fujita Health University School of Medicine Toyoake, Aichi 470-11, Japan*

INTRODUCTION

The principal regulators of the pituitary gonadotropins LH and FSH in mammals are the gonadotropin-releasing hormone (GnRH) from the hypothalamus and the gonadal steroids.

The gonadal steroids exert feedback actions on the brain to attenuate the release of GnRH and on the anterior pituitary gland to inhibit the secretion of LH and FSH. In recent years, several hypophysiotropic protein factors that can suppress or enhance FSH secretion by pituitary cells have been identified from mammalian gonads (Ying, 1988; Linget *et al.*, 1988). These include activin, which stimulates FSH release, and inhibin and follistatin, which inhibit FSH release. These protein factors are part of a novel feedback control mechanism for FSH secretion (Figure 1).

ACTIVIN

Although functionally antagonistics, activin and inhibin are structurally related; inhibin is a heterodimer composed of an α-subunit linked by disulfide bonds to one of two related β-subunits (inhibin A (αβA) or inhibin B (αβB)), whereas activin is a dimer composed of the inhibin β-subunits (activin A (βAβA), activin AB (βAβB) or activin B (βBβB)) (Figure 2) (Burger and Igarashi, 1988; Vale *et al.*, 1988).

Structural studies have shown that these proteins constitute the TGF-β family, with diverse regulatory effects on the growth and differentiation of numerous cell types (Massague, 1987; Vale *et al.*, 1990). There is increasing evidence from various areas of study that activin has multiple functions in a wide variety of biological systems: regulation of the secretion of anterior pituitary hormones (Billestrup *et al.*, 1990; Kitaoka *et al.*, 1988; Ling *et al.*, 1986; Vale *et al.*, 1986), induction of erythropoiesis (Broxmeyer *et al.*, 1988; Eto *et al.*, 1987; Yu *et al.*, 1987), modulation of ovarian and testicular cell functions

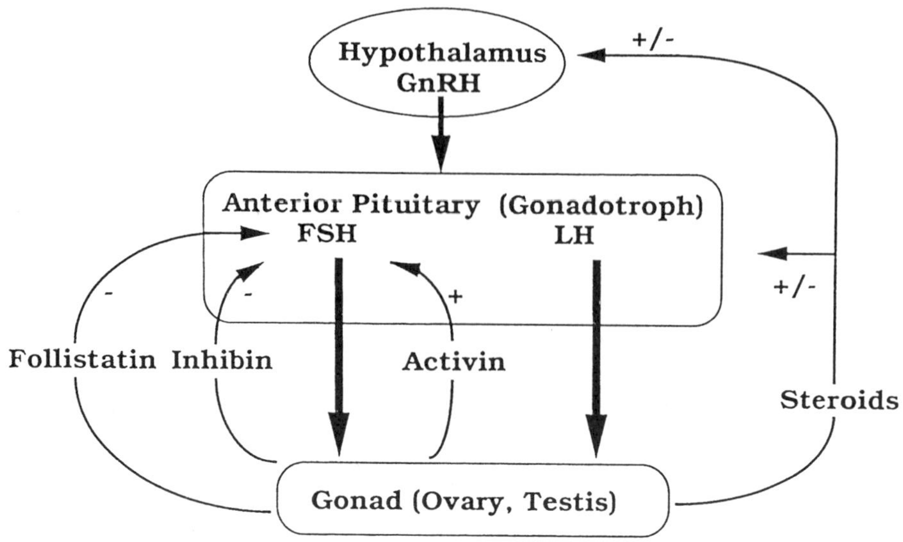

Figure 1. Activin, inhibin and follistatin actions along the hypothalamic-pituitary-ovarian axis. Plus signs indicate augmentation of cell functions and minus sings indicate inhibitory effects.

(Hsueh *et al.*, 1987; Hutchinson *et al.*, 1987; Sugino *et al.*, 1988), promotion of survival of nerve cells (Hashimoto *et al.*, 1990; Schubert *et al.*, 1990) and initiation of early embryonic development of *Xenopus* (Asashima *et al.*, 1990; Green and Smith, 1990; Smith *et al.*, 1990; Thomsen *et al.*, 1990; van den Eijnden-van Raaij *et al.*, 1990).

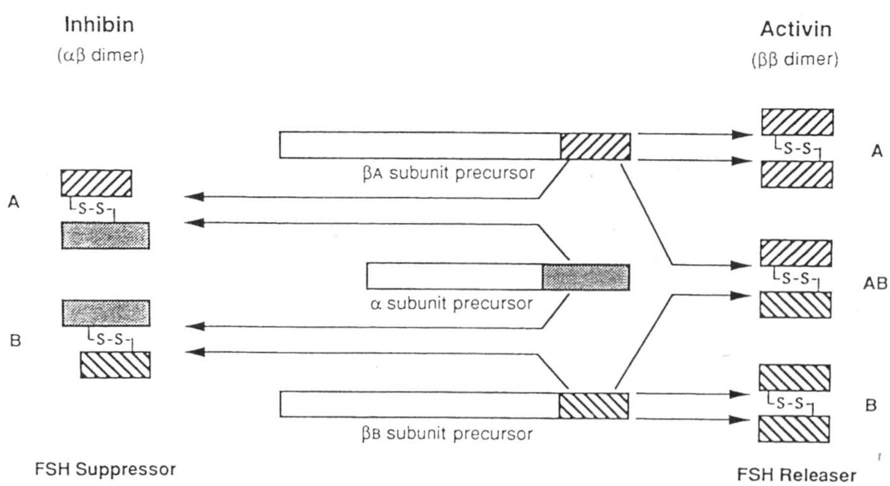

Figure 2. Multiple arrangement of gene products from the α, βA and βB subunit precursors through disulfide bridging to yield FSH releasers (activin) and suppressors (inhibin).

Activin Can Modulate Follicular Granulosa Cell Functions

Similarity of activin to TGF-β led us to focus our interest on autocrine or paracrine actions of activin in the ovary. Thus, we examined the action of activin on the functions of rat granulosa cells cultured in a chemically defined medium. Activin can enhance the FSH-induced LH-receptor expression and progesterone production, while activin itself has no effect on either the basal level in the absence or presence of a low concentration of FSH. In addition to the FSH-dependent actions, activin potentiates the ability of granulosa cells to produce inhibin. Activin also induces the expression of FSH receptors on the granulosa cell surface (Hasegawa *et al.*, 1988). It is widely believed that granulosa cell cytodifferentiation begins with the induction of FSH receptors on the cells. By virtue of the receptors, granulosa cells become competent to respond to the rise in FSH to undergo cytodifferentiation.

Therefore, our findings strongly suggest that activin affects the earlier differentiation stage of granulosa cell and triggers the cytodifferentiation process.

ACTIVIN-BINDING PROTEIN (FOLLISTATIN)

During the process of purification of the activin receptor, we observed that significant activin binding activity was recovered in a soluble fraction from rat ovary homogenates.

The purification of the active entity led us to discover an activin binding protein.

Activin-Binding Protein Is Follistatin

Two cycles of affinity chromatography of the 40,000*g* supernatant on an activin-immobilized column gave a glycoprotein showing several bands at a molecular mass range from 32 to 38 kDa in non-reducing SDS-PAGE (Nakamura *et al.*, 1990). Ligand blotting with the purified preparation established that every protein band does indeed have a specific binding activity for activin. The NH_2-terminal sequence analysis of each band revealed that all the bands have a single and the same sequence. Unexpectedly, this sequence was found to be identical to that of follistatin, which was initially obtained as a factor having the ability to preferentially inhibit the FSH secretion by pituitary cells (Robertson *et al.*, 1987; Ueno *et al.*, 1987).

```
    Deduced protein sequence
   1         95                              259                 303     315
1  GNC........DCSNITW.....................ASDNATY....SCNSISEDTEEEEEDEDQ...LEW
   1         95                              259     288
2  GNC........DCSNITW.....................ASDNATY....SCN

    Determined protein sequence
   1                                         CHO                         315
3  GNC........DCSNITW.....................ASDNATY....SCNSISEDTEEEEEDEDQ...LEW
   1          CHO                            CHO                 303
4  GNC........DCSNITW.....................ASDNATY....SCNSISEDTEEEEEDEDQ
   1                                         CHO                 303
5  GNC........DCSNITW.....................ASDNATY....SCNSISEDTEEEEEDEDQ
   1                                                             303
6  GNC........DCSNITW.....................ASDNATY....SCNSISEDTEEEEEDEDQ
   1                                         CHO     288
7  GNC........DCSNITW.....................ASDNATY....SCN
   1                                                 288
8  GNC........DCSNITW.....................ASDNATY....SCN
```

Figure 3. Structure of isoforms of porcine follistatin/activin-binding protein.
1,2; Predicted from the porcine follistatin cDNA. 3-8; Determined by peptide mapping and amino acid composition and sequence analyses of the resulting peptide. CHO; Asn-linked carbohydrate chain.

Follistatin/Activin-Binding Protein Can Neutralize Activin Activity

The physiological significance of follistatin was considered trivial because its inhibitory potency was only 10% of that of inhibin. To examine the biological roles of follistatin/activin-binding protein in the target cells of activin, we examined the effect of follistatin on activin actions in various systems. Follistatin was found to suppress the activin-stimulated cytodifferentiation of granulosa cells (Nakamura *et al.*, 1992). Further, activin-induced differentiation of mouse Friend leukemic cells (Shiozaki *et al.*, 1992) and *Xenopus* mesoderm induction (Asashima *et al.*, 1991) by activin were also inhibited in a dose-dependent manner. The inhibition patterns suggested the stoichiometric binding of activin to follistatin. This was confirmed by elucidation of the equimolar complex formation between activin and follistatin in a gel filtration experiment (Kogawa *et al.*, 1991a). The resulting complex was shown to have neither stimulatory nor inhibitory activity for FSH release by pituitary cells. These findings indicate that follistatin binds to activin to form an inactive equimolar complex.

Follistatin Can Associate with Heparan Sulfate Chains of Proteoglycan

In the course of a purification study of follistatin, we noticed that it possessed a unique and strong affinity for heparan and dextran sulfate, a property that has greatly facilitated its purification and characterization. Follistatin, which is secreted by cultured rat granulosa cells (Kogawa *et al.*, 1991b; Saito *et al.*, 1991; Shimasaki *et al.*, 1989), was shown to associate with the cell surface by affinity labeling with ^{125}I-activin.

Addition of follistatin to cultured cells demonstrated a typical ligand-binding saturation curve (Nakamura *et al.*, 1991), which suggested that granulosa cells have a specific binding site for follistatin. This binding was markedly inhibited by heparin and heparan sulfate, but not by chondroitin sulfates A and C, keratan sulfate, and dermatan sulfate. When granulosa cells were treated with glycosaminoglycan-degrading enzymes before or after addition of follistatin to the cultures, heparinase and heparitinase treatments significantly suppressed the binding. These findings suggest that follistatin binds to heparan sulfate side chains of proteoglycans on the granulosa cell surface to regulate the various actions of activin. The molecular cloning study proved that follistatin has two types of molecular forms originating from alternatively spliced mRNA with 315 (FS-315) and 288 (FS-288) amino acids (Esch *et al.*, 1987; Shimasaki *et al.*, 1988). However, we purified six molecular forms of follistatin from porcine ovaries. The six forms differed in structure by the truncation of the COOH-terminal region and/or the presence of carbohydrate chains (Figure 3).

All six species have almost the same activin binding activity. By contrast, the COOH-terminal truncated forms of follistatin (FS-288) showed much higher affinity for the rat granulosa cell surface than the longer forms of follistatin, while FS-315 had no affinity. In the rat pituitary culture, FS-288 was found to suppress FSH release more potently. These findings suggest that cell-associated follistatin traps more strongly activin in the matrix and blocks more effectively the activity of activin on the cell surface.

ACTIVIN RECEPTOR

Although activin has various effects on the cell phenotype and physiology, little is knowm about the biochemical mechanisms that enable it to evoke these various responses. Studies on its signal transduction should provide significant clues to our understanding of the molecular basis of the various actions of activin.

Activin Receptor Exists on Rat Ovarian Granulosa Cells

We elucidated that activin has various modulatory actions on follicular granulosa cell functions. This strongly suggested that the specific binding of activin to its granulosa cell surface receptor is the initial event of a process which ultimately elicits the cellular responses. Thus, we attempted to identify the activin receptor on the rat granulosa cells. A cellular receptor for activin was demonstrated by incubation of ^{125}I-activin with rat granulosa cell cultures (Sugino *et al.*, 1988). The specific binding of labeled activin to the cells showed saturation. Scatchard analysis of binding data indicated a single class of receptors having a Kd=340pM. Activin did not share the receptors with the activin-related proteins such as inhibin and TGF-β.

Activin Receptor Is a Transmembrane Serine/Threonine/Tyrosine Protein Kinase

Recently, two types of mouse activin receptors have been cloned successively by an expression cloning method (Attisano *et al.*, 1992; Mathews and Vale, 1991). From the deduced structures, the activin receptor was predicted to be a serine/threonine specific protein kinase. Independently, to clarify the mechanism of transmembrane signaling for activin in the neural differentiation of mouse embryonal carcinoma cell P19 (EC P19) (Hashimoto *et al.*, 1990), we purified and characterized an activin receptor protein from EC P19 (Nakamura *et al.*, 1992). The purified receptor had a specific and high affinity for activins A, AB and B

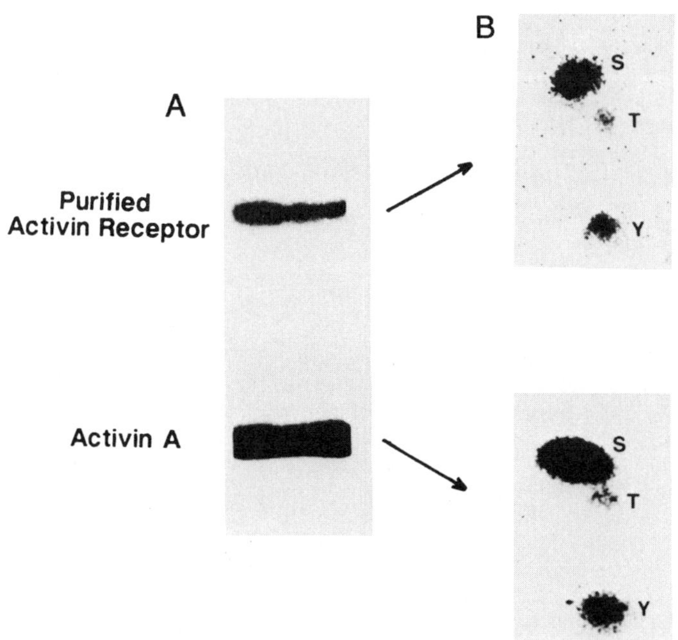

Figure 4. *In vitro* kinase activity of the purified activin receptor (A) and analysis of proteins phosphorylated (B).

(A) The purified receptor was incubated with activin in a buffer containing [γ–^{32}P] ATP. Reaction products were electrophoresed in 10% SDS-gels, electrotransferred to Immobilon membranes and autoradiographed. (B) Regions of the membranes containing the ^{32}P-labeled proteins were incubated with 6N HCl. The hydrolysates were then analyzed by two-dimentional electrophoresis. S, phosphoserine; T, phosphothreonine; Y, phosphotyrosine.

(Kd = 345 pM), but not for TGF-β. The purified activin receptor was identified by SDS-PAGE and ligand blotting analysis as a single protein of 70 kDa. The amino acid sequence of the first 18 NH_2-terminal residues revealed that the receptor is a member of the activin receptor family. The purified receptor phosphorylated itself and exogenous substrate proteins on serine, threonine and tyrosine residues, indicating that the activin receptor is a transmembrane serine/threonine/tyrosine protein kinase (Figure 4). These findings suggest a novel pathway for the signal transduction of activin via a new class of cellular receptor in EC P19 cells.

Acknowledgements: We thank the National Hormone and Pituitary Program, the National Institute of Diabetes and Digestive and Kidney Diseases for supplying the rat FSH assay kit. This work was supported in part by grants from the Ministry of Education, Science and Culture, Japan and the Ryoichi Naito Foundation for Medical Research.

REFERENCES

Asashima, M., Nakano, H., Shimada, K., Kinoshita, K., Ishii, K., Shibai, H. and Ueno, N., 1990 Mesodermal induction in early amphibian embryos by activin A (erythroid differentiation factor), *Roux's Arch. Dev. Biol.* 198:330.

Asashima, M., Nakano, H., Uchiyama, H., Sugino, H., Nakamura, T., Eto, Y., Ejima, D., Davids, M., Plessow, S., Cichocka, I. and Kinoshita, K., 1991 Follistatin inhibits the mesoderm-inducing activity of activin A and the vegetalizing factor from chicken embryo, *Roux's Arch. Dev. Biol.* 200:4.

Attisano, L., Wrana, J. L., Cheifetz, S. and Massagué, J., 1992 Novel activin receptors: Distinct genes and alternative mRNA splicing generate a repertoire of serine/threonine kinase receptors, *Cell* 68:97.

Billestrup, N., González-Manchón, C., Potter, E. and Vale, W., 1990 Inhibition of somatotroph growth and growth hormone biosynthesis by activin *in vitro*, *Mol. Endocrinol.* 4:356.

Broxmeyer, H. E., Lu, L., Cooper, S., Schwall, R. H., Mason, A. J. and Nikolics, K., 1988 Selective and indirect modulation of human multipotential and erythroid hematopoietic progenitor cell proliferation by recombinant human activin and inhibin, *Proc. Natl. Acad. Sci. USA* 85:9052.

Burger, H. and Igarashi, M., 1988 Inhibin: Definition and nomenclature, including related substances, *Endocrinology* 122:1701.

Eto, Y., Tsuji, T., Takezawa, M., Takano, S., Yokogawa, Y. and Shibai, H., 1987 Purification and characterization of erythroid differentiation factor (EDF) isolated from human leukemia cell line THP-1, *Biochem. Biophys. Res. Commun.* 142:1095.

Esch, F., Shimasaki, S., Mercado, M., Cooksey, K., Ling, N., Ying, S.-Y., Ueno, U. and Guillemin, R., 1987 Structural characterization of follistatin: a novel follicle-stimulating hormone release-inhibiting polypeptide from the gonad, *Mol. Endocrinol.* 1: 849.

Green, J. B. A. and Smith, J. C., 1990 Graded changes in dose of a *Xenopus* activin A homologue elicit stepwise transitions in embryonic cell fate, *Nature* 347:391.

Hasegawa, Y., Miyamoto, K., Abe, Y., Nakamura, T., Sugino, H., Eto, Y., Shibai, H. and Igarashi, M., 1988 Induction of follicle stimulating hormone receptor by erythroid differentiation factor on rat granulosa cell, *Biochem. Biophys. Res. Commun.* 156:668.

Hashimoto, M., Kondo, S., Sakurai, T., Etoh, Y., Shibai, H. and Muramatsu, M., 1990 Activin/EDF as an inhibitor of neural differentiation, *Biochem. Biophys. Res. Commun.* 173:193.

Hsueh, A. J. W., Dahl, K. D., Vaughan, J., Tucker, E., Rivier, J. Bardin, C. W. and Vale, W., 1987 Heterodimers and homodimers of inhibin subunits have different paracrine action in the modulation of luteinizing hormone-stimulated androgen biosynthesis, *Proc. Natl. Acad. Sci. USA* 84:5082.

Hutchinson, L. A., Findlay, J. K., de Vos, F. L. and Robertson, D. M., 1987 Effects of bovine inhibin, transforming growth factor-β and bovine activin-A on granulosa cell differentiation, *Biochem. Biophys. Res. Commun.* 146:1405.

Kitaoka, M., Kojima, I. and Ogata, E., 1988 Activin-A: a modulator of multiple types of anterior pituitary cells, *Biochem. Biophys. Res. Commun.* 157:48.

Kogawa, K., Nakamura, T., Sugino, K., Takio, K., Titani, K. and Sugino, H., 1991 Activin-binding protein is present in pituitary, *Endocrinology* 128:1434.

Kogawa, K., Ogawa, K., Hayashi, Y., Nakamura, T., Titani, K. and Sugino, H., 1991 Immunohistochemical localization of follistatin in rat tissues, *Endocrinol. Japon.* 38:383.

Ling, N., Ueno, N., Esch, F., Shimasaki, S., Koga, M., Ying, S.-Y. and Guillemin, R., 1988 Isolation and characterization of gonadal polypeptides that regulate the secretion of follicle stimulating hormone, *in* : Nonsteroidal Gonadal Factors: Physiological Roles and Possibilities in Contraceptive Development, Hodgen, G.D., Rosenwaks, Z. and Spieler, J.M.,eds., Jones Inst. Press, Virginia.

Ling, N., Ying, S.-Y., Ueno, N., Shimasaki, S., Esch, F., Hotta, M. and Guillemin, R., 1986 Pituitary FSH is released by a heterodimer of the β-subunits from the two forms of inhibin, *Nature* 321:779.

Massagué, J., 1987 The TGF-β family of growth and differentiation factors, *Cell* 49: 437.

Mathews, L. S. and Vale, W., 1991 Expression cloning of an activin receptor, a predicted transmembrane serine kinase, *Cell* 65:973.

Nakamura, T., Hasegawa, Y., Sugino, K., Kogawa, K., Titani, K. and Sugino, H., 1992 Follistatin inhibits activin-induced differentiation of rat follicular granulosa cells *in vitro*, *Biochim. Biophys. Acta* 1135:103.

Nakamura, T., Sugino, K., Kurosawa, N., Sawai, M., Takio, K., Eto, Y., Iwashita, S., Muramatsu, M., Titani, K. and Sugino, H., 1992 Isolation and characterization of activin receptor from mouse embryonal carcinoma cells, *J. Biol. Chem.* 267:18942.

Nakamura, T., Sugino, K., Titani, K. and Sugino, H., 1991 Follistatin, an activin-binding protein, associates with heparan sulfate chains of proteoglycans on follicular granulosa cells, *J. Biol. Chem.* 266:19432.

Nakamura, T., Takio, K., Eto, Y., Shibai, H., Titani, K. and Sugino, H., 1990 Activin-binding protein from rat ovary is follistatin, *Science* 247:836.

Robertson, D. M., Klein, R., de Vos, F. L , McLachlan, R. I., Wettenhall, R. E. H., Hearn, M. T. W., Burger, H. G. and de Kretser, D. M., 1987 The isolation of polypeptides with FSH suppressing activity from bovine follicular fluid which are structurally different to inhibin, *Biochem. Biophys. Res. Commun.* 149:744.

Saito, S., Nakamura, T., Titani, K. and Sugino, H., 1991 Production of activin-binding protein by rat granulosa cells *in vitro*, *Biochem. Biophys. Res. Commun.* 176:413.

Schubert, D., Kimura, H., LaCorbiere, M., Vaughan, J., Karr, D. and Fischer, W. H., 1990 Activin is a nerve cell survival molecule, *Nature* 344:868.

Shimasaki, S., Koga, M., Buscaglia, M. L., Simmons, D. M., Bicsak, T. A. and Ling, N., 1989 Follistatin gene expression in the ovary and extragonadal tissues, *Mol. Endocrinol.* 3: 651.

Shimasaki, S., Koga, M., Esch, F., Mercado, M., Cooksey, K., Koba, A. and Ling, N., 1988 Porcine follistatin gene structure supports two forms of mature follistatin produced by alternative splicing, *Biochem. Biophys. Res. Commun.* 152:717.

Shiozaki, M., Sakai, R., Tabuchi, M., Nakamura, T., Sugino, K., Sugino, H. and Eto, Y., 1992 Evidence for the participation of endogenous activin A/erythroid differentiation factor in the regulation of erythropoiesis, *Proc. Natl. Acad. Sci. USA* 89:1553.

Smith, J. C., Price, B. M. J., van Nimmen, K. and Huylebroeck, D., 1990 Identification of a potent *Xenopus* mesoderm inducing factor as a homologue of activin A , *Nature* 345:729.

Sugino, H., Nakamura, T., Hasegawa, Y., Miyamoto, K., Abe, Y., Igarashi, M., Eto, Y., Shibai, H. and Titani, K., 1988 Erythroid differentiation factor can modulate follicular granulosa cell functions, *Biochem. Biophys. Res. Commun.* 153:281.

Sugino, H., Nakamura, T., Hasegawa, Y., Miyamoto, K., Igarashi, M., Eto, Y., Shibai, H. and Titani, K., 1988 Identification of a specific receptor for erythroid differentiation factor on follicular granulosa cell, *J. Biol. Chem.* 263:15249.

Thomsen, G., Woolf, T., Whitman, M., Sokol, S., Vaughan, J., Vale, W. and Melton, D. A., 1990 Activins are expressed early in *Xenopus* embryogenesis and can induce axial mesoderm and anterior structures, *Cell* 63:485.

Ueno, N., Ling, N., Ying, S.-Y., Esch, F., Shimasaki, S. and Guillemin, R., 1987 Isolation and partial characterization of follistatin: a single-chain Mr 35,000 monomeric protein that inhibits the release of follicle-stimulating hormone, *Proc. Natl. Acad. Sci. USA* 84: 8282.

Vale, W., Hsueh, A., Rivier, C. and Yu, J., 1990 The inhibin/activin family of hormones and growth factors, *in* : Handbook of Experimental Pharmacology, Peptide Growth Factors and Their Receptor II, Born, G. V. R., Cuatrecasas, P. and Herken, H., eds., Springer-Verlag, Berlin.

Vale, W., Rivier, C., Hsueh, A., Campen, C., Meunier, H., Bicsak, T., Vaughan, J., Corrigan, A., Bardin, W., Sawchenko, P., Petraglia, F., Yu, J., Plotsky, P., Spiess, J. and Rivier, J., 1988 Chemical and biological characterization of the inhibin family of protein hormones, *in*: Recent Progress in Hormone Research, Clark, J.H., ed., Academic Press, San Diego.

Vale, W., Rivier, J., Vaughan, J., McClintock, R., Corrigagn, A., Woo,.W., Karr, D. and Spiess, J., 1986 Purification and characterization of an FSH releasing protein from porcine ovarian follicular fluid, *Nature* 321:776.

van den Eijnden-van Raaij, A. J. M., van Zoelent, E. J. J., van Nimmen, K., Koster, C. H., Snoek, G. T., Durston, A. J. and Huylebroeck, D., 1990 Activin-like factor from a *Xenopus laevis* cell line responsible for mesoderm induction, *Nature* 345:732.

Ying, S.-Y., 1988 Inhibin, activin and follistatins: gonadal proteins modulating the secretion of follicle- stimulating hormone, *Endoc. Rev.* 9:267.

Yu, J., Shao, L., Lemas, V., Yu, A, L., Vaughan, J., Rivier, J. and Vale, W., 1987 Importance of FSH-releasing protein and inhibin in erythrodifferentiation, *Nature* 330:765.

DO WELL-POPULATED INTERMEDIATES REPRESENT THE PATHWAY OF PROTEIN FOLDING?

Jui-Yoa Chang

Pharmaceuticals Research Laboratories
K-121, 104 Ciba-Geigy Ltd.
Basel CH-4002 Switzerland

INTRODUCTION

The ultimate value of the amino acid sequence is to predict its active three dimensional structure. Protein folding is known to be dictated by the amino acid sequence alone and occurs spontaneously (Anfinsen, 1973). What remain to be fully understood is the mechanism of this spontaneous process. Despite the accumulated efforts, several fundamental issues concerning the folding mechanism of proteins remain hotly debated. For instance, is protein folding a thermodynamically or kinetically governed event (Barker *et al.*, 1992; Creighton, 1990; Dobson, 1990)? is there a preferred pathway or are there multiple pathways (Creighton, 1990; Harrison and Durbin, 1990; Kim and Baldwin, 1990; Weissman and Kim, 1991)? and what are the interplays of "hydrophobic collapse" and "specific tertiary interaction" in driving and guiding the protein folding (Baldwin, 1989; Dill, 1990)? These controversies are caused primarily by our inadequate understanding of the structure and function of protein folding intermediates.

The most extensively investigated case is the disulfide folding pathway of bovine pancreatic trypsin inhibitor (BPTI) (Creighton, 1978; 1990; Weissman and Kim, 1991). These studies provide ample evidence which confirms the existence of partially structured folding intermediates. The original model put forward by Creighton and colleagues (1978; 1990) stresses the involvement of intermediates containing both native and non-native disulfides. A more recently proposed model (Weissman and Kim, 1991), on the other hand, claims that only intermediates containing native disulfide are relevant to the BPTI folding.

Both models of BPTI folding were constructed upon the kinetic analysis of well-populated intermediates accumulated along the folding pathway. This approach is rested upon two major premises: (1) well-populated intermediates are productive species and account for the flow of the folding pathway, and (2) rapid equilibrium among the folding intermediates do not occur. Both premises, however, have not been tested and confirmed in

Methods in Protein Sequence Analysis, Edited by K. Imahori
and F. Sakiyama, Plenum Press, New York, 1993

the case of BPTI. For instance, there is no evidence (Weissman and Kim, 1991) that well-populated l-disulfide intermediates BPTI [Cys^{30}-Cys^{51}] and BPTI[Cys^{5}-Cys^{55}] convert directly to BPTI[Cys^{30}-Cys^{51}, Cys^{14}-Cys^{38}] and BPTI[Cys^{5}-Cys^{55}, Cys^{14}-Cys^{38}] without undergoing additional rearrangements. Furthermore, there are indeed evidences that the folding intermediates of BPTI (Weissman and Kim, 1991) and ribonuclease A (Scheraga *et al.*, 1984) exist in rapid equilibrium.

In the presence of rapid equilibrium of folding intermediates, one can not automatically assume that well-populated intermediates are also productive species even if they do contain the native disulfides. It is necessary that all detectable intermediates, predominant or minor, are trapped in a reversible manner, isolated and allowed to resume the folding *(stop/go folding experiments)* in order to judge and compare their relative productivity along the pathway. Reversible trapping of the disulfide formation can be achieved by acid quenching.

FOLDING OF HIR$^{1-27}$ AND REVERSIBLE TRAPPING OF ITS FOLDING INTERMEDIATES

In order to demonstrate our arguments, a simple two disulfide model is required. In a three disulfide model like BPTI, there are 75 possible disulfide structures (15 l-disulfide isomers, 45 2-disulfide isomers and 15 3-disulfide isomers). A two disulfide model, on the other hand, has only 9 possible species (6 1-disulfide and 3 2-disulfide isomers). There is a reasonable chance that folding intermediates of a 2-disulfide model can be isolated in pure form to pursue the stop/go folding experiments.

Toward this goal, we have examined the folding behavior of soybean trypsin inhibitor and apamin, both containing two disulfides. They were abandoned due to their undesirable patterns of HPLC separation. Eventually, hirudin N-terminal fragment residues 1-27 was selected because of its favorable properties in the reversed phase HPLC, despite the fact that it does not represent the active molecule (Figure 1).

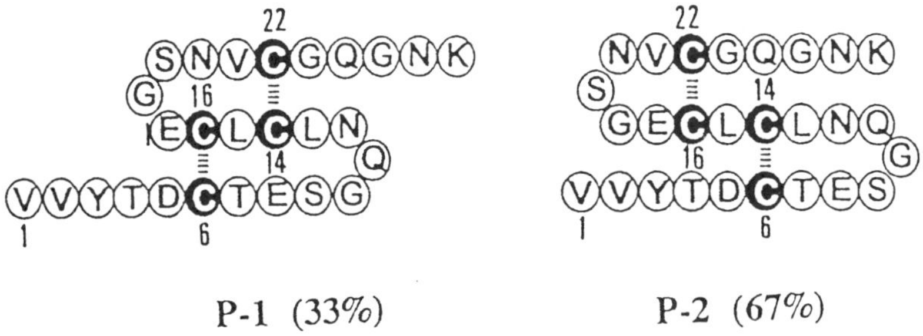

Figure 1. The amino acid sequence and disulfide structures of refolded Hir$^{1-27}$ (P-l and P-2).

Hir$^{1-27}$ was denatured/reduced and allowed to refold in the alkaline buffer (sodium bicarbonate, 50 mM, pH 8.3). Folding intermediates were trapped in a time-course manner by mixing with 4% trifluoroacetic acid. The trapped samples were analysed directly by HPLC (Figure 2). Reduced/denatured Hir$^{1-27}$ refolds via l-disulfide intermediates (I-1, I-2 and I-M) to form 2-disulfide products (P-l and P-2). One important characteristic of the patterns of these intermediates is their constant ratio. Throughout the entire folding process, the ratio of I-1, I-2 and I-M remains unchanged, so do the 2-disulfide products. These results clearly indicate that they all exist in equilibrium along the pathway and the folding proceeds spontaneously via equilibrated 1-disulfide isomers to the equilibrated 2-disulfide isomers.

The 1-disulfide species were isolated and carboxymethylated with iodoacetic acid (1 M) in the Tris-HCl buffer (0.5 M, pH6.5) containing 40% by volume of dimethylformamide. The structures of 1-disulfide and 2-disulfide species were then determined by amino acid analysis and sequence analysis, following thermolysin digestion and HPLC separation of thermolytic peptides. Their structures are summarized in Figure 4. I-2 contains two species and I-M contains three species.

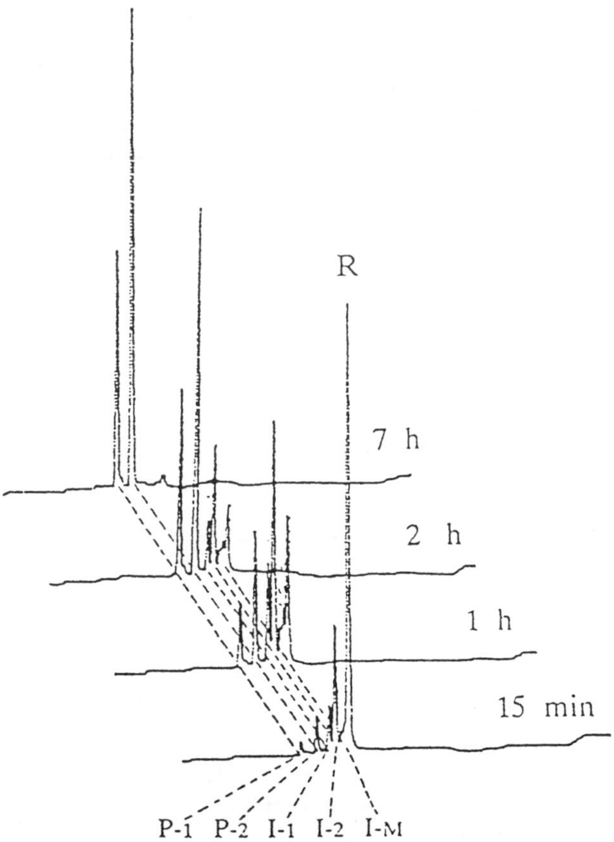

Figure 2. HPLC profiles of the acid-trapped folding intermediates of Hir^{1-27}. "R" stands for fully reduced/denatured Hir^{1-27}. I-1, I-2 and I-M are 1-disulfide intermediates. P-1 and P-2 are 2-disulfide folded products. Note that I-M is partially overlapped with "R" The HPLC column was Vydac C-18 for peptides and proteins. Solvent A was 0.1% trifluoroacetic acid in water. Solvent B was 0.09% trifuoroacetic acid in acetonitrile. The gradient was 13-40% solvent B linear in 25 min. Detector wavelength was 214 nm.

STOP/GO FOLDING EXPERIMENTS OF THE 1-DISULFIDE INTERMEDIATES

Isolated 1-disulfide intermediates (I-1, I-2 and I-M) were dried from the acidic acetonitrile/water solution (HPLC solution) and redissolved in the alkaline buffer (folding buffer) to resume the folding pathway. The folding intermediates were again trapped by

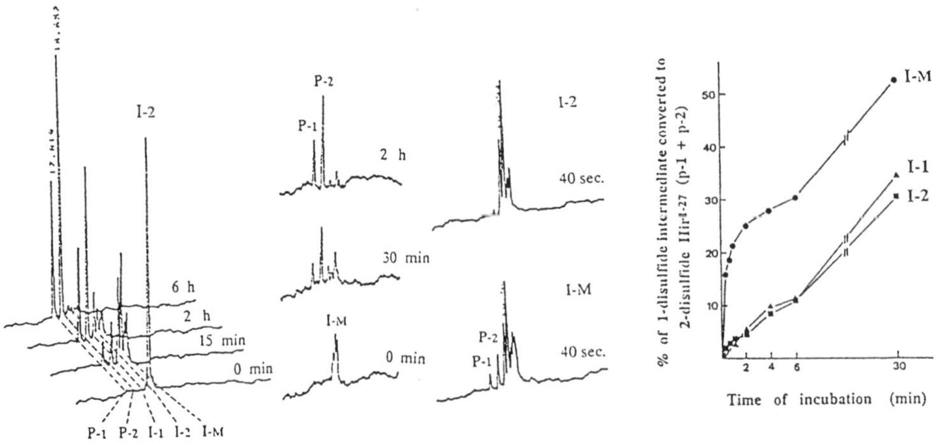

Figure 3. Stop/go folding experiments of isoluted 1-disulfide intermediates.
Purified 1-disulfide intermediates were dried and reconstituted (10 mM) in the folding buffer to resume the folding. (Left) Time-course refolding of purified I-2. (Second from left) Time-course refolding of isolated I-M. (Second from right) Comparison of the productivity of I-2 and I-M during the early phase of resumed folding (40 s). (Right) Relative productivity of I-1, I-2 and I-M.

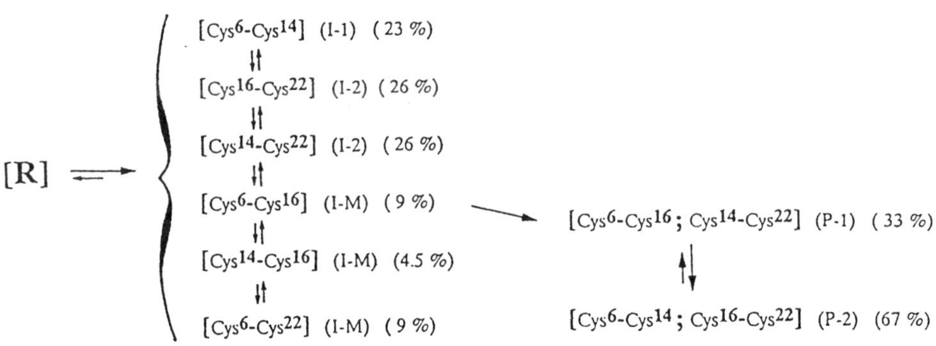

Figure 4. Proposed folding pathway of Hir$^{1-27}$. 1-disulfide species . [Cys^6-Cys^{16}] was identified as the productive intermediate.

4% trifluoroacetic acid and analysed by HPLC (Figure 3). The productivity was judged by their efficiency to form P-1 and P-2.

For I-1 and I-2, rapid equilibrium among 1-disulfide isomers ensued within 20 sec and detectable recovery of P-1 and P-2 was obtained only 1 min after resumed folding. However, I-M appears to contain structures which is ready to form P-1 and P-2. The surge of this productivity apparently took place before full equilibrium among 1-disulfide isomers was established. These findings unambiguously demonstrate that I-M is more productive than the well-populated intermediates (I-1 and I-2). I-M is composed of three 1-disulfide isomers, but only Cys^6-Cys^{16} is a likely candidate to form P-1. The proposed disulfide folding pathway of Hir$^{1-27}$ is shown in Figure 4.

CONCLUSION

The notion that well-populated intermediates account for the flow of protein folding and specify the folding pathway (Creigton, 1978; Creigton, 1990; Weissman and Kim, 1991) can not be taken for granted, particularly when intermediates are presented in rapid equilibrium. One way to identify the productive folding intermediates is to reversibly quench the folding intermediates, isolate all detectable species and allow them to resume the folding. By this approach, we have shown that in the case of Hir^{1-27}, the most productive 1-disulfide intermediate which accounts for the flow of its folding is not a well-populated species.

REFERENCES

Anfinsen, C. B., 1973 Principles that govern the folding of protein chains, *Science* 181:223.

Baker, D., Sohl, J. L. and Agard, D. A., 1992 A protein folding reaction under kinetic control, *Nature* 356:262.

Baldwin, R. L., 1989 How does protein folding get started, *Trends Biochem. Sci.* 14:291.

Creighton, T. E., 1978 Experimental studies of protein folding and unfolding, *Prog. Biophys. Mol. Biol.* 33:231.

Creighton, T. E., 1990 Protein folding, *Biochem. J.* 270:1.

Dill, K. A., 1990 Dominant forces in protein folding, *Biochemistry* 29:7133.

Dobson, C. M., 1992 Resting places on folding pathway, *Current Biol.* 2: 343 .

Harrison, S. C. and Durbin, R., 1985 Is there a single pathway for the folding of a polypeptide chain, *Proc. Natl. Acad. Sci. USA* 82:4028

Kim, P. S. and Baldwin, R. L., 1990 Intermediates in the folding reactions of small proteins, *Annu. Rev. Biochem.* 59: 631.

Scheraga, H. A., Konishi, Y. and Ooi, T., 1984 Multiple pathway for regenerating ribonuclease A, *Adv. Biophys.* 18:21

Weissman, J. S. and Kim, P. S., 1991 Reexamination of the folding of BPTI: Predominance of native intermediates, *Science* 253:1386.

ALCOHOL DEHYDROGENASES: PATTERNS OF PROTEIN EVOLUTION

Hans Jörnvall, Olle Danielsson, Jan-Olov Höög and Bengt Persson

Department of Chemistry I
Karolinska Institutet
S-104 01 Stockholm, Sweden

INTRODUCTION

Alcohol dehydrogenases and related enzymes constitute a large group of proteins belonging to several super-families, which also encompass many other enzyme activities. The enzymes, in turn, exhibit further multiplicity in the form of separate classes, separate isozymes, and different allelozymes. These overall relationships are exemplified in Table 1 and have been summarized recently (Jörnvall *et al.*, 1992). The different levels of complexity reflect general principles of protein relationships. A feature observed in many families is the modular build-up of proteins, showing separate organizations but still individual units in common, and an extensive divergence. Another such feature is the frequent examples of insertions/deletions.

Later levels of complexity, however, represent properties where alcohol dehydrogenases and related enzymes exceptionally well illustrate protein patterns. One is a functional convergence, which is seen repeatedly in this enzyme group. Another is a remarkable conservation of glycine residues and its molecular explanation. A third property to note is the metalloprotein relationships. Finally, alcohol dehydrogenases have multiplicity at the level of classes (Vallee and Bazzone, 1983) between the levels of enzymes and isozymes (Kaiser *el al.*, 1988). Each of these special properties, particularly pronounced in alcohol dehydrogenases but typical of protein relationships in general, are detailed below, after consideration of the two most studied alcohol dehydrogenase super-families that correspond to different types of the enzyme.

Different Types of Enzyme

Two types are well-known and widely represented, the medium-chain alcohol dehydrogenases and the short-chain dehydrogenases. The different characteristics of these two families are summarized in Table 2. However, other forms also exist, a long-chain type

Methods in Protein Sequence Analysis, Edited by K. Imahori
and F. Sakiyama, Plenum Press, New York, 1993

(Inoue *et al.*, 1989) and additional types (Scopes, 1983; Williamson and Paquin, 1987). The number of known enzymes of each type differ. Of the medium-chain alcohol dehydrogenases a limited number (~5) have now been detected, whereas of the short-chain dehydrogenases over several tens of different enzymes are known (Persson *et al.*, 1991; Jörnvall *et al.*, 1992).

Table 1. Different levels of multiplicity of alcohol dehydrogenases and related enzymes.

Level	Examples		Structural divergence (percent residue identity)
Different families	Medium-chain families Short-chain families Additional families		<20
Different enzymes	Alcohol dehydrogenases Polyol dehydrogenases Other dehydrogenases ζ-Crystallin/Quinone reductases		~25
Different classes[1]	M-I M-II M-III M-IV M-V		~60
Different isozymes[2]	I-α I-β I-γ		~90
Different allelic variants[3]	β_1 β_2 β_3	γ_1 γ_2	>95

[1]M-I to M-V denote five classes (Vallee and Bazzone, 1983; Parés *et al.*, 1992) for the medium-chain alcohol dehydrogenases.
[2]I-α to I-γ denote 3 isozyme subunits (Smith *et al.*, 1971) for the class I medium-chain alcohol dehydrogenases.
[3]β_1, β_2, β_3 and γ_1, γ_2 denote known allelic variants of the β and γ subunits, respectively.

Classes have been detected for the medium-chain alchol dehydrogenases in particular (Vallee and Bazzone, 1983), where thus far five classes are defined (Parés *et al.*, 1992) and given the designations I-V (Table 1). Similarly, isozymes have been detected mainly in some enzymes and classes, in particular the class I enzyme, for which three genes and the corresponding subunit types exist (Smith *et al.*, 1971). However, isozymes have recently been detected also for a class III enzyme (Danielsson and Jörnvall, 1992). Known allelic variants give further variabilities to the class I enzyme forms (Table 1).

Functional Convergence

Alcohol dehydrogenases and related enzymes repeatedly show examples of functional convergence. Thus, a single type of activity like the ethanol dehydrogenase activity, appears to have arisen several times and from different evolutionary lines. The separate origins of the ethanol dehydrogenase activities in the short-chain enzyme of *Drosophila* and the

Table 2. Characteristics of two different alcohol dehydrogenase families.
The sizes are given in residues per protein chain; the organization with lines and bars denote the medium-chain family (top) and the short-chain family (bottom), the bars representing the coenzyme-binding domains which are differently situated in the linear architecture of the whole subunits, and with stippling to indicate variations between different enzymes within the family (Krook *et al.*, 1992); the heading catalysis shows the characteristic structures in each family, with a catalytic zinc atom in the medium-chain alcohol dehydrogenases, and a conserved tyrosine residue (and closely spaced Lys residue) in the short-chain family; the heading second zinc indicates that four ligands are typical of a structural zinc atom in the medium-chain family, while structural zinc atom has not been reported in short-chain alcohol dehydrogenases (hence designated as -) although suggested in other short-chain dehydrogenases (designated by(+)); deletions/insertions occur in both families; and the variability is either quite large (+) or small (-), differing by a factor of about 3 (denoted by x3) in the medium-chain family, but thus far largely unknown (denoted by ?) in species variation for the short-chain family.

Property	Families: Medium-chain alcohol dehydrogenases	Short-chain dehydrogenases
Size	~350	~250
Organization	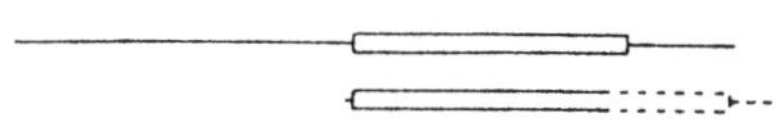	
Catalysis	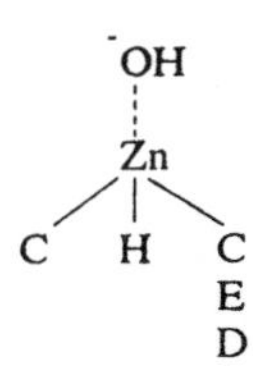	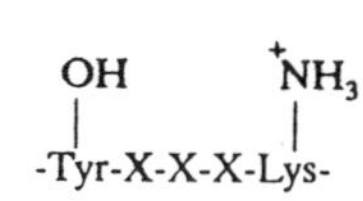
Second Zn	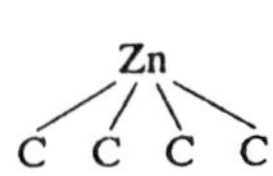 –	– (+)
Deletions/Insertions	+	+
Variability	+ – (x 3)	?

medium-chain enzyme of mammals was early pointed out (Schwartz and Jörnvall, 1976; Thatcher and Sawyer, 1980; Jörnvall *et al.*, 1981). Later divergence in structure within the medium-chain family gave dimeric and tetrameric enzymes (Jörnvall *et al.*, 1987), and more recent divergence of the class I and III enzymes gave the liver ethanol dehydrogenase of class I type (Danielsson and Jörnvall, 1992). Hence, three separate origins have converged in function toward ethanol dehydrogenase activity as shown in Figure 1, at the same time as the corresponding structures have diverged in the usual manner, and additional forms have appeared.

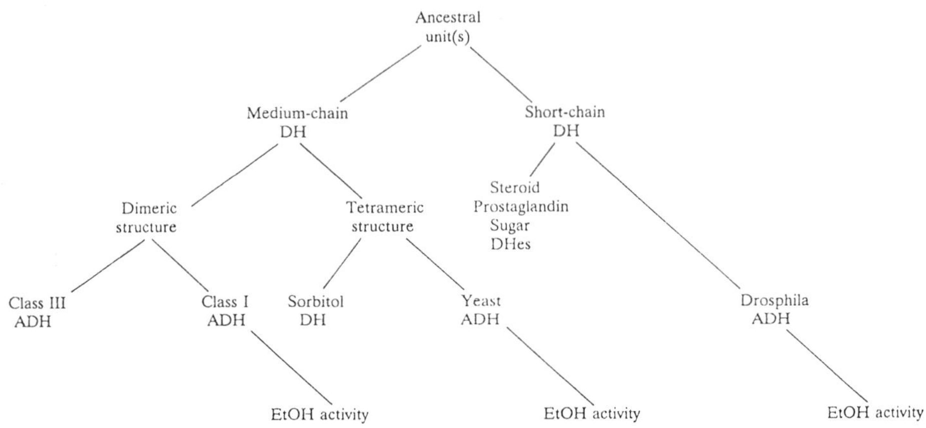

Figure 1. Conventional divergence and functional convergence in alcohol dehydrogenases. The functional convergence is exemplified by separate emergences of ethanol dehydrogenase activity, but is not limited to this activity and is also present (not shown) regarding polyol dehydrogenase activity or xylitol dehydrogenase activity. Each cross point indicates a gene duplication. Bottom right-hand lines indicate the functional convergence, remaining lines conventional divergence.

This pattern suggests a peculiar feature of alcohol dehydrogenase. Apparently, its ethanol dehydrogenase activity is not crucial *per se* in the sense that it is not constantly present in all life forms but has separate duplicatory origins. At the same time, the repeated emergence suggests that the biological activity has a positive value and presumably participates in a wide range of detoxication reactions, more or less in the same manner as the enzymes of other multi-member protein families, such as cytochrome P450 and glutathione transferase, the latter with an organizational pattern largely identical to that of alcohol dehydrogenases (cf. Figure 2 in Jörnvall *et al.*, 1992). In addition to this functional conclusion for the whole enzyme system, the pattern constitutes an example of substrate convergence from separate structural origins in a protein family.

Glycine Conservation

The medium-chain alcohol dehydrogeneases and related enzymes exhibit a pronounced pattern of residue conservation. By far the most conserved residue is glycine (Table 3). This was early detected in alcohol dehydrogenases (quoted in Jörnvall *et al.*, 1987) and was then also observed in other well-studied protein families such as cytochorome c (Smith and Margoliash, 1964; Dickerson, 1970). Additions of the ζ-crystallin structure (Borrás *et al.*, 1989) and further variants make the pattern of Gly conservation in the alcohol dehydrogenase family even more pronounced. Most of the conserved glycine residues correspond to reverse turn elements in the secondary structure and therefore suggest that all these enzymes are closely related in conformation (Eklund *et al.*, 1990) and that the conformational conservation forces the Gly conservation. Interestingly, these enzymes correspond in the primary structure to residue identities at the 25% level (Table 1). In fact, at the same time as almost no residue, not even several of the functionally important ones at the active site, are conserved, close to half of all glycine residues are non-exchanged.

Table 3. Residue conservation in the alcohol dehydrogenase families.
The glycine conservation is most pronounced in the medium-chain family (left column, showing the extent of conservation in over half of all variants of each of the dehydrogenases), less so in the short chain family (right column, showing the extent of conservation in at least all but two of all the dehydrogenases characterized at that time).

Residue conserved	Medium-chain family (Borrás *et al.*, 1989)	Short-chain family (Persson *et al.*, 1991)
Asp	1	2
Asn	-	1
Thr	-	1
Ser	-	1
Pro	2	-
Gly	13	5
Ala	1	1
Val	3	-
Tyr	-	1
Lys	-	1

Among proteins in general this extreme glycine conservation is especially clear in medium-chain alcohol dehydrogenases. In protein families with still larger variation, like the short-chain dehydrogenases, glycine conservation in less pronounced and just fairly visible (Persson *et al.*, 1991), as is also other conservations (Table 3). Obviously, the proteins within the short-chain family have diverged to such an extent that the conformations are not as much conserved as in the medium-chain family, and therefore also not the reverse turns or glycine residues. Thus, extreme Gly conservation seems to correspond to an interval above a lower limit in variability. In the case of alcohol dehydrogenases this border appears to be at about 25% residue identity. Similarly, an upper limit of identities also exists for the Gly conservation to be visible. Class variants or species variants, which exhibit large identities in primary structure, do also not reveal the glycine conservation equally clearly. Apparently, in closely related structures, many other residues are conserved and the glycine pattern therefore not equally unique.

In conclusion, it would appear that a pronounced glycine conservation is associated with conserved conformations and functional requirements. In closely related proteins high overall identities fail to reveal the unique nature of the glycine residues, whereas proteins with very low extents of identity have already accepted so many conformational changes that the glycine conservation is also not evident. Medium-chain alcohol dehydrogenases appear to be exactly at the correct level for maximal observation of glycine conservation in proteins.

Metalloprotein Variability

Medium-chain alcohol dehydrogenases have another property of general interest: they exhibit a wide range of differences in metalloprotein relationships. The classical liver enzyme has two zinc atoms per subunit and of course the corresponding zinc ligands, one Cys, one His, and a second Cys to the catalytic zinc atom, and four cysteine residues to the structural zinc atom (Eklund *et al.*, 1976). The related sorbitol dehydrogenase does not have the structural zinc atom (Jeffery *et al.*, 1884) and probably another residue for one of the ligands to the catalytic zinc atom (the second Cys residue, in all likelihood replaced by a Glu residue, Eklund *et al.*, 1985). With these changes the structural zinc atom is absent and the metalloprotein characteristics just one zinc per subunit (Table 2). Continuing this pattern, further missing segments and ligands (Borrás *et al.*, 1989), explain the absence of also the second zinc atom per subunit in ζ-crystallin, making this an enzyme lacking both the sites and structures typical of the zinc atom in classical alcohol dehydrogenases. Thus, alcohol dehydrogenase and related enzymes exhibit a whole pattern with gradual changes from zero zinc and zero ligands to one zinc and variable ligands, and two zincs with all ligands. Few other protein families have this successively altered metalloprotein pattern equally evident.

Insertions/Deletions

Both medium-chain and short-chain alcohol dehydrogenases exhibit insertions/deletions of long segments, in addition to many single-residue gaps. In the medium-chain dehydrogenases, a long loop, superficial in the subunit, is absent in sorbitol dehydrogenase (Eklund *et al.*, 1985), and in ζ-crystallin this loop and an additional, adjacent loop are absent (Borrás *et al.*, 1989), making the polypepteide size differences even larger (Figure 2). Both the one-loop and the two-loop differences involve a subunit surface structure and affect the

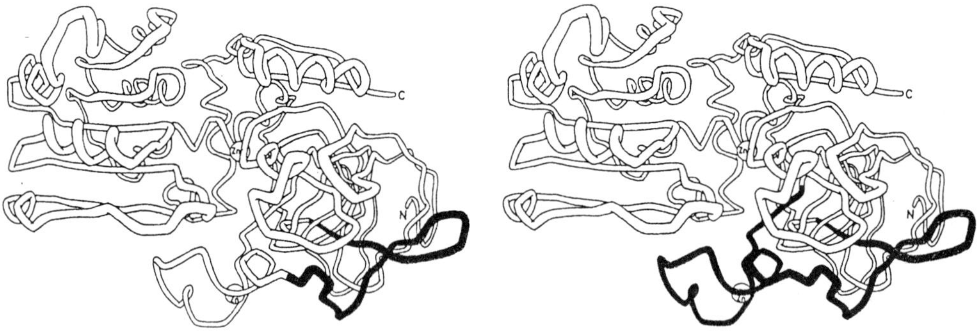

Figure 2. Insertions/deletions in medium-chain dehydrogenases.
The continuous fold represents liver alcohol dehydrogenase. The black segment in the left model shows the missing part in yeast alcohol dehydrogenase and liver sorbitol dehydrogenase (two tetrameric dehydrogenases), the black segment to the right the missing part in ζ-crystallin/quinone reductase. Black therefore corresponds to deletions in those enzymes, or to insertions in liver alcohol dehydrogenase, and as shown has superficial positions in the subunits. The situation with insertion/delections and superficial regions in the subunit is similar in short-chain dehydrogenases as shown from computer drawings (Figure 5 in Krook *et al.*, 1992). Original drawing by Hans Furugren, modifications by Carina Palmberg.

likely area for subunit interactions (Rouimi *et al.*, 1992) in these dimeric molecules, explaining the different quaternary structures.

Similarly, in the short-chain dehydrogenases, comparisons of $NADP^+$-dependent 15-hydroxyprostaglandin dehydrogenase and $3\alpha/20\beta$-hydroxysteroid dehydrogenase reveal an insertion (in the former) or deletion (in the latter) in the C-terminal part. This conformational change is possible to adjust into the rest of the tertiary structure (Krook *et al.*, 1992).

In conclusion, both dehydrogenase families exhibit insertions/deletions that in common involve superficial segments of the subunit tertiary structures.

Classes - In Between Enzymes and Isozymes

The property of classes is perhaps the one most clearly outlined in alcohol dehydrogenases. Five classes are now known for medium-chain alcohol dehydrogenases. All have, as far as in known, different genetic control and different organ experssions. The classes represent intermediate forms, in between those of truly different enzymes and typical isozymes. The classes have at least one substrate in common, like isozymes, but the corresponding structures are highly divergent and approach a relationship like that for different proteins. Hence, the class distinction is especially clear in alcohol dehydrogenases. The corresponding levels in other families have been pointed out (Figure 2 in Jörnvall *et al.*, 1992).

CONCLUSIONS

Alcohol dehydrogenase constitutes a complex enzyme system with separate enzymes, separate classes, separate isozymes, and separate alleles. The system is derived from a series of gene duplications and has been gradually revealed. Because of the separate levels of relationship both those typical of distantly related proteins and those typical of closely related proteins are illustrated within a single family. In particular, the functional convergence, the glycine conservation, and the metalloprotein variability are properties where the alcohol dehydrogenase family exceptionally well illustrates properties typical of proteins in general. Functional convergence coupled with structural divergence give proteins with hybrid properties in some species (Danielsson *et al.*, 1992), glycine conservation suggests a most clear pattern at the extent of residue conservation around 25%, and the metalloprotein variability shows a successive difference in zinc content.

Acknowledgements. This study was supported by the Swedish Medical Research Council (project 03X-3532) and the Swedish Alcohol Research Fund (project 88/12:4). Collaboration in different aspects of the separate structural studies are gratefully acknowledged and evident from the references quoted.

REFERENCES

Borrás, T., Persson, B. and Jörnvall, H., 1989 Eye lens ζ-crystallin relationships to the family of "long-chain" alcohol/polyol dehydrogenases. Protein trimming and conservation of stable parts, *Biochemistry* 28:6133.

Danielsson, O. and Jörnvall, H., 1992 "Enzymogenesis": Classical liver alcohol dehydrogenase origin from the glutathione-dependent formaldehyde dehydrogenase line, *Proc. Natl. Acad. Sci. USA* 89:9247.

Dickerson, R. E., 1971 The structure of cytochrome *c* and the rates of molecular evolution, *J. Mol. Evol.* 1:26.

Eklund, H., Nordstrom, B., Zeppezauer, E., Soderlund, G., Ohlsson, I., Boiwe, T., Soderberg, B.-O., Tapia, O., Branden, C.-I. and Akeson, A., 1976 Three-dimensional structure of horse liver alcohol dehydrogenase at 2.4 Å resolution, *J. Mol. Biol.* 102:27.

Eklund, H., Horjales, E., Jornvall, H., Branden, C. I. and Jeffery, J., 1985 Molecular aspects of functional differences between alcohol and sorbitol dehydrogenases, *Biochemistry* 24:8005.

Inoue, H., Sunagawa, M., Mori, A., Imai, C., Fukuda, M., Takagi, M. and Yano, K., 1989 Cloning and sequencing of the gene encoding the 72-kilodalton dehydorgenase subunit of alcohol dehydrogenase from *Aceobacter aceti*, *J. Bacteriol.* 171:3115.

Jeffery, J., Chester, J., Mills, C., Sadler, P. J. and Jörnvall, H., 1984 Sorbitol dehydrogenase is a zinc enzyme, *EMBO J.* 3:357.

Jörnvall, H., Persson, M. and Jeffery, J., 1981 Alcohol and polyol dehydrogenases are both divided into two protein types, and structural properties cross-relate the different enzyme activities within each type, *Proc. Natl. Acad. Sci. USA* 78:4226.

Jörnvall, H., Persson, B. and Jeffery, J., 1987 Characteristics of alcohol/polyol dehydrogenases. The zinc-containing long-chain alcohol dehydorgenases, *Eur. J. Biochem.* 167:195.

Jörnvall, H., Danielsson, O., Eklund, H., Hjelmqvist, L., Höög, J.-O., Pares, X. and Shafqat, J., 1992 Enzyme and isozyme developments within the medium-chain alcohol dehydorgenase family, *in*: Enzymology and Molecular Biology of Carbonyl Metabolism, Weiner, H., Flynn, T.G. and Crabb, D.W., eds., Plenum Press, in press.

Kaiser, R., Holmquist, B., Hempel, J., Vallee, B. L. and Jörnvall, H., 1988 Class III human liver alcohol dehydrogenase: a novel structural type equidistantly related to the class I and class II enzymes, *Biochemistry* 27:1132.

Krook, M., Ghosh, D., Stromberg, R., Carlquist, M. and Jörnvall, H., 1992 Native carbonyl reductase/$NADP^+$-dependent prostaglandin dehydrogenase. Multiplicity and reductive alkylation in a short-chain dehydrogenase, *Proc. Natl. Acad. Sci. USA*, in press.

Parés, X., Cederlund, E., Moreno, A., Saubi, N., Höög, J.-O. and Jörnvall, H., 1992 Class IV alcohol dehydrogenase (the gastric enzyme). Structural analysis of human σσ-ADH reveals class IV to be variable and confirms the presence of a fifth mammalian alcohol dehydrogenase class, *FEBS Lett.* 30:69.

Persson, B., Krook, M. and Jörnvall, H., 1991 Characteristics of short-chain alcohol dehydrogenases and related enzymes, *Eur. J. Biochem.* 200:537.

Rouimi, P., Loomes, K. and Jörnvall, H., 1992 Comparative proteolysis of sorbitol and alcohol dehydrogenases, *Eur. J. Biochem.*, in press.

Scopes, R. K., 1983 An iron-activiated alcohol dehydrigenase, *FEBS Lett.* 156:303.

Smith, E. L. and Margoliash, E., 1964 Evolution of cytocrome c, *Fed. Proc.* 23:1243.

Smith, M., Hopkinson, D. A. and Harris, H., 1971 Developmental changes and polymorphism in human alcohol dehydrogenase, *Ann. Hum. Genet.* 34:251.

Schwartz, M. F. and Jörnvall, H., 1976 Structural analyses of mutant and wild-type alcohol dehydrogenases from *Drosophila melanogaster*, *Eur. J. Biochem.* 68:159.

Thatcher, D.R. and Sawyer, L., 1980 Secondary-structure prediction from the sequence of *Drosophila melanogaster* (fruitfly) alcohol dehydrogenase, *Biochem. J.* 187:884.

Vallee, B. L. and Bazzone, T. J., 1983 Isozymes of human liver alcohol dehydrogenases, *Isozymes: Curr. Top. Biol. Med. Res.* 8:219.

Williamson, V. M. and Paquin, C. E., 1987 Homology of *Saccharomyces cerevisiae* ADH4 to an iron-activated alcohol dehydrogenase form *Zymonomas mibilis*, *Mol. Gen. Genet.* 209:374.

STRUCTURE AND POST-TRANSLATIONAL MODIFICATION OF THE LIPOYL DOMAIN OF 2-OXO ACID DEHYDROGENASE COMPLEXES: A NEW FAMILY OF PROTEIN DOMAINS

Richard N. Perham, Nicola G. Wallis, Simon M. Brocklehurst, Frederic Dardel[1], Adrian L. Davis and Ernest D. Laue

Cambridge Centre for Molecular Recognition
Department of Biochemistry, University of Cambridge
Tennis Court Road, Cambridge CB2 1QW, England, U.K.

INTRODUCTION

The 2-oxo acid dehydrogenase multienzyme complexes are among the best examples of multifunctional proteins fashioned from independently-folded protein domains joined by linker sequences of various lengths and degrees of conformational flexibility (Patel and Roche, 1990; Perham, 1991). The structural core of these complexes is provided by the dihydrolipoyl acyltransferase (E2) component, an aggregate of 24 or 60 polypeptide chains arranged with octahedral or icosahedral symmetry, respectively, according to the source. The E2 chains themselves are highly segmented: they comprise, from the N-terminus, up to three lipoyl domains, a peripheral subunit-binding domain, and a large core-forming acyltransferase domain, all linked together by long (25-30 residue) segments of polypeptide chain rich in alanine, proline and charged/hydrophilic amino acids (Reed and Hackert, 1990; Perham, 1991).

The structure of the peripheral subunit-binding (approx. 40 residues) of the E2o chain of the 2-oxoglutarate dehydrogenase complex of *Escherichia coli* has been determined by means of NMR spectroscopy (Robien *et al.*, 1992) and that of the acetyltransferase domain of the assembled E2p core of the pyruvate dehydrogenase (PDH) complex of *Azotobacter vinelandii* has been determined at 2.6Å resolution by means of X-ray crystallography (Mattevi *et al.*, 1992). The structure of the lipoyl domain is of particular interest for two reasons: first, despite the fact that the dithiolane ring of the lipoyl group is at the end of a

---

[1]Permanent address: Laboratorie de Biochimie, Ecole Polytechnique, 91128 Palaiseau Cedex, France.

lipoyl-lysine side chain 1.4 nm long, and is free to rotate with respect to the bulk of the protein, the lipoyl group must be attached to a lipoyl domain if it is to serve as an effective substrate for the 2-oxo acid decarboxylase (E1) component of the complex; and secondly, the lipoyl domain confers specificity on the pendant dithiolane ring for reductive acetylation only by the E1 component of the parent 2-oxo acid dehydrogenase complex (Graham *et al.*, 1989), a beautiful mechanism for substrate channelling (Perham, 1991).

In the present paper, we describe the determination of the structure of the lipoyl domain of the PDH complex of *Bacillus stearothermophilus* by means of NMR spectroscopy (Dardel *et al.*, 1991; 1992). This has enabled us to initiate experiments designed to unravel the molecular mechanism by which the E1 component selects its cognate lipoyl domain for reductive acetylation of the attached lipoyl group, and also to analyse the specificity of the enzyme(s) in the *E.coli* cell that selectively lipoylate the N^6-amino group of the target lysine residue in the lipoyl domain. In addition, we have devised a new method for predicting the three-dimensional structures of proteins from homologous template structures and used it to show that a putative biotinyl domain of yeast pyruvate carboxylase (Lim *et al.*, 1988) and the lipoylated H-protein of the pea leaf glycine cleavage system (Kim and Oliver, 1990) are likely closely to resemble the lipoyl domain. These structures unexpectedly appear to be representative of a new class of protein domains.

STRUCTURE OF THE LIPOYL DOMAIN

A sub-gene encoding the single lipoyl domain (residues 1-79) of the E2p chain of the PDH complex of *B. stearothermophilus* has been expressed in *E. coli* and lipoylated and non-lipoylated forms of the domain have been purified (Dardel *et al.*, 1990). It was apparent from the initial analysis by NMR spectroscopy that the post-translational modification was without detectable effect on the conformation of the polypeptide chain (Dardel *et al.*, 1990) and that the predominant secondary structure was β-sheet (Dardel *et al.*, 1991).

A full structural analysis (Dardel *et al.*, 1992) has now revealed that the domain is composed largely of two four-stranded β-sheets forming a flattened β-barrel around a well-defined hydrophobic core (Figure 1). The polypeptide chain weaves backwards and forwards from one sheet to the other, with the exception of a single type-I turn on the corner of one sheet. Significantly, the lipoyl-lysine residue (Lys-42) is prominently displayed in this tight turn and is thus readily accessible to the active site of the E1 component. The N- and C-terminal residues of the domain are located close together in the other β-sheet. A further important feature of the domain is its two-fold quasi-symmetry. The Cα-positions of residues 15-39 can be superposed on those of residues 52-76 after a rotation of 180° about the barrel axis, a symmetry reflected in a weak similarity between the N- and C-terminal halves of the amino acid sequence.

Comparison of the amino acid sequences of lipoyl domains from a wide range of E2 chains from pyruvate, 2-oxoglutarate and branched-chain 2-oxo acid dehydrogenase complexes (Perham, 1991; Russell and Guest, 1991) indicates that amino acids occupying key positions in the lipoyl domain fold are highly conserved. This in turn suggests that all lipoyl domains will exhibit similar structures.

MOLECULAR RECOGNITION AND THE LIPOYL DOMAIN

As described above, the structure of the lipoyl domain must hold the key to two important questions of molecular recognition: how does the domain interact

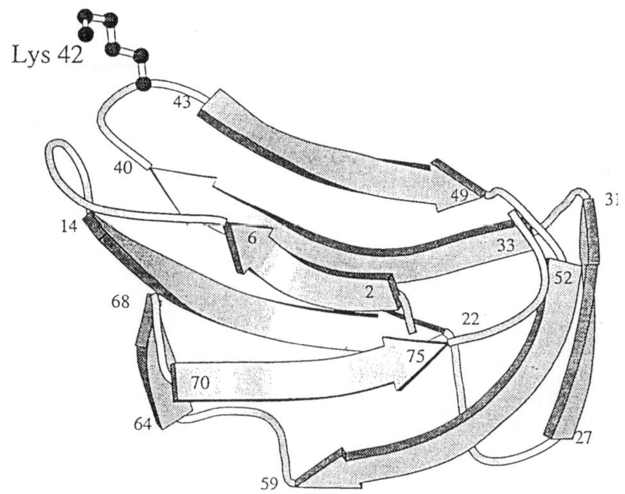

Figure 1. Schematic ribbon drawing (Kraulis, 1991) of the structure of the lipoyl domain of *B. stearothermophilus* PDH complex. The lipoylated lysine residue (Lys-42) is indicated. The numbers refer to the position of the residue within the domain, which is N-terminal in the intact E2p polypeptide chain.

productively with its cognate E1 component and what dictates selective lipoylation of the appropriate lysine side-chain? In the prominent β-turn containing the lipoyl-lysine residue, there is a sequence -D[K]A- (where [K] represents the lipoylated lysine residue), which is highly conserved in the primary structures of all lipoyl domains determined to date (Perham, 1991; Russell and Guest, 1991). To investigate its potential involvement in these processes of molecular recognition, we have subjected the sub-gene encoding the *B. stearothermophilus* domain (Dardel *et al.*, 1990) to directed mutagenesis. Thus, the aspartic acid residue has been replaced with lysine (D41K) and the alanine residue with methionine (A43M).

Electrospray mass spectrometry of both mutant domains purified after expression of the sub-gene in *E. coli* revealed that they were generated in two forms differing in mass by 188 Da; these were identified as the lipoylated and non-lipoylated domains, as found previously for the native domain (Dardel *et al.*, 1990). We infer that neither mutation affected the ability of the *E. coli* lipoylating enzyme(s) to recognise the domain. The D41K mutant domain showed no evidence of a version carrying two lipoyl groups. Proteolysis (*S. aureus* V8 proteinase) of the lipoylated domains after reduction with DTT and alkylation with vinyl pyridine (Dardel *et al.*, 1990) appeared to release only one lipoylated peptide, the sequence of which was found to be VQNKXAVVE. No peptide with the sequence VQNXKAVVE has yet been found. We tentatively infer that the D41K mutant domain had been correctly lipoylated on Lys-42 and that the lipoylating enzyme(s) can distinguish between the N^6-amino groups of the neighbouring lysine residues. Further tests of this conclusion are in progress.

To examine the efficacy of the mutant lipoyl domains as substrates for E1, their ability to become reductively acetylated by the PDH complex in the presence of [2-^{14}C]-pyruvate (Packman *et al.*, 1984) was analysed. The results for the A43M mutant show that it behaves similarly to the wild-type domain, whereas the D41K mutant is reductively acetylated much more slowly, if at all, under comparable conditions. We infer that one of these mutations, A43M, in which a bulky aliphatic side-chain is placed next to the lipoyl-lysine residue, does not affect recognition by E1. On the other hand, the D41K mutation appears to have greatly inhibited reductive acetylation and thus to be a recognition signal for E1.

A comparison of 2D NOESY NMR spectra of the D41K mutant domain and unlipoylated wild-type domain indicates that the mutant domain was correctly folded. These experiments have set the stage for a systematic attack on the structural code that dictates the correct post-translational modification of the lipoyl domain and its subsequent recognition by its cognate E1 component.

A BIOTINYL DOMAIN IN BIOTIN-DEPENDENT CARBOXYLASES

The biotin cofactor in biotin-dependent carboxylases is attached in amide linkage to the N^6-amino group of a lysine residue in an amino acid sequence that is widely conserved among such enzymes. In yeast pyruvate carboxylase, the biotinylated region is thought to form a protein domain of about 70 amino acids located at the C-terminal end of each of the four constitutent polypeptide chains. There is some sequence similarity between this biotinyl domain and the lipoyl domains of the *E. coli* PDH complex (Lim *et al.*, 1988) and the potential resemblance is underlined by the similarity of the interaction of avidin with the biotinyl domain of carboxylases and the lipoyl domains of PDH complexes (Hale *et al.*, 1992).

By means of a new, largely automated procedure, we have identified potential key residues in the folding of the lipoyl domain and used the lipoyl domain structure as a template to predict the three-dimensional structure of the biotinyl domain of yeast pyruvate carboxylase from its amino acid sequence. Dihedral angles, inter-Cα distances and hydrogen bond distances and angles were among the structural features that were restrained. Surprisingly, despite the low level of sequence identity between the target and template structures, a plausible three-dimensional structure for the biotinyl domain can be deduced (Figure 2), closely related to that of the lipoyl domain. The biotinyl-lysine residue is located in a comparable β-turn. As with the lipoyl domain (see above), there is some evidence that replacement of the residue on the N-terminal side of the biotinyl-lysine need

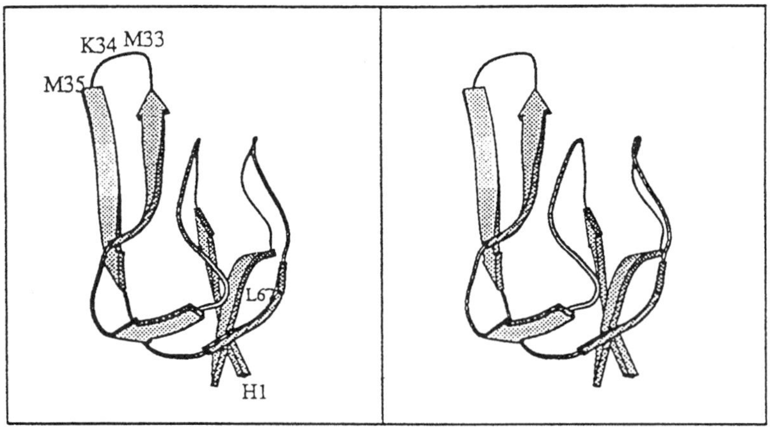

Figure 2. Schematic stereo ribbon drawing (Kraulis, 1991) of the predicted three-dimensional structure of the biotinyl domain from yeast pyruvate carboxylase. The numbers refer to the position of the residue within the domain (His-1 is position 1102 in the native sequence; K34 is the biotinylated lysine residue).

not affect the post-translational modification but does affect the intermediate carboxylation of the biotinyl group (Samols *et al.*, 1988). Thus, the parallel is very close.

THE H-PROTEIN OF THE GLYCINE CLEAVAGE SYSTEM

In the reaction catalysed by the glycine cleavage system, the aminomethyl group derived from the decarboxylated glycine is transferred to the lipoyl-lysine residue of the H-protein (Hiraga and Kikuchi, 1980). The H-protein is about 125 amino acids in length and, as for the biotinyl domain just described, there is evidence of some sequence similarity to the lipoyl domain (Fujiwara *et al.*, 1991).

Using the same methods as we developed for the biotinyl domain, we have found it possible to build a plausible three-dimensional structure for approximately 80 of the 131 amino acid residues of the pea leaf H-protein (Kim and Oliver, 1990), based on the structure of the lipoyl domain as a template. Once again, the lipoyl-lysine residue is located in a tight β-turn, although secondary structure prediction has previously indicated that this region of the H-protein is α-helical (Fujiwara *et al.*, 1991). The pea leaf H-protein has been crystallized (Sieker *et al.*, 1991) and it will be of interest to compare our predicted structure with that ultimately determined by means of X-ray crystallography.

CONCLUSIONS

The determination of the structure of the lipoyl domain of the *B. stearothermophilus* PDH complex paves the way to understanding the unusual part played by the domain in the catalytic mechanism of 2-oxo acid dehydrogenase complexes. Based on this structure, we have been able to build plausible models of the biotinyl domain in biotin-dependent carboxylases and the H-protein of the glycine cleavage system. The models are necessarily speculative, given the low (less than 15%) level of sequence identity with the template lipoyl domain, but they indicate that the three structures are likely to be closely similar and to be representative of a new protein module incorporated in a range of multifunctional proteins.

Acknowledgements: We are grateful to the Science and Engineering Research Council, The Wellcome Trust and Meiko Limited for financial support and The Alliance British-French Joint Research Programme for a grant to facilitate collaboration. Many of the facilities were provided by the Cambridge Centre for Molecular Recognition. We thank Drs J.C.J. Barna, W. Boucher, P.J. Kraulis and A.R.C. Raine for help with computing, 3D NMR data processing and the structure calculations. F.D. was an I.C.I. Fellow of Churchill College, Cambridge; N.G.W. is the holder of a Wellcome Trust Prize Studentship: and S.M.B. is the holder of a Research Studentship awarded by the Protein Engineering Initiative of the S.E.R.C.

REFERENCES

Dardel, F., Packman, L. C. and Perham, R. N., 1990 Expression in *Escherichia coli* of a subgene encoding the lipoyl domain of the pyruvate dehydrogenase complex of *Bacillus stearothermophilus*, *FEBS Lett.* 264:206.

Dardel, F., Laue, E. D. and Perham, R. N., 1991 Sequence-specific assignments and secondary structure of the lipoyl domain of the *Bacillus stearothermophilus* pyruvate dehydrogenase multienzyme complex, *Eur. J. Biochem.* 201:203.

Dardel, F., Davis, A. L., Laue, E. D. and Perham, R. N., 1992 The three-dimensional structure of the lipoyl domain from *Bacillus stearothermophilus* pyruvate dehydrogenase multienzyme complex, *J. Mol. Biol.* in the press.

Fujiwara, K., Okamura-Ikeda, K. and Motokawa, Y., 1991 Lipoylation of the H-protein of the glycine cleavage system, *FEBS Lett.* 293:115.

Graham, L. D., Packman, L. C. and Perham, R. N., 1989 Kinetics and specificity of reductive acetylation of lipoyl domains from 2-oxo acid dehydrogenase complexes, *Biochemistry* 28:1574.

Hale, G., Wallis, N. G. and Perham, R. N., 1992 Interaction of avidin with the lipoyl domains in the pyruvate dehydrogenase multienzyme complex: three-dimensional location and similarity to biotinyl domains in carboxylases, *Proc. R. Soc. Lond. B.* 248:247.

Hiraga, K. and Kikuchi, G., 1980 The mitochondrial glycine cleavage system: functional association of glycine decarboxylase and aminomethyl carrier proteins, *J. Biol. Chem.* 255:11671.

Kim, Y. H. and Oliver, D. J., 1990 Molecular cloning, transcriptional characterization, and sequence of cDNA encoding the H-protein of the mitochondrial glycine decarboxylase complex in peas, *J. Biol. Chem.* 265:848.

Kraulis, P. J., 1991 MOLSCRIPT: a program to produce both detailed and schematic plots of protein structures, *J. Appl. Crystallogr.* 24:946.

Lim, F., Morris, C. P., Occhiodoro, F. and Wallace, J.C., 1988 Sequence and domain structure of yeast pyruvate carboxylase, *J. Biol. Chem.* 263:11493.

Mattevi, A., Obmolova, G., Schulze, E., Kalk, K. H., Westphal, A., de Kok, A. and Hol, W. G. J., 1992 Atomic structure of the cubic core of the pyruvate dehydrogenase multienzyme complex, *Science* 255:1544.

Packman, L. C., Perham, R. N. and Roberts, G. C. K., 1984 Domain structure and ^{1}H-n.m.r. spectroscopy of the pyruvate dehydrogenase complex of *Bacillus stearothermophilus, Biochem. J.* 217:219.

Patel, M. S. and Roche, T. E., 1990 Molecular biology and biochemistry of pyruvate dehydrogenase complexes, *FASEB J.* 4:3224.

Perham, R. N., 1991 Domains, motifs and linkers in 2-oxo acid dehydrogenase multienzyme complexes: a paradigm in the design of a multifunctional protein, *Biochemistry* 30:8501.

Reed, L. J. and Hackert, M. L., 1990 Structure-function relationships in dihydrolipoamide acyltransferases, *J. Biol. Chem.* 265:8971.

Robien, M. A., Clore, G. M., Omichinski, J. G., Perham, R. N., Appella, E., Sakaguchi, K. and Gronenborn, A., 1992 Three-dimensional solution structure of the E3-binding domain of the dihydrolipoamide succinyltransferase core from the 2-oxoglutarate dehydrogenase multienzyme complex of *Escherichia coli, Biochemistry* 31:3463.

Russell, G. C and Guest, J. R., 1991 Sequence similarity within the family of dihydrolipoamide acyltransferases and discovery of a previously unidentified fungal enzyme, *Biochim. Biophys. Acta* 1076:225.

Samols, D., Thornton, C. G., Murtif, V. L., Kumar, G. K., Haase, F. C. and Wood, H. G., 1988 Evolutionary conservation among biotin enzymes, *J. Biol. Chem.* 263:6461.

Sieker, L., Cohen-Addad, C., Neuberger, M. and Douce, R., 1991 Crystallographic data for H-protein from the glycine decarboxylase complex, *J. Mol. Biol.* 220:223.

STRUCTURE ANALYSIS OF THE TOPOGRAPHY AND MOLECULAR ORGANIZATION OF PROTEIN-RNA COMPLEXES AS REVEALED IN RIBOSOMES

Brigitte Wittmann-Liebold

Max-Delbrueck-Centrum fuer Molekulare Medizin
Department Protein Chemistry
Robert-Roessle-Str. 10, 0-1115 Berlin-Buch, Germany

INTRODUCTION

Molecular structure analysis of multi-protein complexes or RNP-particles is still a difficult task. Although performing essential functions in the cell the nature of the interactions of protein and nucleic acids of such complexes at message transcription and translation or at cell differentiation and regulation is yet not understood. In order to gain information whether common structural elements or typical domain structures of the protein-protein or RNA-protein complexes may be preserved to assist these processes we study the topography of proteins and RNA in ribosomes and try to sheet light on conserved structural domains in ribosomes of various organisms.

Ribosomes are complex organelles constituted of many different proteins and 3 to 4 RNA strains. They vary considerably in different organisms and contain between 50 to 70 proteins of different sizes and net charges (Wittmann, 1986). Since ribosomes occur in all organisms their RNAs and proteins are ideal objects for studying evolutionary aspects as well as RNA-protein binding principles and structural-functional correlations.

Prerequisite for these studies is the knowledge of the primary structures of all constituents of the ribosome. To date a complete set of known ribosomal compounds is only available for the gram-negative eubacterium *E. coli* (Brosius *et al.*, 1978; Brosius *et al.*, 1980; Giri *et al.*, 1984; Wittmann-Liebold, 1986). Further, more than 40 protein sequences of the eubacterial gram-positive organism *B. stearothermophilus* and for the archaebacterial halophilic *Haloarcula marismortui* (Arndt *et al.*, 1991; Wittmann-Liebold *et al.*, 1990) and the thermophile *Methanococcus vannielii* (Koepke and Wittmann-Liebold, 1989; Matheson *et al.*, 1990; Wittmann-Liebold, 1988) are available. Among the eukaryotes most r-protein sequences are known for yeast (Warner, 1989; Planta *et al.*, 1986) and rat liver ribosomes (Wool *et al.*, 1990). Also chloroplast and mitochondrial ribosomes are under investigation (Grohmann *et al.*, 1992).

Methods in Protein Sequence Analysis, Edited by K. Imahori
and F. Sakiyama, Plenum Press, New York, 1993

Table 1. Approaches for topographical and three-dimensional structure analysis of the ribosome.

ARCHITECTURE
X-ray Analysis of Microcrystals
Immune Electron Microscopy
Neutron Scattering
Protein Binding Sites on RNA
Crosslinking
RNA-RNA
RNA-Protein
Protein-Protein
Surface Peptides
Antibody Approach
Limited Proteolysis

Various approaches have been undertaken to study the architecture of the ribosome as listed in Table 1. Detailed data on the secondary and tertiary structures of the constituents and of the quarternary structure of the particles would be requested for constructing a three-dimenional model of the entire ribosome on a molecular level. Crystals of individual compounds diffracting to sufficient resolution for X-ray analysis are available only for a few proteins and the 5S RNA (Lorenz, *et al.*, 1991; Liljas, 1991; Ramakrishnan and White, 1992; Wilson *et al.*, 1986). Secondary structure elements of the 5S and 16S RNA of *E. coli* were derived by biochemical and crosslinking experiments which allowed to construct models of the 5S RNA (Wolters and Erdmann, 1988; Barciszewska, *et al.*, 1992) and the 16S RNA fold within the 30S subunit (Brimacombe *et al.*, 1990; Stern *et al.*, 1988); yet, no model is available for the 23S RNA of the 50S subunit. Furthermore, individual proteins were localized at the RNAs by analysing crosslinks between RNA and protein obtained by iminothiolane treatment followed by UV-irradiation (Brimacombe *et al.*, 1986). These experiments identified the nucleotide site of the crosslink but did not allow to assign the peptides or amino acids involved in the hetero-oligomer due to the minute amounts of the crosslink except for two contact points (Ehresmann *et al.*, 1976; Maly *et al.*, 1980; Moeller *et al.*, 1978).

Immunological methods revealed binding sites of antibodies directed against individual proteins at the surface of the 30S and 50S subunits and at the 70S particle of *E. coli* and for some proteins of *B. stearothermophilus* (Stoeffler, G. and Stoeffler-Meilicke, 1986; Stoeffler-Meilicke and Stoeffler, 1990). Localization of proteins within the subunits was also done by triangulations in neutron scattering experiments of 30S and 50S ribosomes in *E. coli* (Capel *et al.*, 1988, May *et al.*, 1992; Moore *et al.*, 1986; Nowotny *et al.*, 1986). Suitable models of the subunits became available on the basis of these different approaches. The proteins were placed as globular molecules at the RNA site in the 30S subunit (Brimacombe *et al.*, 1990; Schueler and Brimacombe, 1988) and yet without RNA's, in the 50S subunit (Walleczek *et al.*, 1988). However, the present models do not reach the necessary resolution to explain the translationary machinery on a molecular level. Furthermore, for some of the r-proteins elongated shapes were measured (Giri *et al.*, 1984) and more than one antibody-binding site was found for some proteins, thus complicating any location of such a protein within the particle. The protein antibody approach only allows to assign the proteins at the surface of the particle with a deviation of +- 20 to 30 Å due to the large space of the FAB-antibody. These experiments yield no details on the individual structures nor their peptide regions residing at the surface or at the inside of the particle. Neutron scattering measures the distances of the center of the masses of the proteins but does not yield parameters on the extensions of the peptide backbone. RNA crosslinking data allow to construct helices of the RNA; however, assigning directions of these helices and details of the loop structures is hardly possible. Interestingly, it was suggested that the RNA loop structures may be involved in binding of the individual proteins. Thus, the present 30S and 50S models are still limited in respect to three-dimensional resolution of the constituents and their relative arrangements but may serve as working hypothesis for designing further experimental approaches.

In order to arrive at a quarternary model of the ribosome and its subparticles crystals of the entire ribosomes and their subunits were grown the best of which were obtained from halophilic and thermophilic ribosomes which resolve to 3-5 Å resolution (Wittmann *et al.*, 1982; Yonath *et al.*, 1980; Yonath *et al.*, 1988; Yonath *et al.*, 1990). The image reconstruction of two-dimensional crystal sheets allowed to prove the predicted tunnel of the 50S subunit through which the growing peptide chain is leaving the ribosome after translation of the message (Wittmann *et al.*, 1988). However, approaches to obtain good three-dimensional crystals failed for the *E. coli* ribosome; yet, for this ribosome most biochemical and biophysical data are available. Obviously, *E. coli* ribosomes are not stable enough to derive suitable crystals. On the other hand, ribosomes and their subunits from archaebacteria are more tight and resulted in suitable crystals for collecting full data sets by X-ray measurements at liquid nitrogen temperature. These crystals allow to establish the three-dimensional shapes of ribosomal subunits and their relative arrangements, once the phase problem is solved. They may facilitate to trace the fold of the RNAs within the particles and may also yield informations about the locations of predominant protein helices or beta-strands. Solving of the phase problem in such a complex crystal is rather difficult; not only have several heavy atom derivatives of the crystal to be made, but clusters of heavy atoms have to be introduced covalently bound to specific sites into the entire ribosome. Such a cluster containing 11 gold atoms has been successfully bound to protein L_{11} and introduced into L_{11}-deficient mutant ribosomes (see Boeckh and Wittmann, 1991; Makowski *et al.*, 1987; Yonath *et al.*, 1988; Yonath *et al.*, 1990). However, since the entire ribosome structure is so complex one cannot expect to arrive at a resolution to the atomic level with the present crystals and data collection techniques in opposite to X-ray analysis of individual protein crystals, a comparable easy task! Therefore, many biochemical and biophysical data of contact points, secondary structure elements, conserved regions, RNA binding sites, protein-protein interactions and fine structure analysis of functional important domains within the ribosomal subunits and their interphase in addition to X-ray analysis of suitable crystals will be demanded in unravelling the quarternay structure of the entire ribosome.

In our laboratory we have undertaken, by proteinchemical means in combination with mass spectrometry and nucleic acid methods, to identify close contact regions by crosslinking between neighboring proteins and RNA-protein contacts within the ribosomal 30S and 50S subunits of various organisms on a molecular level and have employed new strategies in order to gain information about surface peptide regions and peptide epitopes accessible by anti-peptide antibodies on intact ribosomes and their subunits. By limited proteolysis we cut off exposed peptides at the surface of the particles and investigated the remaining core structures for the presence of unaffected protein and peptide regions still anchored in the complex. By this means differentiation between surface exposed and inner regions of the particles became possible. The data obtained and the methods employed are dealt with in this paper.

CROSSLINKING EXPERIMENTS

Suitable Reagents and Identification of Crosslinks

Crosslinking of proteins was done on ribosomal subunits of *E. coli*, *B. stearothermophilus* and *H. marismortui* in order to obtain topographical and evolutionary data on the r-proteins at the amino acid level. Crosslinking was performed with homobifunctional reagents, such as suberimidates, diepoxybutane, or bifunctional reagents that are bridged by S-S-groups (Baeumert and Fasold, 1989). It proved to be important to employ cleavable reagents which allow to split the crosslinked protein pair at the various purification steps in order to detect the traces of crosslinked material in the bulk of protein moiety which remains unmodified or is monovalently derivatized. Furthermore, it is

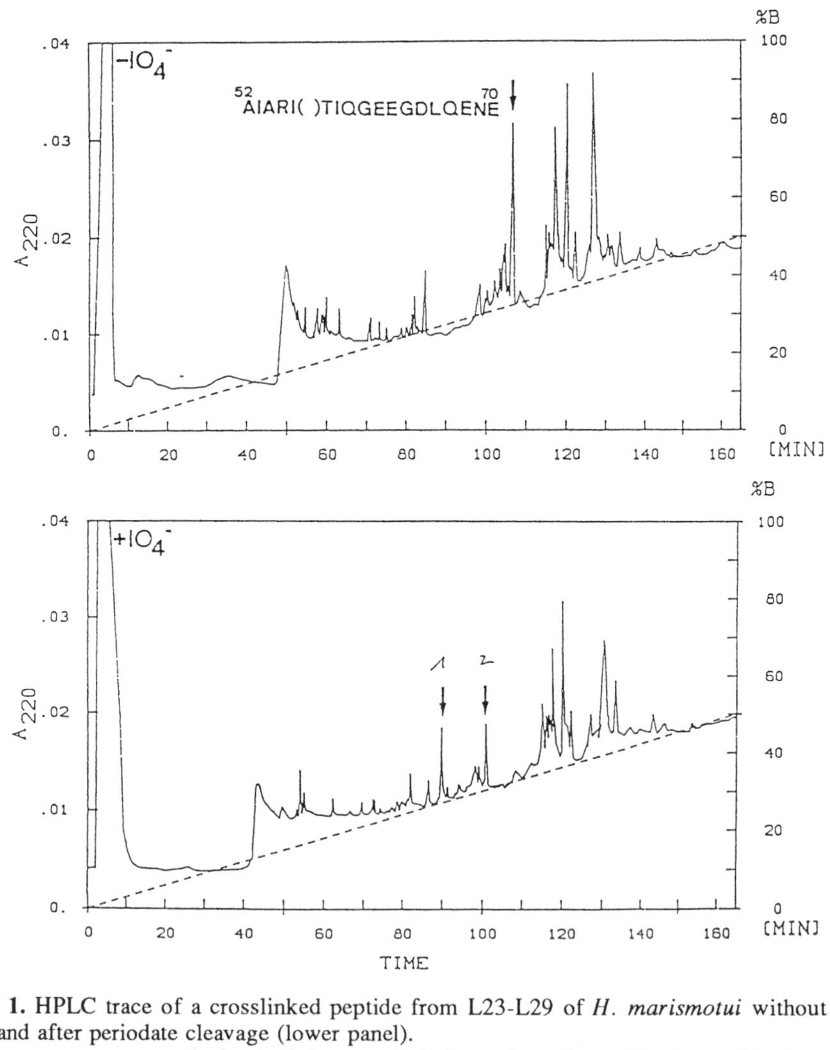

Figure 1. HPLC trace of a crosslinked peptide from L23-L29 of *H. marismotui* without (upper panel) and after periodate cleavage (lower panel).
Only one sequence was obtained of the crosslinked, uncleaved peptide due to blockage of the second peptide chain. Arrows indicate the master peptide and its daughter peptides after cleavage. In backets: crosslinked residue not identified in the sequencer.

necessary to use watersoluble reagents that allow to perform the reaction under mild and native buffer conditions in order to avoid any conformational changes of the particles (Kamp, 1988). By employing dilute concentrations of particles and low excess of the reagent (we used up to 30 fold reagent excess over modificable groups) low yields of crosslinked material in the order of less than 5% per protein were obtained. Since monovalent modifications may occur upon the reaction which obscure the identification of crosslinked protein pairs at the purification steps it turned out not to be useful to employ radiolabeled reagents since thereby labeled peptide fractions occured anywhere in the chromatogram but not necessarily contained crosslinked material (Brockmoeller and Kamp, 1986).

By the use of dimethylsuberimidate (DMS) the crosslinked protein pairs can be cleaved by ammonia as done in case of protein pair S5-S8 from *E. coli* (Allen *et al.*, 1979). However, the cleavage is rather incomplete.

In case of diepoxybutane which forms vicinal hydroxyl-groups at the crosslinking reaction with proteins the cleavage was done by mild treatment with periodate solution at ambient temperature, and depending on the time of reaction releases the parent proteins in reasonable yields. Further, the released proteins showed unmodified amino acids which can be easily determined by amino acid analysis or identified by N-terminal microsequencing. When purifications of the crosslinked proteins and the peptide fragments thereof were done by HPLC the purified crosslinked pair was twice reinjected under identical conditions, once after treatment with periodate and once without any further treatment: in this way one peak in the HPLC trace had to disappear to form two new ones after cleavage, see Figure 1. But in case of crosslinked peptides sometimes one peak disappeared and only one new was detected in the HPLC trace due to the generation of one short peptide migrating in the front peak or due to the formation of a very hydrophobic fragment retained. On the other hand, also more than

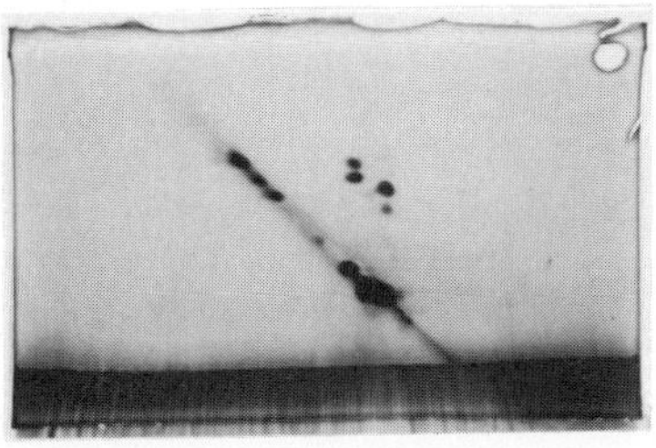

Figure 2. Diagonal 2D PAGE of crosslinked protein mixture. Identical conditions in the first and second gel electrophoresis; however, prior to the second dimension sucking of the gel in periodate solution was done.

one peptide might disappear resulting in several new peaks due to incomplete enzymatic cleavage of the crosslinked master peptide. By applying mass spectrometry (measurements were kindly provided by Dr. Klaus Eckart, MPI fuer Experimentelle Medizin, Abteilung Neuroendokrinologie, Goettingen, in a BioIon mass spectrometer) all HPLC peptide fractions of one run were examined for the presence of crosslinked peptides with and without cleavage.

Alternatively, the cleavage was done on gels: applying a few nanograms of the HPLC fractions 2D micro PAGE (3 x 3 cm gels, according the Brockmoeller and Kamp, 1986) was performed without and with cleavage. After the ID PAGE the gel was sucked in periodate solution and the second dimension PAGE performed under identical conditions. In this manner the cleaved proteins separate out of the non-crosslinked protein moiety found in the diagonal; the cleaved proteins are visible as two separate spots migrating below the diagonal (diagonal PAGE, see Figure 2).

Figure 3. 2D PAGE of the total 30S protein mixture of *E. coli* prior to (a) and after (b) crosslinking: purified S13-S19 corrlinked protein pair XL prior to (c) and after cleavage in periodate (d).

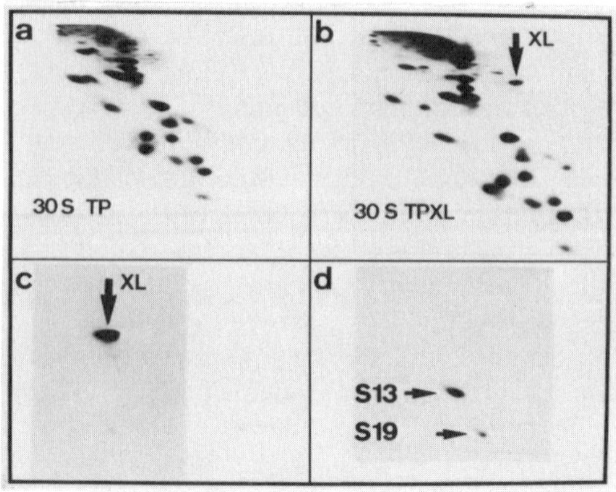

The use of DEB was very successfully applied for the topographical analysis of ribosomal particles although only few crosslinks were detected in each of the various ribosomes studied obviously due to the short span (4 Å) of the reagent (Figure 3). In case of the halophilic ribosome the crosslinking protocoll had to be adapted to the high salts under which these ribosomes are native. This caused some solubility problems of the reagent which were overcome by the addition of DMF to the reaction mixture (Bergmann and Wittmann-Liebold, 1992).

When applying S-S-bridged crosslinkers, e.g. DSP, dithio-bis-(succinimidyl propionate) or DTBP, dimethyl-3,3'-dithio-bis-(propioimidate), the crosslinked proteins or peptides were cleaved by treatment with dithioerythritol (DTT). Hence, the isolation of the protein and peptide pairs had to be performed without any addition of DTT and the ribosomal particles had to be freed of any traces of mercaptanes prior to the crosslinking reaction. The same techniques as described above were used to detect the protein and peptide pairs.

Depending on the pH of the crosslinking reaction (usually native buffers of pH 7.5 to 7.9 were employed) and the type of reagent used mostly histidines and cysteines were found covalently linked by the bifunctional reagent, but also side chain lysines and the N-terminal free amino group of the entire protein chain served as crosslinking sites (see below).

Purification of Crosslinked Protein Pairs and Application of Antibodies

The *E. coli* protein pair S5-S8 had been isolated in large scale in oder to purify material sufficient for conventional peptide purification applying a combination of open column chromatography and thin-layer techniques; sequence analysis of the crosslinked peptide pair was done by the DABITC/PITC double coupling method (Allen *et al.*, 1979). Another example employing conventional techniques was the identification of the intracrosslink of two molecules of protein L12 that form the stalk pertuberance of the 50S subunit in *E. coli* (Maassen *et al.*, 1981).

Other investigators employed 2D-PAGE for the separation and identification of the crosslinked proteins (Traut *et al.*, 1980). However, since the proteins are modified under the crosslinking reaction identification by 2D-techniques of the parent proteins is ambiguous due to changes in the migration pattern of the proteins after the reaction. Later, this technique was modified by employing immunological methods to identify the proteins of the crosslink in the 2D-pattern using antibodies raised against individual proteins and isolated peptide fragments (Traut *et al.*, 1986; Walleczek, 1990). Under these conditions reliable topographical data can be obtained provided that the antibodies are of high purity.

Further limitations are due to the fact that on the peptide level crosslinked material usually is not detectable by the anti-protein-antibody. This problem led us to try whether it is possible to use an antibody directed against a hapten, e.g. the dansyl group, to modify the ribosomes with this reagent and to perform binding studies with the anti-dansyl antibody (Bergmann and Wittmann-Liebold, 1990). Alternatively, monoclonal antibodies were used which imply the characterization of the antigenic epitopes recognized by the different antibodies (Nag *et al.*, 1987; Olson *et al.*, 1991; Syu *et al.*, 1990).

We isolated the crosslinked protein pairs after acetic acid or salt extraction of the total protein mixture by a combination of ion exchange chromatography and one- to two-step HPLC purification (Bergmann and Wittmann-Liebold, 1992; Brockmoeller and Kamp, 1988; Herwig, 1990). In some cases we blotted the protein pair after 2D PAGE onto PVDF and extracted the protein pair for further studies (Herwig *et al.*, 1992). The crosslinked peptides were exclusively separated by HPLC on an analytical scale.

Microsequence Analysis of the Crosslinked Proteins and Peptides

Microsequence analysis by the automated Edman technique was performed on the pulsed-liquid gas-phase sequencer (model 477 from Applied Biosystems, Foster City) or in the flow-through reactor of the Knauer sequencer (Dr. Knauer GMBH, Berlin). Crosslinked peptides with and without cleavage were usually sequenced in the Knauer sequencer employing covalent attachment of the carboxyl-end of the peptide to arylamino-PVDF (Millipore, Burlington, MA.) without any addition of polybrene (Herfurth *et al.*, 1991a and b). The crosslinked peptides gave a negative result at the position of the crosslinked amino acid residue due to the fixation to the second peptide chain. In all these cases the crosslinking site was verified by mass spectrometric analysis (see above) of the smallest crosslinked peptide pair available.

The entire procedures to arrive at the crosslinked residues of two adjacent proteins are given in Table 2.

Protein-Protein Contact Points in Diverse Ribosomes

Fast and sensitive isolation methods such as a combination of salt extraction and size exclusion chromatography (see Figure 4) of the entire protein mixture after crosslinking,

Table 2. Protein-protein crosslinking scheme.

1. Isolation of particles under native conditions
2. Crosslinking reaction employing a homobifunctional reagent
3. Extraction of total protein mixture (TP 30 or TP50)
4. 2D PAGE of crosslinked total protein mixture with and without cleavage
5. Isolation of crosslinked protein pairs
6. Identification of proteins in the crosslink (N-terminal sequence analysis; treatment with antiprotein antibodies)
7. Identification of parent proteins after cleavage (2D PAGE, N-terminal sequencing, anti-protein antibody approach)
8. Preparative purification of protein pair
9. Isolation of larger sized crosslinked peptide pair
10. Sequencing of peptide pair without cleavage and after cleavage and isolation of parent peptides.
11. Isolation of shorter crosslinked peptide pair
12. Identification of crosslinked amino acids (N-terminal sequencing and mass spectrometry)

Table 3. Investigated crosslinked protein pairs in various ribosomes.

Reagent	Protein Pair	References
DMS	EcoS5-EcoS8	Allen *et al.*, 1979
DMS	EcoL12-EcoL12	Maassen *et al.*, 1981
DMS	BstL4-BstL15	V. Kruft, unpublished results
DTBP	BstL4-BstL15	A.-W. Geissler *et. al.*, unpublished results
DTBP	EcoL14-EcoL32	T. Pohl, Ph.D. thesis Free University Berlin, 1988
DEB	EcoS13-EcoS19	Pohl and Wittmann-Liebold, 1988
DEB	BstS13-BstS19	Brockmoeller and Kamp, 1988
DEB	EcoL3-EcoL19	T. Pohl, unpublished results
DEB	BstL3-BstL19	Herwig *et al.*, 1992
DEB	BstL23-BstL29	Herwig *et al.*, 1992
DEB	HmaL23-HmaL29	Bergmann and Wittmann-Liebold, 1992
o-PDM	EcoL23-EcoL29	Walleczek *et al.*, 1990
DSP	HmaL23-HmaL29	Bergmann and Wittmann-Liebold, 1992
DSP	HmaL4-HL29	U. Bergmann, unpublished results
DSP	HmaL18-HL31	U. Bergmann, unpublished results

symbols: DMS, dimethylsuberimidate; DTBP, dimethyl-3,3-dithio-bis-(propioimidate); DEB,1,4-diepoxybutane; DSP,dithiobis-(succinimidylpropionate);*o*-PDM,ortho-phenylenedimaleimide. Eco, *Escherichia coli*; Bst, *Bacillus stearothermophilus*; Hma, *Haloarcula* (*Halobacterium*) *marismortui*, HL, *H. marismortui* proteins that have no equivalent protein in the eubacterial ribosome

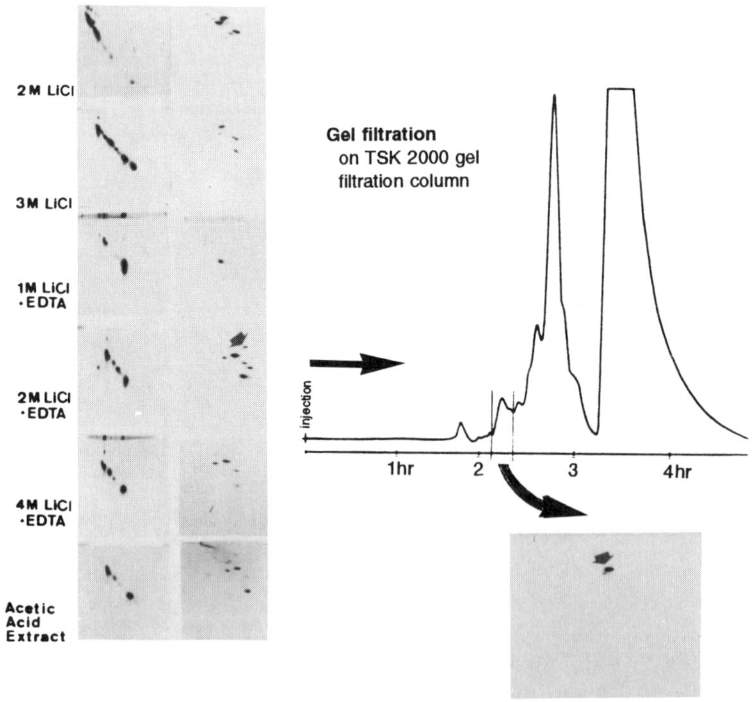

Figure 4. Purification of a 30S crosslink (BstS13-S19) after extraction with increasing amounts of LiCl by gel filtration on a TSK 2000 SW column. The arrows indicate the fractions of the protein pair and its migration in 2D PAGE (diagonal and common 2D-PAGE).

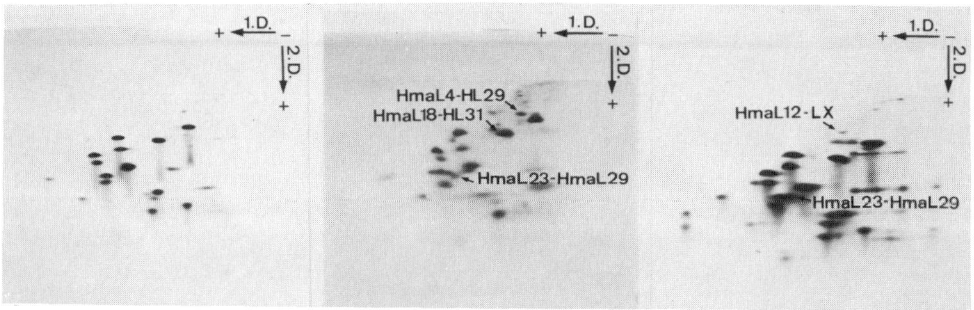

TP50 TP50-DSP TP50-DEB

Figure 5. Crosslink formation in H. marismortui 50S ribosomes. 2D PAGE of total 50S protein mixture without crosslinking (left figure) and after reaction by DSP (middle) and by DEB (right).

or one conventional ion exchange column chromatography (see Figure 4) combined with a one-step HPLC purification step allowed us to investigate several crosslinked ribosomal protein pairs as indicated in Table 3 and Figure 5. The results facilitated a comparison of structural domains of the *E. coli* ribosome with that of *B. stearothermophilus* or the archaebacterial *H. marismortui* ribosome.

Protein pair S13-S19 was found crosslinked in *E. coli* and in *B. stearothermophilus* 30S particles but not in the halophilic organism, likely due to the different native conditions under which the crosslinking reaction had to be performed. As stated above the *H. marismortui* ribosome is only stable under high salt conditions (3M NaCl) under which the eubacterial ribosome falls apart. Interestingly, the crosslinked amino acids, namely Cys-84 in protein S13 and His-68 in protein S19 both were found crosslinked in the two eubacterial ribosomes and are located in the most conserved sequence regions of these proteins (Brockmoeller and Kamp, 1988; Pohl and Wittmann-Liebold, 1988).

Protein pair Bst L3-L19 was also formed in both eubacterial ribosomes but not in *H. marismortui* whereas the pair L23-L29 could be isolated only from *B. stearothermophilus* (Bst L23-BstL29, see Figure 6) and *H. marismortui* (Hma L23-HmaL29, see Figure 7). But in *E. coli* the latter crosslink was identified by reaction with *o*-phenylenedimaleimide (*o*-PDI) (Walleczek *et al.*, 1990). In *B. stearothermophilus* Lys-4 of protein L29 was crosslinked to Met-l in protein L23 as proved by N-terminal sequencing and mass spectrometry (Figure 6, Herwig *et al.*, 1992). However, in the halophilic ribosome Lys-57 in the C-terminal region of protein L29 was linked to the amino-terminal group, Ser-l of protein L23 (Figure 7). These latter contacts were confirmed applying both reagents, DEB and DSP, which span 4 Å and 12 Å, respectively (Bergmann and Wittmann-Liebold, 1992).

In the pair Bst L3-L19 His-28 in L3 and the three N-terminal amino acids, Met-l, His-2 and His-3, of protein L19 share the crosslinking site (Figure 8, Herwig *et al.*, 1992). Obviously the N-terminal residues in protein L19 are so flexible that all three residues, potent for the crosslinking reaction, reacted partially and were crosslinked to some degree. An unambiguous result in these cases was only possible by applying mass spectrometric analysis of each individual crosslinked peptide pair obtained. The above results show that, although the overall configuration of these r-protein domains are conserved between eubacteria and archaebacteria, the fine structure might be slightly variable, since different amino acid residues were covalently attached in case of protein L29. It should be noted that the results obtained were reproducible in several isolation procedures of the crosslinks.

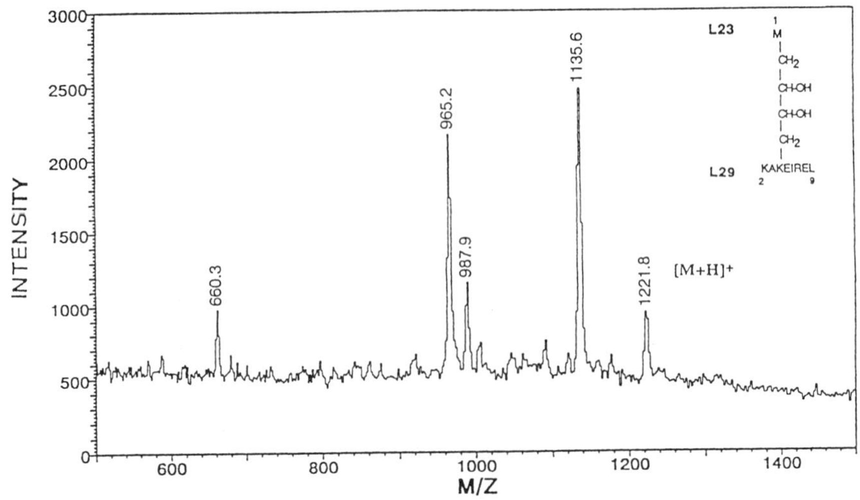

Figure 6. Mass spectrometric analysis of protein pair L23-L29 derived from *B. stearothermophilus* after treatment with DEB.
Measurements were done in a time-of-flight mass spectrometer by Dr. K. Eckart in Goettingen.

RNA-Protein Crosslinks

Crosslinking of RNA and protein in ribosomal subunits is usually performed by employing 2-aminothiolane (Baeumert and Fasold, 1989) in order to attach the reagent

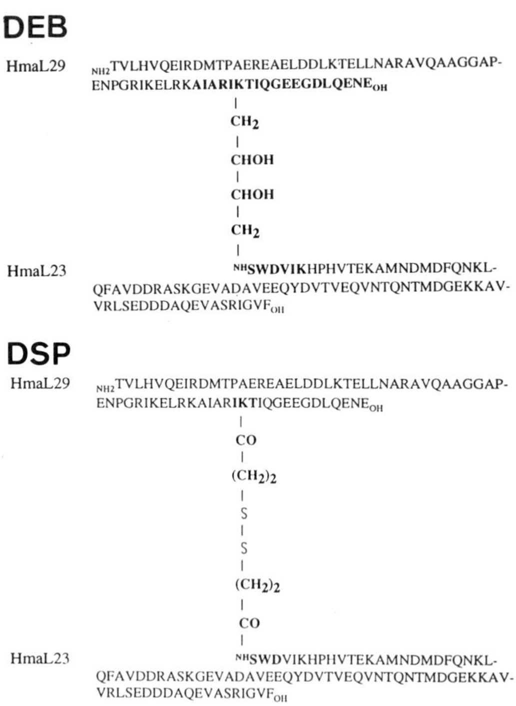

Figure 7. Sequence analysis of protein pair L23-L29 derived from *H. marismortui* using DEB and DSP.
The crosslinked amino acids were established by N-terminal sequencing and mass spectrometry.

monovalently to a lysine group of the protein and secondly, to perform the crosslinking to the RNA by UV irradiation. Further, experiments were performed applying UV irradiation only (Brimacombe, 1986; Bergmann, Ph.D. thesis, Free University Berlin, 1992). Several proteins were localized at their respective nucleotide sites of the 16S and 23S RNA's using nucleic acid purification schemes (Brimacombe *et al.*, 1986).

```
L3        TKGILGRKIGMTQIFAENGDLIPVTVIHATPNVVLQKKTIENDGYEAIQLGFEDISEKRA
          NKPQIGHAAKANTAPKRFIREIRGANINEYEVGQEVKVDIFSEGDIVDVTGISKGKGFQ
          GAIKRHGQSRGPMAHGSRYHRRPGSMGAIAPNRVFKTKNLPGRMGGERVTIQNLKI
          VKVDPERNLLLIKGNVPGPRKGLVIVKSAVKAKAKAK
                         18                      41
L3-GluC                   NGDLIPVTVIHATPNVVLQKKTIE

                                    CH2
                                    |
                                    CH-OH
                                    |
                                    CH-OH
                                    |
                                    CH2
                                    |\
L1 9-GluC                          MHHLIQE

L19       MHHLIQEITKEQLRTDLPDFRPGDTVRVHVKVVEGNRERIQVFEGVVIKRRGAGISE~
          VRKVSYGVGVERTFPVHTPKIAKLEVIRRGKVRRAKLYYLRELRGKAARIKEKTAQ
```

Figure 8. Sequence analsis of protein pair L3-Ll9 derived from *B. stearothermophilus* ribosomes.

As stated above protein-RNA crosslinks are much more difficult to analyse on the amino acid level, since by radioactive labeling of the RNA the nucleotide involved in the crosslink can be identified relatively easily whereas the crosslinked amino acid could not be detected due to the rather low crosslinking yield. So far, only two crosslinks were established on the molecular level, namely Met-l of protein S7 to U-1239 of the 16S RNA and Tyr-35 of protein L4 to U-615 of the 23S RNA, both in *E. coli* (Ehresmann *et al.*, 1976; Maly *et al.*, 1980; Moeller *et al.*, 1978). Here, we designed new approaches which confirmed these results and enhanced the data collection of RNA-protein interactions in ribosomes considerably (V. Kruft and Ulrike Bergmann, unpublished experiments).

PEPTIDE REGIONS AT THE SURFACE OF RIBOSOMES AND THEIR SUBUNITS

Limited Proteolysis

Limited proteolysis under controlled conditions using different proteolytic enzymes, such as trypsin, thermolysin, chymotrypsin, Lys-C and Glu-C enzyme was applied to 30S, 50S and 70S tight couples of ribosomes from *E. coli*, *B. stearothermophilus*, *H. marismortui* and *S. solfataricus* in order to determine the primary cleavage points at the surface of the particles and to compare their accessibilities to the enzymes. Valuable informations about the fine structure of the particles and their functional important domains were obtained (Kruft and Wittmann-Liebold, 1991; Kruft, Bischoff and Wittmann-Liebold, maniscript in preparation).

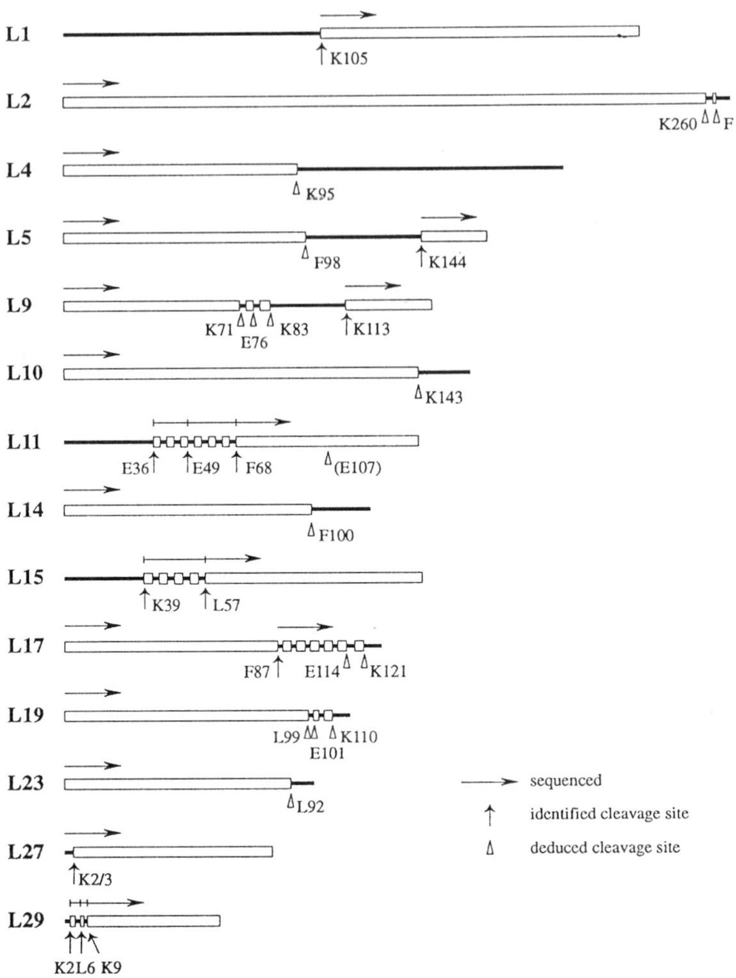

Figure 9. Limited proteolysis of intact 50S ribosomes from *E coli.*
Peptides exposed at the surface were cleaved off (thin bars); peptide regions still anchored in the remaining core particles were analysed by sequence analysis (thick bars). Cleavage sites were detected with trypsin (K), chymotrypsin (F, L), and Glu-C-enzyme (E).

The methodical approach is rather straight forward: the particles are incubated with different amouts of enzymes in a time course experiment and freed from cleaved peptides and enzymes by gradient centrifugation. The core particles are extracted for protein and anchored fragments and the mixture is applied to 2D PAGE. The gels are blotted onto PVDF membranes and the newly occurring spots not present in untreated 30S or 50S total protein mixtures are sequenced. By comparison with the RIBO data base (B. Wittmann-Liebold, M. Dzionara, A. Beck, A.K.E. Koepke, T. Boeckh, unpublished) the sequences obtained are identified and the cleavage sites assigned (Figure 9). By this means many surface-located peptide regions could be determined. Therefore, this approach provides many new data on the surface and the inner parts of the ribosome, respectively, which was not possible before. Further, the cores can be examined for peptide regions that are still in close contact with the RNA strands. Therefore, this approach allows to investigate interactions between r-proteins and their RNA counterparts on a molecular level.

The data collected for the various 50S particles (Kruft and Wittmann-Liebold, 1991) and 30S ribosomes (V. Kruft, O. Bischoff and B. Wittmann-Liebold, manuscript in preparation) were compared and conserved domain structures detected among eubacterial and archaebacterial sources. Yet, also differences had been noticed due to larger sized r-proteins from archaee or to changes in the fine structure of these ribosomes. In general, proteins that were localized at the surface of the particles by immunological methods were attacked by the enzymes; on the other hand, proteins that could not be detected by anti-protein antibodies were stable under limited proteolytic digestion. Therefore, the new data are in good agreement with previous approaches to localize r-proteins at the surface of the particles.

Anti-Peptide Antibodies

Predictive methods, such as turn structure predictions according to Chou, Adler and Fassman, calculations of hydrophilic peptide areas and predictions of surface structures have been employed to calculate antigenic sites of various r-proteins. The predicted peptide regions might be located at the surface of the proteins and could also contribute to the surface structure of the ribosomal subunits with some probability. According to these predictive methods several peptides deriving from r-proteins with lengths of 8 to 20 residues were selected for synthesis. Frequently N- or C-terminal sequences of the proteins were surface-predicted. These peptides were compared by computer programs to the entire RIBO-protein data bank (see above) in order to assure that these peptides were not found in other r-protein sequences. Then, these peptides were synthesized, usually by adding a cysteine to one of the peptide's ends, then purified by HPLC chromatography and covalently attached to serum albumin via these cysteines. Antibodies directed against these conjugates were raised in rabbits and purified by gradient centrifugations. In this way several anti-peptide antibodies were obtained (Herfurth and Wittmann-Liebold, unpublished) as listed in Table 4.

The anti-peptide antibodies were subjected to different ELISA techniques and their specific binding to 30S and 50S subunits, respectively, could be demonstrated in competition experiments (E. Herfurth, Ph.D. Thesis, Free University Berlin, 1992). In case of anti-peptide antibodies raised against proteins L2, L5 and L25 the antibody reacted specifically

Table 4. Synthesis of peptides derived from r-protein sequences for anti-peptide antibody production

Protein	Total no. residues	Synthetized Peptide	Position
EcoL2	272	CDKFIVRRRSK	261-272
BstL2	275	CKTRKKKNKSDKFIIRRRKK	257-275
EcoL5	177	CIFPEIDYDKVDRVRG	136-150
EcoL25	94	CQDVQRHPYKPKLQHID	75- 90
EcoS3	232	CQPKKQQRKGRK	222-232
BstS3	218	CRRAQKQAIQRAMRPGRKGVKT	130-150
BstS20	83	RAKTSEKRRAC	9- 18
BstS20	83	CAASRQKSRLAK	70- 80

with 50S subunits but not with 70S particles (Figure 10). This is in accordance with other findings that these proteins are located at the inner surface of the 50S subunit hidden under the addition of 30S particles in 70S ribosomes. On the other hand, the anti-S3-peptide antibody did bind to both, the 30S and 70S ribosomes as this protein is located at the opposite site to the interphase between both subunits (Stoeffler-Meilicke and Stoeffler, 1990). Therefore, the described approach via anti-peptide antibodies is well suited to locate peptides at the surface of the ribosomal particles. Further work is under way.

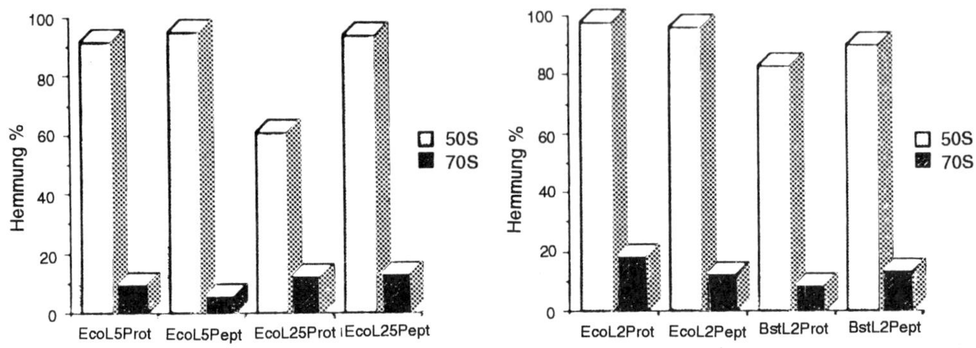

Figure 10. Decrease of the reactivity (in %) of anti-peptide antibodies against their parent protein in competition to the entire 50S and 70S particles.
The reactivity of *E. coli* anti-L5-peptide antibody with protein EcoL5 is almost decreased (rate of decrease indicated in %) when the antibody is preincubated with 50S ribosomes, but not when incubated with 70S particles since the binding site resides at the inside surface of the 50S subunit covered by the 30S particle in the 70S complex (upper left panel). The same effect is noticed with anti-EcoL25-peptide antibody (upper right figure), and with anti-EcoL2 peptide antibody and anti-BstL2 antibody (lower figures), but not with anti-EcoS3 peptide antibody which is located outside of the 50S/30S interphase (not shown).

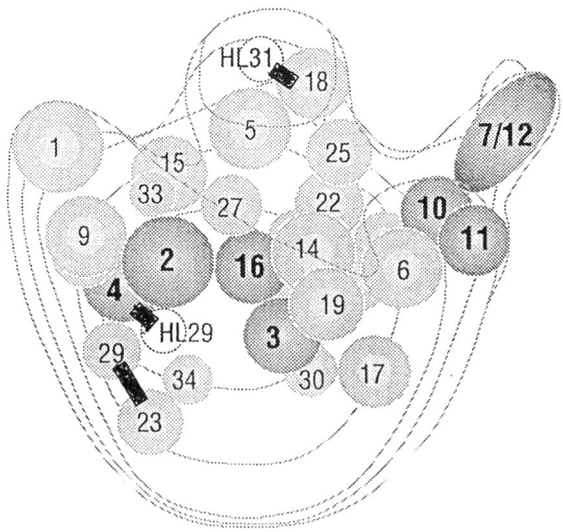

Figure 11. Crosslinked protein pairs analysed in *H. marismortui* (indicated in bars) were placed into the 50S model of *E. coli* ribosomes (Walleczek *et al.*, 1988). Hma-proteins having no counterpart in *E. coli* could be localized by this method, e.g. HL31 near protein L18, and HL29 in close neighborhood to protein L4.

CONCLUSIONS

The topography of the proteins and RNA in ribosomes deriving from eubacteria and archaee was studied by crosslinking studies and limited proteolysis of entire particles as well as by binding studies of antibodies directed against individual peptide sequences predicted at the surface of r-proteins. These approaches allowed us to assign contact points between the proteins on the amino acid level, they defined RNA-protein interactions of the complex and allowed to compare functional important domains in different ribosomes. The techniques employed led to many new data on peptide areas residing inside and at the surface of the particles. The present models of the 30S and 50S ribosome can be refined on the basis of these data. Finally, conclusions about the localization of r-proteins in archaebacteria which have no counterpart in the eubacterial ribosome can be drawn (Figure 11).

REFERENCES

Allen, G., Capasso, R. and Gualerzi, C., 1979 Identification of the amino acid residues of proteins S5 and S8 adjacent to each other in the 30S ribosomal subunit of *Escherichia coli*, *J. Biol. Chem.* 254:9800.

Arndt, E., Scholzen, T., Kroemer, W., Hatakeyama, T. and Kimura, M., 1991 Primary structures of ribosomal proteins from the archaebacterium *Halobacterium marismortui* and the eubacterium *Bacillus stearothermophilus*, *Biochimie* 73:657.

Baeumert, H. G. and Fasold, H., 1989 Crosslinking techniques, *Methods Enzymol.* 172:584.

Barciszewska, M. Z., Huang, H.-W., Marshall, A. G., Erdmann, V. A. and Barciszewski, J., 1992 Biochemical and NMR spectroscopy evidence for a new tertiary A-U base pair in lupin 5S RNA structure, *J. Biol. Chem.* 267:16691.

Bergmann, U. and Wittmann-Liebold, B., 1992 Identification of crosslinked amino acids of protein pair HmaL23-HmaL29 from the 50S ribosomal subunit of the archaebacterium *Haloarcula marismortui*, paper submitted.

Bergmann, U. and Wittmann-Liebold, B., 1990 Use of a hapten specific anti-dansyl antibody for the localization of ribosomal proteins by immune electron microscopy, *Biochem. Intern. J.* 21:741.

Boeckh, T. and Wittmann, H. G., 1991 Synthesis of a radiolabeled undecagoldcluster for application on X-ray structure analysis of ribosomes, *Biochim. Biophys. Acta* 1075:50.

Brimacombe, R., Atmadja, J., Kyriatsoulis and Stiege, W., 1986 RNA-structure and RNA-protein neighborhoods in the ribosome *in*: Structure, Function and Genetics of Ribosomes, Hardesty, B.and Kramer, G., eds., Springer Verlag, Heidelberg, New York, pp.184-202.

Brimacombe, R., Gornicki, P., Greuer, B., Mitchell, P., Osswald, M., Rinke-Appel, J., Schueler, D. and Stade, K., 1990 The 3-dimensional structure and function of *Escherichia coli* ribosomal-RNA, as studied by crosslinking techniques, *Biochim. Biophys. Acta* 1050:8.

Brockmoeller, J. and Kamp, R. M., 1986 Isolation and identification of two protein-protein crosslinks within the two subunits of *Bacillus stearothermophilus* ribosomes, *Biol. Chem. Hoppe-Seyler* 367:925.

Brockmoeller, J. and Kamp, R. M., 1988 Cross-linked amino acids in the protein pair S13-S19 and sequence analysis of protein S13 of *Bacillus stearothermophilus* ribosomes, *Biochemistry* 27:3372.

Brosius, J., Dull, T. J. and Noller, H. F., 1980 Complete nucleotide sequence of a 23S ribosomal RNA gene from *Escherichia coli*, *Proc. Natl. Acad. Sci. USA* 77:201.

Brosius, J., Palmer, M. L., Kennedy, P. L. and Noller, H. F., 1978 Complete nucleotide sequence of a 16S ribosomal RNA gene from *Escherichia coli*, *Proc. Natl. Acad. Sci. USA* 75:4801.

Capel, M. J., Kjelgaard, M., Engelmann, D. M. and Moore, P. B., 1988 Positions of S2, S13, S16, S17, S19 and S21 in the 30S ribosomal subunit of *Escherichia coli*, *J. Mol. Biol.* 200:65.

Ehresmann, B., Reinbolt, J., Backendorf, C., Tritsch, D. and Ebel, J. P., 1976 Studies on the binding sites of *Escherichia coli* ribosomal protein S7 with the 16S RNA by ultraviolet irradiation, *FEBS Lett.* 67:316.

Giri, L., Hill, W. E., Wittmann, H. G. and Wittmann-Liebold, B., 1984 Ribosomal proteins: Their structure and spatial arrangement in prokaryotic ribosomes, *Adv. Prot. Chem.* 36:1.

Grohmann, L., Graack, H.-R. and Subramanian, A.R., 1992 Ribosomal RNAs and proteins and their genes *in*: The Molecular Biology of Mitochondria, Academic Press, Inc., New York, in press.

Herfurth, E., Hirano, H. and Wittmann-Liebold, B., 1991a The amino acid sequences of the *Bacillus stearothermophilus* ribosomal proteins S17 and S21 and comparison to homologous proteins of other ribosomes, *Biol. Chem. Hoppe-Seyler* 372:955.

Herfurth, E., Pilling, U. and Wittmann-Liebold, B., 1991b Microsequencing of proteins and peptides in the Knauer sequencer with and without covalent attachment to polyvinylidene difluoride membranes by the wet-phase degradation technique, *Biol. Chem. Hoppe-Seyler* 372:351.

Herwig, S., 1990 Preparation and comparison of RP-HPLC of crosslinked ribosomal protein pairs by chromatography on PLRP- and Vydac-columns, *Chromatographia* 30:696.

Herwig, S., Kruft, V., Eckart, K. and Wittmann-Liebold, B., 1992 Amino acids in the protein pairs L3-L19 and L23-L29 of *Bacillus stearothermophilus* ribosomes after treatment with diepoxybutane, *J. Biol. Chem.*, in press.

Kamp, R. M., 1988 Application of bifunctional reagents for topological investigation, *in*: Modern Methods in Protein Chemistry, Tschesche, H. ed., vol. 3, Walter deGruyter & Co., Berlin, New York, pp. 275-298.

Koepke, A. K. E. and Wittmann-Liebold, B, 1989 Comparative studies of ribosomal proteins and their genes from *Methanococcus vannielii* and other organisms, *Can. J. Microbiol.* 35:10.

Kruft, V. and Wittmann-Liebold, B., 1991 Determination of peptide regions on the surphase of the eubacterial and archaebacterial ribosome by limited proteolytic digestion, *Biochemistry* 30:11781.

Liljas, A, 1991 Comparative biochemistry and biophysics of ribosomal proteins, *Int. Rev. Cytol.* 124:103.

Lorenz, S., Betzel, C., Raderschall, E.,.Dauter, Z., Wilson, K. S. and Erdmann, V.A., 1991 Crystallization and preliminary diffraction studies of 5S ribosomal RNA from the thermophilic bacterium *Thermus flavus*, *J. Mol. Biol.* 219:399

Maassen, J. A., Schop, E. N. and Moeller, W., 1981 Structural analysis of ribosomal protein L7/L12 by the hetero-bifunctional crosslinker 4-(6-Formyl-3-azidophenoxy)butyrimidate, *Biochemitry* 20:1020.

Makowski, I., Frolow, F., Saper, M. A., Shoham, M., Wittmann, H. G. and Yonath, A, 1987 Single crystals of large ribosomal particles from *Halobacterium marismortui* diffract to 6Å, *J. Mol. Biol.* 193:819.

Maly, P., Rinke, J., Ulmer, E., Zwieb, C. and Brimacombe, R., 1980 Precise location of the site of crosslinking between protein L4 and 23S RNA induced by mild ultraviolet irradiation of *Escherichia coli* 50S ribosomal subunits, *Biochemistry* 19:4179.

Matheson, A. T., Auer, J., Ramirez, C. and Boeck, A., 1990 Structure and evolution of archaebacterial ribosomal proteins, *in*: The Ribosome: Structure, Function and Evolution, Hill, W. E., Dahlberg, A., Garrett, R. A., Moore, P. B., Schlessinger, D. and Warner, J. R., eds., American Society for Microbiology, Washington, D.C., pp.617-635.

May, R. P., Nowotny, V., Nowotny, P. and Nierhaus, K. H., 1992 Inter-protein distances within the large subunit from *Escherichia coli* ribosomes, *EMBO J.* 11:373.

Moeller, K., Zwieb, C. and Brimacombe, R., 1978 Identification of the oligonucleotide and oligopeptide involved in an RNA-protein crosslink induced by UV irradiation of *E. coli* 30S, *J. Mol. Biol.* 126:489.

Moore, P. B., Capel, M., Kjeldgaard, M. and Engelman, D. M., 1986 A 19 protein map of the 30S ribosomal subunit of *Escherichia coli*, *in*: Structure, Function and Genetics of Ribosomes, Hardesty, B. and Kramer, G., eds., Springer Verlag, Heidelberg, New York, pp. 87-100.

Nag, B. Tewari, D. S. and Traut, R. R., 1987 Monoclonal antibodies to epitopes in both C-terminal and N-terminal domains of *Escherichia coli* ribosomal protein L7/L12 inhibiting elongation factor binding but not peptidyl transferase activity, *Biochemistry* 26:461.

Nowotny, V., May, R. P. and Nierhaus, K., 1986 Neutron-scattering analysis of structural and functional aspects of the ribosome: The strategy of the glassy ribosome, *in*: Structure Function and Genetics of Ribosomes, Hardesty, B. and Kramer, G., eds., Springer Verlag, Heidelberg, New York, pp. 101-111.

Olson, H. M., Nag, B., Etchison, J. R., Traut, R. R. and Glitz, D. G., 1991 Differential localization of two epitopes of *Escherichia coli* ribosomal protein L2 on the large ribosomal subunit by monoclonal anti-bodies, *J. Biol. Chem.* 266:1898.

Planta, R. J., Mager, W. H., Leer, R. J., Woudt, L. P., Raue' an El-Baradi, T. T. A. L., 1986 Structure and expression of ribosomal protein genes in yeast, *in*: Structure, Function and Genetic of Ribosomes, Hardesty, B. and Kramer, G., eds., Springer Verlag Heidelberg, New York, pp. 699-718.

Pohl, T. and Wittmann-Liebold, B., 1988 Identification of a crosslink in *Escherichia coli*

ribosomal protein pair S13-Sl9 at the amino acid level, *J. Biol. Chem.* 263:4293.
Ramakrishnan, V. and White, S. W., 1992. The structure of ribosomal protein S5 reveals sites of interaction with 16S rRNA, *Nature* 358:768.
Schueler, D. and Brimacombe, R., 1988 The *Escherichia coli* 30S ribosomal subunit; an optimized three-dimensional fit between the ribosomal proteins and the 16S RNA, *EMBO J.* 7:1509.
Stern, S., Weiser, B. and Noller, H. F., 1988 Model for the three-dimensional folding of 16S RNA, *J. Mol. Biol.* 204:447.
Stoeffler, G. and Stoeffler-Meilicke, M., 1986 Immuno electron microscopy on *Escherichia coli* ribosomes, *in*: Structure, Function and Genetics of Ribosomes, Hardesty, B.and. Kramer, G., eds., Springer Verlag, Heidelberg, New York, pp. 28-46.
Stoeffler-Meilicke, M. and Stoeffler, G., 1990 Topography of the ribosomal proteins from *Escherichia coli* within the intact subunits as determined by immuno electron microscopy, *in*: The Ribosome, Structure, Function and Evolution, Hill, W. E., Dahlberg, A., Garrett, R. A., Moore, P. B., Schlessinger, D. and Warner, J. R., eds., Am. Soc. Microbiol., Washington, D.C., pp. 123-133.
Syu, W.-J., Kahan, B. and Kahan, L., 1990 Epitope mapping of monoclonal antibodies to *Escherichia coli* ribosomal protein S3, *J. Protein Chem.* 9:159.
Traut, R. R., Lambert, J. M., Boileau, G. and Kenny, J. W., 1980 Protein topography of *Escherichia coli* ribosomal subunits as inferred from protein crosslinking *in*: Structure, Function and Genetics of Ribosomes, Chambliss, G., Craven, G. R., Davies, J., Kahan, L. and Nomura, M., eds., Universit Park Press, Baltimore, pp. 89-110.
Traut, R. R., Tewari, D. S., Sommer, A., Gavino, G. R., Olson, H. M and Glitz, D. G., 1986 Protein topography of ribosomal functional domains: Effect of monoclonal antibodies to different epitopes in *E. coli* protein L7/L12 on ribosome function and structure, *in*: Structure, Function and Genetics of Ribosomes, Hardesty, B. and Kramer, G., eds., Springer Verlag, Heidelberg, New York, pp. 286-308.
Walleczek, J., Albrecht-Ehrlich, R., Stoeffler, G. and Stoeffler-Meilicke, M., 1990 Three-dimensional localization of the NH_2- and carboxyl-terminal domain of ribosomal protein Sl on the surface of the 30S subunit from *Escherichia coli*, *J. Biol. Chem.* 265:11338.
Walleczek, J., Schueler, D., Stoeffler-Meilicke, M., Brimacombe, R. and Stoeffler, G., 1988 A model for the spatial arrangement of the proteins in the large subunit of the *Escherichia coli* ribosome, *EMBO J.* 7:3571.
Warner, J. R., 1989 Synthesis of ribosomes in *Saccharomyces cerevisiae*, *Microbiol. Rev.* 53:256.
Wilson, K. S., Appelt, K., Badger, J., Tanaka, I. and White, S. W., 1986 Crystal structure of a prokaryotic ribosomal protein, *Proc. Natl. Acad. Sci. USA* 83:7251.
Wittmann, H. G., 1986 Structure of ribosomes, *in*: Structure, Function and Genetics of Ribosomes, Hardesty, B. and Kramer, G., eds., Springer Verlag, Heidelberg, New York, pp. 1-27.
Wittmann, H. G., Muessig, J., Gewitz, H. S., Piefke, J., Rheinberger, H. J. and Yonath, A., 1982 Crystallization of *E. coli* ribosomes, *FEBS Lett.* 146:217.
Wittmann-Liebold, B., 1986 Ribosomal proteins: Their structure and evolution, *in*: Structure, Function and Genetics of Ribosomes, Hardesty, B. and Kramer, G., eds., Springer Verlag, Heidelberg, New York, pp. 326-361
Wittmann-Liebold, B., 1988 Structure and evolution of ribosomal proteins and their genes, *Endocytobiosis & Cell Res.* 5:259.
Wittmann-Liebold, B., Koepke, A. K. E., Arndt, E., Kroemer, W., Hatakeyama, R. T. and Wittmann, H. G., 1990 Sequence comparison and evolution of ribosomal proteins and their genes, *in*: The Ribosome, Structure, Function and Evolution, Hill, W. E., Dahlberg, A., Garrett, R. A., Moore, P. B., Schlessinger, D. and Warner, J. E., eds., Am. Soc. Microbiol., Washington, D.C., pp. 598-616.
Wolters, J. and Erdmann, V. A., 1988 Compilation of 5S rRNA and 5S rRNA gene sequences, *Nucl. Acids Res.* 16:rl.
Wool, I. G., Yaeta, E., Chan, Y.-L. and Glueck, A., 1990 Structure, function, and evolution of mammalian ribosomes, *in*: The Ribosome, Structure, Function and Evolution, Hill, W.E., Dahlberg, A., Garrett, R. A., Moore, P. B., Schlessinger, D. and J. R. Warner, eds., Am. Soc. Microbiol., Washington, D.C., pp. 203-214.
Yonath, A., Frolow, F., Shoham, M., Muessig, J., Makowski, I., Glotz, C., Jahn, W., Weinstein, S. and Wittmann, H. G., 1988 Crystallography of ribosomal particles, *Journal Crystal Growth* 90:231.
Yonath, A., Bennett, W., Weinstein, S. and Wittmann, H. G., 1990 Crystallography and image

reconstructions of ribosom, *in*: The Ribosome, Structure, Function and Evolution, Hill, W. E., Dahlberg, A., Garrett, R. A., Moore, P. B., Schlessinger, D. and Warner, J. E., eds., Am. Soc. Microbiol., Washington, D.C., pp. 134-147.

Yonath, A., Muessig, J., Tesche, B., Lorenz, S., Erdmann, V. A., Wittmann, H. G., 1980 Crystals of the large ribosomal subunits from *Bacillus stearothermophilus*, *Biochem. Intern. J.* 1:428.

INDEX